T0189030

Signals and Systems
A One Semester Modular Course

Synthesis Lectures on Signal Processing

Editor
José Moura, *Carnegie Mellon University*

Synthesis Lectures in Signal Processing publishes 80- to 150-page books on topics of interest to signal processing engineers and researchers. The Lectures exploit in detail a focused topic. They can be at different levels of exposition-from a basic introductory tutorial to an advanced monograph-depending on the subject and the goals of the author. Over time, the Lectures will provide a comprehensive treatment of signal processing. Because of its format, the Lectures will also provide current coverage of signal processing, and existing Lectures will be updated by authors when justified.
Lectures in Signal Processing are open to all relevant areas in signal processing. They will cover theory and theoretical methods, algorithms, performance analysis, and applications. Some Lectures will provide a new look at a well established area or problem, while others will venture into a brand new topic in signal processing. By careful reviewing the manuscripts we will strive for quality both in the Lectures' contents and exposition.

Signals and Systems: A One Semester Modular Course
Khalid Sayood
2020

Smartphone-Based Real-Time Digital Signal Processing, Third Edition
Nasser Kehtarnavaz, Abhishek Sehgal, Shane Parris, and Arian Azarang
2020

Anywhere-Anytime Signals and Systems Laboratory: from MATLAB to Smartphones, Third Edition
Nasser Kehtarnavaz, Fatemeh Saki, Adrian Duran, and Arian Azarang
2020

Reconstructive-Free Compressive Vision for Surveillance Applications
Henry Braun, Pavan Turaga, Andreas Spanias, Sameeksha Katoch, Suren Jayasuriya, and Cihan Tepedelenlioglu
2019

Smartphone-Based Real-Time Digital Signal Processing, Second Edition
Nasser Kehtarnavaz, Abhishek Sehgal, Shane Parris
2018

Anywhere-Anytime Signals and Systems Laboratory: from MATLAB to Smartphones, Second Edition
Nasser Kehtarnavaz, Fatemeh Saki, and Adrian Duran
2018

Anywhere-Anytime Signals and Systems Laboratory: from MATLAB to Smartphones
Nasser Kehtarnavaz and Fatemeh Saki
2017

Smartphone-Based Real-Time Digital Signal Processing
Nasser Kehtarnavaz, Shane Parris, and Abhishek Sehgal
2015

An Introduction to Kalman Filtering with MATLAB Examples
Narayan Kovvali, Mahesh Banavar, and Andreas Spanias
2013

Sequential Monte Carlo Methods for Nonlinear Discrete-Time Filtering
Marcelo G.S. Bruno
2013

Processing of Seismic Reflection Data Using MATLAB™
Wail A. Mousa and Abdullatif A. Al-Shuhail
2011

Fixed-Point Signal Processing
Wayne T. Padgett and David V. Anderson
2009

Advanced Radar Detection Schemes Under Mismatched Signal Models
Francesco Bandiera, Danilo Orlando, and Giuseppe Ricci
2009

DSP for MATLAB™ and LabVIEW™ IV: LMS Adaptive Filtering
Forester W. Isen
2009

DSP for MATLAB™ and LabVIEW™ III: Digital Filter Design
Forester W. Isen
2008

DSP for MATLAB™ and LabVIEW™ II: Discrete Frequency Transforms
Forester W. Isen
2008

DSP for MATLAB™ and LabVIEW™ I: Fundamentals of Discrete Signal Processing
Forester W. Isen
2008

The Theory of Linear Prediction
P. P. Vaidyanathan
2007

Nonlinear Source Separation
Luis B. Almeida
2006

Spectral Analysis of Signals: The Missing Data Case
Yanwei Wang, Jian Li, and Petre Stoica
2006

Signals and Systems: A One Semester Modular Course

Khalid Sayood

ISBN: 978-3-031-01417-8 paperback
ISBN: 978-3-031-02545-7 ebook
ISBN: 978-3-031-00338-7 hardcover

DOI 10.1007/978-3-031-02545-7

A Publication in the Springer series
SYNTHESIS LECTURES ON SIGNAL PROCESSING

Lecture #20
Series Editor: José Moura, *Carnegie Mellon University*
Series ISSN
Print 1932-1236 Electronic 1932-1694

Signals and Systems

A One Semester Modular Course

Khalid Sayood
University of Nebraska–Lincoln

SYNTHESIS LECTURES ON SIGNAL PROCESSING #20

ABSTRACT

This book is designed for use as a textbook for a one semester Signals and Systems class. It is sufficiently user friendly to be used for self study as well. It begins with a gentle introduction to the idea of abstraction by looking at numbers—the one highly abstract concept we use all the time. It then introduces some special functions that are useful for analyzing signals and systems. It then spends some time discussing some of the properties of systems; the goal being to introduce the idea of a linear time-invariant system which is the focus of the rest of the book. Fourier series, discrete and continuous time Fourier transforms are introduced as tools for the analysis of signals. The concepts of sampling and modulation which are very much a part of everyday life are discussed as applications of the these tools. Laplace transform and Z transform are then introduced as tools to analyze systems. The notions of stability of systems and feedback are analyzed using these tools.

The book is divided into thirty bite-sized modules. Each module also links up with a video lecture through a QR code in each module. The video lectures are approximately thirty minutes long. There are a set of self study questions at the end of each module along with answers to help the reader reinforce the concepts in the module.

KEYWORDS

linear time-invariant systems, impulse response, convolution, Fourier series, Fourier transform, discrete Fourier transform, Laplace transform, Z transform, feedback systems

To Füsun

Contents

Preface

There are a number of very nice books on Signals and Systems. The book by Oppenheim, Willsky and Nawab, the one by Lathi and Green, and the one by Mitra, to name just three. So why another book. The motivation behind this book, as reflected in the title, was to provide a textbook for a modular one semester class which could also be used for self study. Therefore, the scope of this book is both more restricted and focused than most Signals and Systems books. For each of the modules there is also an approximately thirty minute long video available. I recorded those videos for my Signals and Systems class using Open Broadcaster Software (OBS) which I would strongly recommend to novices like me for video recording and streaming. I apologize for the less than professional video editing. The modules in the text are linked to the videos through a QR code.

The order of topics is a familiar one. We begin with three modules devoted to preliminaries: complex numbers, functions, and special functions. The next six modules are devoted to understanding properties of systems, in particular linearity and time invariance, which are then used to develop the convolution relationship for linear time-invariant systems. The following eight modules are devoted to Fourier analysis and its applications in sampling and modulation. The sampling application is used to motivate the discrete Fourier transform (DFT). We use the next six modules to look at the Laplace transform and explore its use on the analysis and design of feedback systems. The final six modules are focused on the Z transform and the analysis and design of discrete systems.

In my teaching of this class I go through three modules a week. Our semesters are fifteen weeks long so this leaves me with five weeks for reviews and tests.

Khalid Sayood
May 2021

Acknowledgments

There are a number of people who helped me through this process and to whom I am very grateful. As with all my writing Pat Worster tried to make me follow the rules of the English language. Mike Hoffman, my long suffering colleague who would really like me to stop writing, read through several of these modules and provided pungent criticism. Kadri Özçaldıran at Bogazici University spent many hours pointing out the errors of my ways and discovered some nasty typos. Jerry Hudgins generously agreed to do a test run of the material for his class—who could ask for more. My friend John Beck was one of the initial inspirations for this modular form and has always been very gentle in his critiques. My partner Füsun Sayood who has supported my many wanderings this time suffered through days of my amateur video productions which slowly took over the whole house. She was my reality check on various aspects of the video production.

My thanks to Joel Claypool who has always been very gracious and patient with my various writing adventures (except one) and who came up with the idea of linking the modules to the corresponding videos. And of course my thanks to Dr. C.L. Tondo for his able shepherding of this manuscript through the publication process.

Khalid Sayood
May 2021

MODULE 0

What is This Course About

A signal is anything that carries information. A smile or a wink is a signal, the flash of lightening is a signal as is thunder, the song of a bird is a signal, a speech is a signal, as is a song, or a snarl. A picture is a signal, as is a movie. A strand of DNA is a signal, a protein is a signal, a hormone is most definitely a signal. Signals surround us so completely that we could if we wanted to classify the entirety of our world in terms of signals. In these modules we will be more modest and look only at signals which are one dimensional functions of time. Speech signals fall into this category, as does music, or temperature variations or the price of a particular stock. In Figure 1 I am saying the word "test." You can see the plosive *t* sound in the beginning and end of the waveform. The sound of the *e* is the large signal which seems to dominate the plot and the sound of *s* is the low amplitude almost noiselike signal before the final plosive *t*.

Figure 2 contains a signal which shows the concentration of CO_2 in the environment over the last decade or so. This plot is from the NASA website https://climate.nasa.gov/. (This is an interesting website with all kinds of information about climate change.)

The global change in temperature over the last century and a half is shown in Figure 3.

Figure 4 shows the progression of the Spanish Flu in the United Kingdoms from June of 1918 to the end of April 1919. You can see the first, second, and third waves in a one year period.

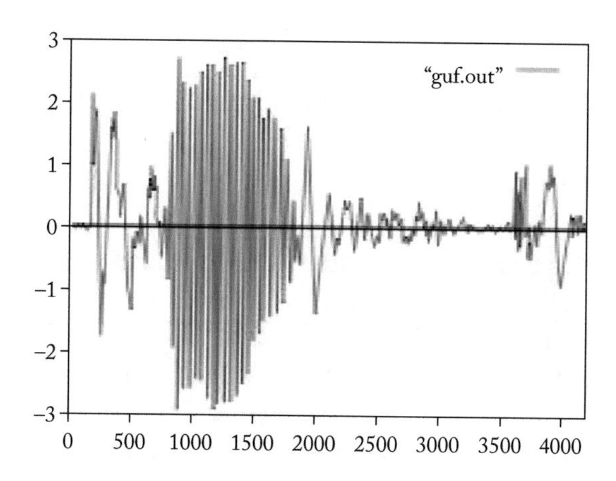

Figure 1: "Test."

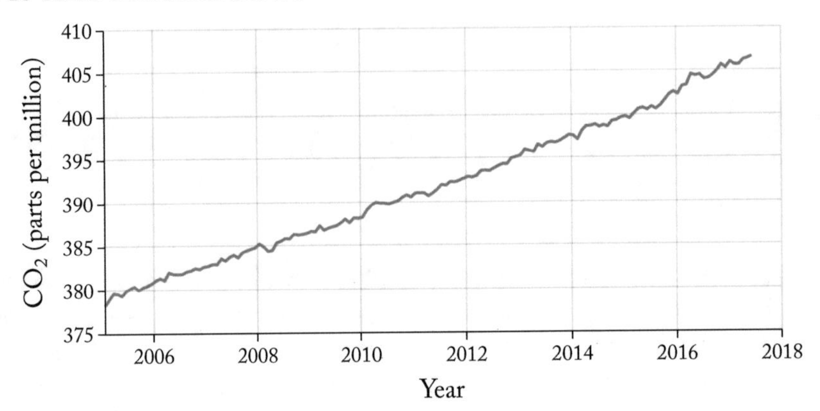

Figure 2: Concentration of CO_2 in the environment. From https://climate.nasa.gov/vital-signs/carbon-dioxide/.

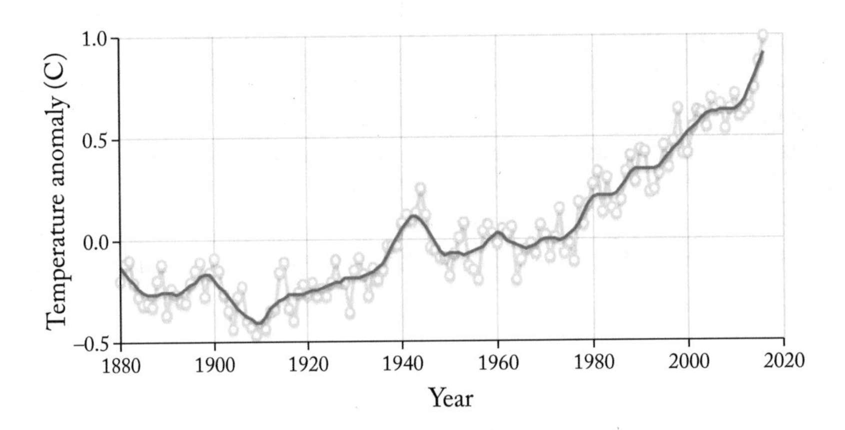

Figure 3: Global change in temperature. From https://climate.nasa.gov/vital-signs/global-temperature/.

Each of these signals contains information, and information is something that will rule your professional life. As electrical and computer engineers you will be designing and working with signals regardless of your particular specialization. You will work with broadcast signals, power signals, signals from instruments, measurements of control signals, or signals for processing in embedded systems. Regardless of the engineering specialty you end up embracing, signals will be a part of your life. That being the case it makes sense to become proficient in as many tools as possible which will allow you to understand a signal. In this book we will look at a number of ways to view signals.

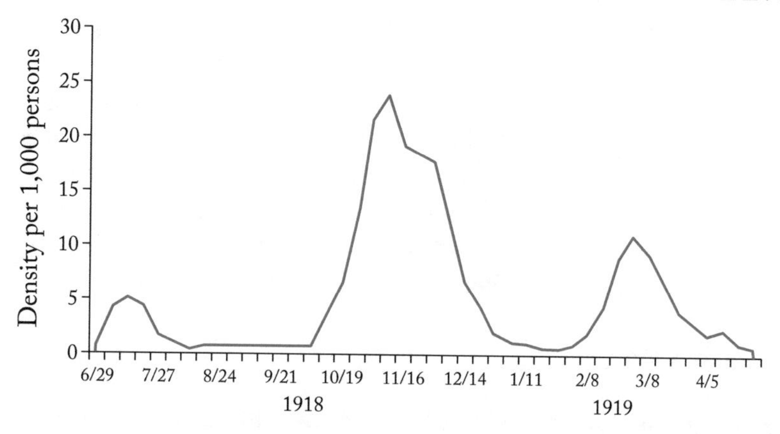

Figure 4: 1918 Spanish Flu progression in the United Kingdom (from *1918 Influenza: the Mother of All Pandemics*, Emerging Infectious Diseases, Jan. 2006).

Once you have a signal you will want to do a number of things with it. You may want to enhance it, distort it, or extract some particular piece of information from it. Something which does any of this is a system. As in the case of signals we can look at the world as a collection of systems. Your body is a system of systems, as is the world. There is the system in the body which generates speech. There is the system of processes in the world that generate CO_2. There is the system we call the atmosphere which responds to the signals the sun is putting out by heating up or cooling down. Our economic well being is effected by the stock market which is a system that processes various signals from political and economic systems. To understand how all these systems work is a tall order indeed. And we will not come anywhere close to doing so. In these modules we will limit our attention to a very small class of systems called linear time invariant (LTI) systems. However, you will find that through creative approaches for partitioning or approximating systems which do not fit into this small class we will be able to use the methods of analysis we develop for linear time-invariant systems more widely.

0.1 WHAT DO WE PLAN TO COVER

Our plan in this book is to learn some tools that can be used to analyze both discrete time and continuous time signals and systems. For analyzing continuous time signals and systems we will use the Fourier transform and the Laplace transform. For discrete time signals and systems we will use the discrete Fourier transform and the Z transform. This is most emphatically not a mathematics class; our goal is to learn some tools that happen to be mathematical tools. As with any tool practice makes perfect so the more you can practice the more comfortable you will be with using these tools. As our goal is to study these as tools we will try to include as many applications as we can. This is unfortunately only a single semester course so our opportunities are somewhat limited but we will try our best.

Having said that, there are some mathematical concepts we will use a lot and which you should be comfortable with. You should be reasonably comfortable with complex numbers. You should be able to do simple integrals. These include integrals of x^n, e^{ax}, and xe^{ax}. You should understand what a function is and finally you should be comfortable with summations, in particular geometric sums. These are concepts that you should be familiar with but just to be on the safe side we will go through some of them in the first modules.

0.2 ABSTRACTIONS

A Signals and Systems class is usually considered to be difficult because it deals with abstractions. However you have been dealing with abstractions all of your life. Consider numbers for a moment. You deal with numbers daily without really thinking about it. You probably don't consciously think of numbers as abstract. Yet to understand how much of an abstraction numbers are consider how long it has taken human beings to become comfortable with ideas about numbers that you would consider obvious.

The first practical use of numbers was to keep count. Tally sticks dating back tens of thousands of years have been discovered. Going from notches to different symbolic representations for different numbers also occurred many thousands of years ago. So the idea of numbers has been around for a long time. Yet to get to the present day understanding of numbers took thousands of years. The paths to modern numbers are numerous with one path taken in the Indian subcontinent and what is now called the Middle East, another path (or several) in China and yet other paths in Africa and the Americas. Let's look at the path taken in western Asia and Europe.

As we said the first use of numbers was probably for counting. We can keep track of things, like the number of sheep, by making a one-to-one equivalence with pebbles or tokens. But as the number of sheep grows the number of tokens might become cumbersome. We can then use different kinds of tokens for ten sheep and for a single sheep which would be a primitive positional number system. The earliest documented positional number system was that used by people in Mesopotamia (modern day Iraq and Syria) close to 6000 years ago. They used base 60 numbers—which seems a bit odd—until you remember that we still use the base 60 numbering system to measure time. The first documented use of the number zero (as opposed to the concept of zero) is in India in the work of the 7th century mathematician Brahmagupta in a work entitled *Brahmasputasiddhanta*. In around 825 a mathematician by the name of Muhammad ibn Musa Al-Khwarizmi who worked at the House of Wisdom in Baghdad wrote a book on Indian mathematics. Three centuries later his book was translated into Latin by Adelard of Bath finally introducing the number zero to Europe. This idea of zero being a number was then popularized by the Italian mathematician Leonardo of Pisa, better known as Fibonacci, in 1202—a centuries long journey for an abstraction even a child knows today. (Al-Khwarizmi gave several words to the English language. He wrote a book describing a procedural approach to solving equations called *Kitab al-mukhtasar fi hisab al-jabr wal-muqabala*—Book on Calculations by Completion and Balancing. Procedural approaches became known as the method of Al-Khwarizmi which in time became algoritni which then became algorithm. *Kitab al-mukhtasar fi hisab* **al-jabr** *wal-muqabala* was translated into Italian by Robert Chester with the title *Liber* **algebrae** *et almucabala* giving us the word *Algebra*.)

While zero came to Europe in the 12th century, negative numbers were resisted for another few centuries. (In China on the other hand negative numbers had been accepted since the second century BC.) The first real treatment of negative numbers in Europe appears in *Ars Magna* (The Great Art) in 1545 by Gerolamo Cardano who will show up again in just a bit. At this point if we have the integers we can also get fractions—ratios of integers. However numbers which are not the ratio of integers, $\sqrt{2}$ and π to name a couple, were more difficult to accept. Once you have all of these you can represent numbers as points along a line—the real number line.

Numbers are abstractions; and as you can see, abstractions are difficult to comprehend—until they are not. One thing that helps the transition from incomprehensible to obvious is repeated exposure. The concepts we will study in these modules are nowhere near as abstract as numbers so the transition to obvious should be relatively swift. You just need to be willing to expose yourself to these concepts repeatedly.

MODULE 1

Complex Numbers

I doubt if you spend too much mental energy trying to comprehend a negative number. But, for the longest time in the history of the world west of India, a negative number was considered, "absurd," or"false," or as late as the 19th century "imaginary." Somebody who didn't like negative numbers was the Egyptian (or Greek if you are talking to a Greek) mathematician Hero (or Heron) of Alexandria. Hero was looking for the volume of slices of a pyramid and ended up with a result of $\sqrt{81-144}$ or $\sqrt{-63}$. Instead of marching down the road to fame and fortune he let his aversion for negative numbers take over and rewrote that as $\sqrt{63}$. This aversion to negative numbers in Europe continued even to the late 17th century. In an 1897 paper W. W. Beman refers to the book *A Treatise of Algebra* by John Wallis (Figure 1.1) in which Wallis while promoting the idea of imaginary numbers says

> But it is also Impossible that any Quantity (though not a Supposed Square) can be Negative. Since that it is not possible that any Magnitude can be Less than Nothing or any Number Fewer than None.

This, by the way, is the same John Wallis who deciphered coded messages from Charles I which at least tangentially lead to the beheading of Charles I. Mathematics is a bloody field.

Given that you have accepted a negative number since the days when your age was in the single digits you really shouldn't get uptight about complex numbers. The concept of complex numbers first comes up in Gerolamo Cardano's *Ars Magna* as the solution of the problem of finding two numbers whose sum is equal to 10 and whose product is 40 (these would be $5 \pm \sqrt{-15}$). "As subtle as it is useless" was how Cardano saw the imaginary number. His contempt for these numbers was shared by Renee Descartes who coined the term "imaginary," which shows everyone is capable of saying dumb things. There is nothing imaginary, or *more* imaginary about imaginary numbers. They are as imagined as real numbers. The effect of the name imaginary probably did have some effect in keeping complex numbers somewhat obscure. Gauss in a paper in 1831 alludes to this:

> If $+1$, -1, $\sqrt{-1}$, had not been called positive, negative, imaginary (or impossible) unity, but perhaps direct, inverse, lateral unity, such obscurity could hardly have been suggested.

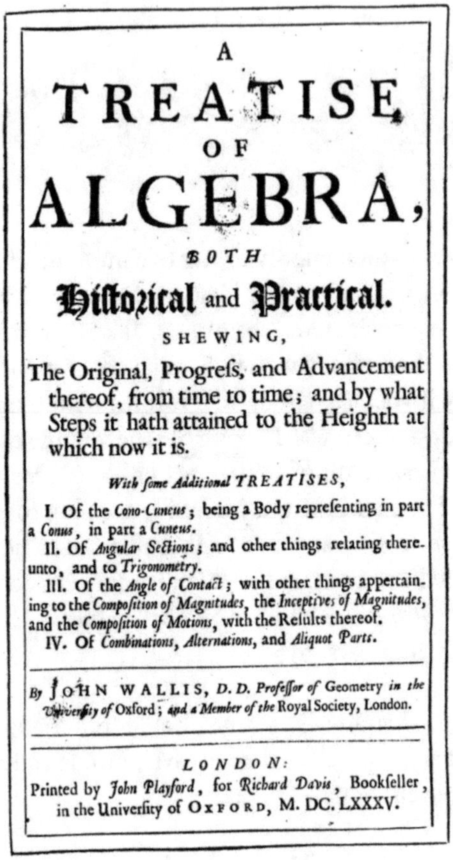

A
TREATISE
OF
ALGEBRA,
BOTH
Historical and Practical.
SHEWING,
The Original, Progrefs, and Advancement thereof, from time to time; and by what Steps it hath attained to the Heighth at which now it is.

With fome Additional TREATISES,

I. Of the *Cono-Cuneus*; being a Body reprefenting in part a *Conus*, in part a *Cuneus*.
II. Of *Angular Sections*; and other things relating thereunto, and to *Trigonometry*.
III. Of the *Angle of Contact*; with other things appertaining to the *Compofition of Magnitudes*, the *Inceptives of Magnitudes*, and the *Compofition of Motions*, with the Refults thereof.
IV. Of *Combinations*, *Alternations*, and *Aliquot Parts*.

By JOHN WALLIS, *D. D. Profeffor of* Geometry *in the Univerfity of* Oxford; *and a Member of the* Royal Society, London.

LONDON:
Printed by *John Playford*, for *Richard Davis*, Bookfeller, in the Univerfity of OXFORD, M. DC. LXXXV.

Figure 1.1: The title page of *A Treatise of Algebra*.

Wallis along with his role in the beheading of Charles I (though to be fair he did argue against the execution) also came close to coming up with a geometric way of representing complex numbers called the *polar representation*. While Wallis came close, the Danish (or Norwegian if you are talking to a Norwegian) cartographer Caspar Wessel in 1797 presented the first exposition of the Cartesian representation of complex numbers. While he was the first, the representation is known as the *Argand diagram* after the Swiss amateur mathematician Jean-Robert Argand who independently came up with the same representation. We don't know much about Argand except that at the time he came up with this he was working in Paris as a bookkeeper. He published his work in a pamphlet entitled *Essay on the Geometric Interpretation of Imaginary Quantities*. Someone sent a copy of the pamphlet to the French mathematician Adrian–Marie Legendre who mentioned it in a letter to another mathematician Francois Francais (and if you didn't figure out that he was French you need to go get some coffee) without mentioning Ar-

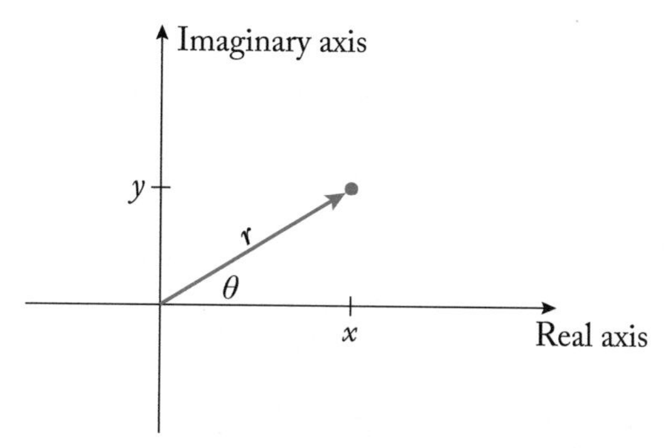

Figure 1.2: A point in the complex plane.

gand's name. As it happens to all mortals Francois died and his brother Jacques found the letter among his papers. Jacques submitted this idea of a geometric representation of complex numbers for publication, mentioning that he got the idea from Legendre's letter and requesting the author of the pamphlet mentioned by Legendre contact him. Argand actually saw the paper, contacted Jacques, and got credit for his work. Wessel got his credit posthumously. For more heart stopping stories like this I recommend *An Imaginary Tale: The Story of* $\sqrt{-1}$.

In this geometric representation of Wessel, Argand, and almost Wallis, complex numbers are points on the complex plane whose two orthogonal axes are the real number line and the imaginary number line. A point in two dimensional space can be represented in a number of ways. We can represent it by its coordinates—the real and imaginary values. Or we can see it as a vector in two dimensional space with a magnitude and an angle. The first representation is nowadays called the Cartesian representation while the second is called the polar representation. The Cartesian representation of the complex number z shown in Figure 1.2 is $x + jy$ where $j = \sqrt{-1}$. The polar representation of the same number is $re^{j\theta}$. Converting between polar and Cartesian coordinates is made very simple because of an identity known as Euler's formula.

1.1 EULER'S FORMULA

There are some weird and wonderful things in the world. Three of them are e, π and our recent friend j. We already know about j (or i as people not blessed to be engineers refer to it). The other two, e and π, are irrational numbers—that is they cannot be written as a ratio of two integers (the ir**ratio**nal comes from *ratio* not some psychotic tendency of the number), which show up often when we try to describe relationships in the world. The number π is the value we get when we divide the circumference (or **p**eriphery) of a circle by its diameter (despite the best efforts of some members of the Indiana legislature). While π has been known in one way or another since at least the Egyptian and Babylonian civilizations, the number e has a more recent history and was first discovered by Bernoulli in 1683 when he was trying to solve a compound interest problem. Both constants show up in all kinds of places. One is in an equation that a fifteen year old Richard Feynman in 1933 called "**the most remarkable formula in math.**"

$$e^{j\pi} + 1 = 0$$

This equation, called Euler's identity, is a special case of Euler's formula:

$$e^{jx} = \cos(x) + j\sin(x)$$

developed by the Swiss mathematician (and everything else) Leonhard Euler who was a student of Bernoulli and one of the greatest (known) mathematicians of all times. (He loses a few points for coining the $i = \sqrt{-1}$ notation.) This is a relationship we will use often so let's take some time and convince ourselves of its validity. We can easily derive this relationship using a power series expansion of all terms around $x = 0$. Recall the power series expansion of $f(x)$ around $x = 0$ is given by

$$f(x) = f(0) + \frac{f'(0)}{1!}x^1 + \frac{f''(0)}{2!}x^2 + \frac{f'''(0)}{3!}x^3 + \frac{f'''(0)}{4!}x^4 + \cdots$$

where prime (') corresponds to derivative.

If $f(x) = \cos(x)$ then

$$\begin{aligned} & & f(0) &= \cos(0) = 1 \\ f'(x) &= -\sin(x) & f'(0) &= -\sin(0) = 0 \\ f''(x) &= -\cos(x) & f''(0) &= -\cos(0) = -1 \\ f'''(x) &= \sin(x) & f'''(0) &= \sin(0) = 0 \\ f''''(x) &= \cos(x) & f''''(0) &= \cos(0) = 1 \end{aligned}$$

Therefore,

$$\begin{aligned} \cos(x) &= 1 + \frac{0}{1}x + \frac{-1}{2!}x^2 + \frac{0}{3!}x^3 + \frac{1}{4!}x^4 + \cdots \\ &= 1 - \frac{x^2}{2!} + \frac{x^4}{4!} + \cdots \end{aligned}$$

Similarly if $f(x) = \sin(x)$

$$\begin{aligned} & & f(0) &= \sin(0) = 0 \\ f'(x) &= \cos(x) & f'(0) &= \cos(0) = 1 \\ f''(x) &= -\sin(x) & f''(0) &= -\sin(0) = 0 \\ f'''(x) &= -\cos(x) & f'''(0) &= -\cos(0) = -1 \\ f''''(x) &= \sin(x) & f''''(0) &= \sin(0) = 0 \end{aligned}$$

Therefore,

$$\begin{aligned}\sin(x) &= 0 + \frac{1}{1}x + \frac{0}{2}x^2 + \frac{-1}{3!}x^3 + \frac{0}{4!}x^4 + \cdots \\ &= x - \frac{x^3}{3!} + \frac{x^5}{5!} + \cdots\end{aligned}$$

Finally, let $f(x) = e^{jx}$.

$$\begin{aligned} & f(0) = e^0 = 1 \\ f'(x) = je^{jx} \quad & f'(0) = je^0 = j \\ f''(x) = j^2e^{jx} \quad & f''(0) = j^2e^0 = -1 \\ f'''(x) = j^3e^{jx} \quad & f'''(0) = j^3e^0 = -j \\ f''''(x) = j^4e^{jx} \quad & f''''(0) = j^4e^0 = 1 \\ f^{(5)}(x) = j^5e^{jx} \quad & f^{(5)}(0) = j^5e^0 = j\end{aligned}$$

Therefore,

$$\begin{aligned}e^{jx} &= 1 + \frac{j}{1}x + \frac{-1}{2!}x^2 + \frac{-j}{3!}x^3 + \frac{1}{4!}x^4 + \frac{j}{5!}x^5 + \cdots \\ &= 1 - \frac{x^2}{2!} + \frac{x^4}{4!} - \cdots + j\left(x - \frac{x^3}{3!} + \frac{x^5}{5!} - \cdots\right) \\ &= \cos(x) + j\sin(x)\end{aligned}$$

This is Euler's formula! Something we will use often in many different ways. One way we use this identity is to write $\cos(x)$ and $\sin(x)$ in terms of e^{jx}. In order to see how to do this we begin with the expression for e^{-jx}. We can write this as $e^{j(-x)}$ and use Euler's formula

$$\begin{aligned}e^{j(-x)} &= \cos(-x) + j\sin(-x) \\ &= \cos(x) - j\sin(x)\end{aligned}$$

where we have used the fact that $\cos(-x) = \cos(x)$ and $\sin(-x) = -\sin(x)$; $\cos(x)$ is an even function and $\sin(x)$ is an odd function. So we have

$$\begin{aligned}e^{jx} &= \cos(x) + j\sin(x) \\ e^{-jx} &= \cos(x) - j\sin(x)\end{aligned}$$

Adding both equations we get

$$e^{jx} + e^{-jx} = \cos(x) + j\sin(x) + \cos(x) - j\sin(x) = 2\cos(x)$$

from which we get

$$\cos(x) = \frac{e^{jx} + e^{-jx}}{2}$$

If we subtract the second equation from the first

$$e^{jx} - e^{-jx} = \cos(x) + j\sin(x) - [\cos(x) - j\sin(x)] = 2j\sin(x)$$

which leads to

$$\sin(x) = \frac{e^{jx} - e^{-jx}}{2j}$$

1.2 CARTESIAN REPRESENTATION

In the cartesian representation the complex number, which is a point in the complex plane, is written in terms of the two coordinates—the real part and the imaginary part. If the number is $4 + j5$ it's easy to see which is the real part and which is the imaginary part. But what about je^{j3} or $\frac{3j}{3+4j}$? What is the real part and what is the imaginary part of each of these numbers? To answer that we need two relationships, Euler's formula (of course) and the concept of the complex conjugate. The conjugate of a complex number is simply the complex number with j replaced by $-j$. So the complex conjugate of $4 + j5$ is $4 - j5$, the complex conjugate of je^{j3} is $-je^{-j3}$ and the complex conjugate of $\frac{3j}{3+4j}$ is $\frac{-3j}{3-4j}$. Multiplying a number with its complex conjugate results in a positive real number. If we view the complex number as a vector in the complex plane, multiplying a number with its complex conjugate gives us the magnitude squared of that vector. Armed with these facts lets see how to find the real and imaginary parts of je^{j3} and $\frac{3j}{3+4j}$.

$$je^{j3} = j(\cos(3) + j\sin(3)) = j\cos(3) + j^2\sin(3) = j\cos(3) - \sin(3)$$

$$Re\left\{je^{j3}\right\} = -\sin(3)$$

$$Im\left\{je^{j3}\right\} = \cos(3)$$

For $\frac{3j}{3+4j}$, the first thing we want to do is remove the complex number from the denominator. We can do that by multiplying the denominator with its (the denominator's) complex conjugate. If we just did that, that would change the number. So we multiply both the numerator and the denominator with the complex conjugate of the denominator,

$$\begin{aligned}
\frac{3j}{3+4j} &= \frac{3j}{3+4j} \cdot \frac{3-4j}{3-4j} \\
&= \frac{3j(3-4j)}{(3+4j)(3-4j)} \\
&= \frac{9j - 12j^2}{9 - 12j + 12j - 16j^2} \\
&= \frac{9j + 12}{9 + 16} \\
&= \frac{12}{25} + j\frac{9}{25}
\end{aligned}$$

1.3 SUMMARY

In this module we looked at complex numbers and their representations. Given a complex number z

- we can represent it in polar coordinates as

$$z = re^{j\theta}$$

- we can represent it in Cartesian coordinates as

$$z = x + jy$$

We also introduced and proved Euler's excellent formula

$$e^{jx} = \cos(x) + j\sin(x)$$

1.4 EXERCISES

(Answers on the following page)

1. Using Euler's formula evaluate the following complex numbers:

   (a) $z = e^{j\pi/2}$

   (b) $z = e^{-j\pi/2}$

   (c) $z = e^{j\pi}$

   (d) $z = e^{-j\pi}$

   (e) $z = e^{j2\pi}$

   (f) $z = e^{-j2\pi}$

   (g) $z = e^{jn\pi}$ where n is an integer

2. Find the Cartesian representation of the following complex numbers

   (a) $z = 1/(1 + e^{jx})$ where x is a real number

   (b) $z = 1/j$

3. Write z in rectangular form, $x + jy$, where x and y are real valued and $j = \sqrt{-1}$

   (a) $z = je^{j}$

   (b) $z = \frac{1+j}{4+j3}$

   (c) $z = \frac{j}{2e^{3j}+1}$

   (d) $z = \frac{2e^{j3}+1}{j}$

   (e) $z = \frac{1+j}{1-j}$

4. Using Euler's formula derive the following:

   $\cos^2(\theta) = \frac{1}{2}(1 + \cos(2\theta))$

1.5 ANSWERS

1. (a) j
   (b) $-j$
   (c) -1
   (d) -1
   (e) 1
   (f) 1
   (g) $(-1)^n$
2. (a) $\frac{1}{2} - j\frac{\sin(x)}{2+2\cos(x)}$
   (b) $-j$
3. (a) $-\sin(1) + j\cos(1)$
   (b) $\frac{7}{25} + j\frac{1}{25}$
   (c) $\frac{2\sin(3)}{4\cos(3)+5} + j\frac{2\cos(3)+1}{4\cos(3)+5}$
   (d) $2\sin(3) - j(2\cos(3) + 1)$
   (e) j
4.

$$\begin{aligned}
\cos^2(\theta) &= \left(\frac{e^{j\theta} + e^{-j\theta}}{2}\right)^2 \\
&= \frac{e^{j2\theta} + e^{-j2\theta} + 2}{4} \\
&= \frac{2}{4} + \frac{1}{2}\left(\frac{e^{j2\theta} + e^{-j2\theta}}{2}\right) \\
&= \frac{1}{2}(1 + \cos(2\theta))
\end{aligned}$$

MODULE 2

Functions

While numbers are abstractions you have been familiar with from childhood, functions are a different matter altogether. These are abstractions that you might think you understand (and maybe you do) but there is a good chance that you are missing some essential aspect of it. Formally, a function is a mapping from one set, generally called the domain, to another set, called the range.

Functions will be important to us because we will model, or abstract, signals as functions. If the signal is an image it will be a function of the spatial coordinates of the pixels, $i(x, y)$. If the signal is a video it will be a function of the two spatial coordinates and time, $v(x, y, t)$. However, most of the signals we study in this class will be functions of only one variable—time. Pictorially we could draw it as shown in Figure 2.1

Or we can think of a function as a machine which given an *argument* generates a *value* (Figure 2.2).

The function operates on its arguments to generate a value. You can specify a function in a number of ways. You can set up a table which gives a value from the range for each value from the domain, or you can describe a rule which contains a set of operations to be performed on the value from the domain to generate a value from the range. An easy way of describing the set of operations is often a formula. For example, consider the function x defined as

$$x(t) = t^2 + 2$$

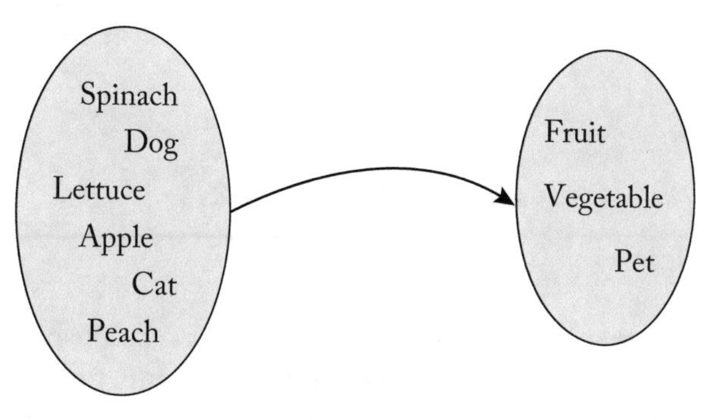

Figure 2.1: A function.

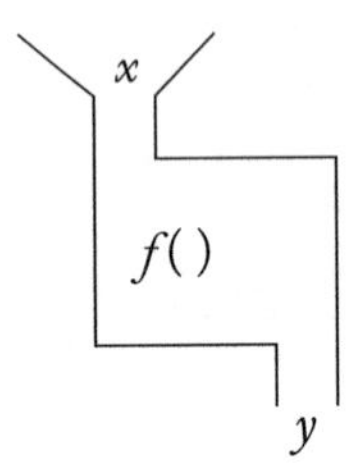

Figure 2.2: A function imagined as a machine.

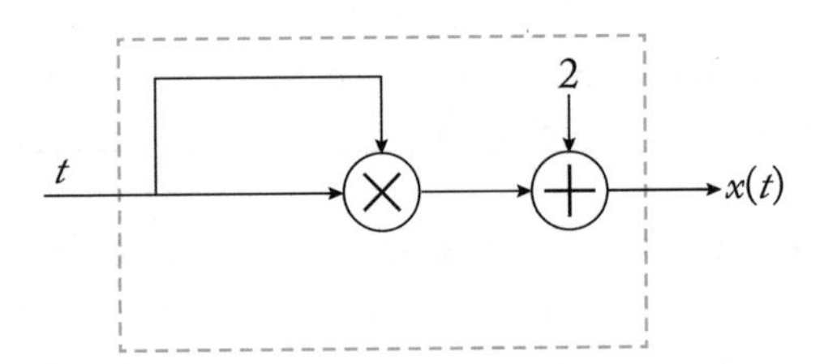

Figure 2.3: The function $t^2 + 2$ imagined as a machine.

if the domain is the set containing the entire real number line, the range will be the set of non-negative real numbers greater than or equal to 2. The function maps each real number by following the rule above to generate a positive real number. Or we can think of the function as a machine, as shown in Figure 2.3, that will take in a number, multiply it with itself and then add 2 to the result.

Let's see what happens when we put in different values of t. If we put in 1 in place of t into this machine we will get $1 \times 1 + 2$ or 3. If we put in -2 instead of t we will get $-2 \times -2 + 2$ or 6. What if we put in $t - 1$ instead of t? Well, the rule hasn't changed, only the input. So, let's follow the rule which is to multiply the input with itself and then add 2 and we get $(t - 1) \times (t - 1) + 2$, or $(t - 1)^2 + 2$.

Often we will be working with functions that are piece-wise continuous. An example of such a function is

$$x(t) = \begin{cases} 1 & 0 \leq t < 1 \\ 2 & 1 \leq t < 2 \\ 0 & \text{otherwise} \end{cases}$$

graphed in Figure 2.4. A graph expresses the input output relationship of a function so it is a valid representation of a function. Let's examine how this function looks as we modify the argument—change the input. Let's begin with a simple input. Let's look at what happens if we replace t in the argument with $t - 1$. If what is going into the machine is $t - 1$ instead of t then

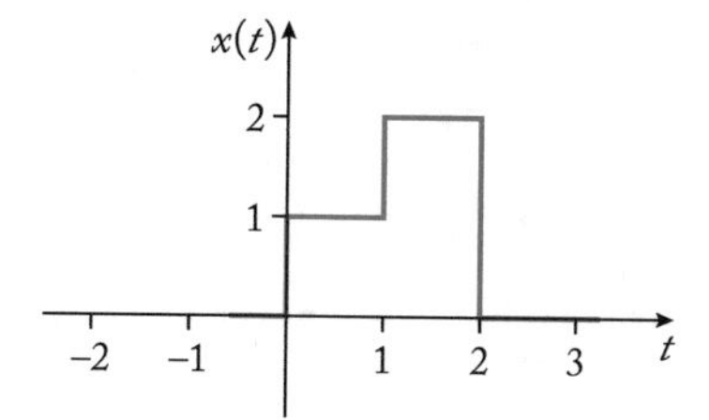

Figure 2.4: A piecewise continuous function.

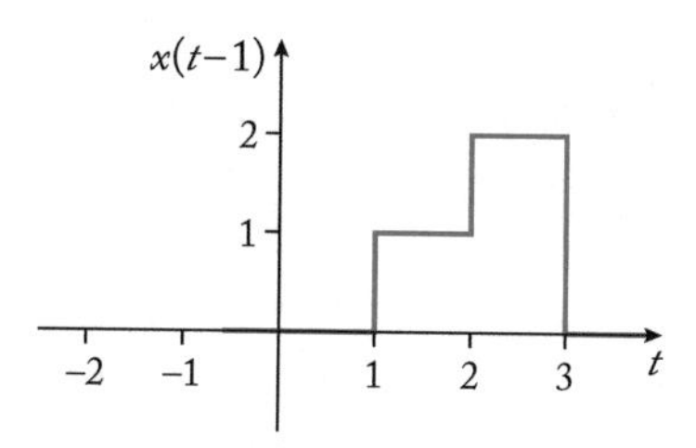

Figure 2.5: A piecewise continuous function.

we replace t with $t-1$ in the description of the function.

$$x(t-1) = \begin{cases} 1 & 0 \leq t-1 < 1 \\ 2 & 1 \leq t-1 < 2 \\ 0 & \text{otherwise} \end{cases}$$

Notice that we replaced t with $t-1$ wherever t appears in the original function description. We can now simplify the inequalities on the right hand side of the equation.

$$x(t-1) = \begin{cases} 1 & 1 \leq t < 2 \\ 2 & 2 \leq t < 3 \\ 0 & \text{otherwise} \end{cases}$$

If we look at a graph of $x(t-1)$ we can see that replacing t with $t-1$ results in a shift of the function to the right by 1 (Figure 2.5). If you did the same thing replacing t with $t+1$ instead of $t-1$ you would get a shift of the graph of the function to the left by 1.

What if we change the sign of t. Let's look at what happens with the simplest of examples $x(-t)$. Again how we find out the effect of this change is to feed $-t$ instead of t into the machine which is our function.

$$x(-t) = \begin{cases} 1 & 0 \leq -t < 1 \\ 2 & 1 \leq -t < 2 \\ 0 & \text{otherwise} \end{cases}$$

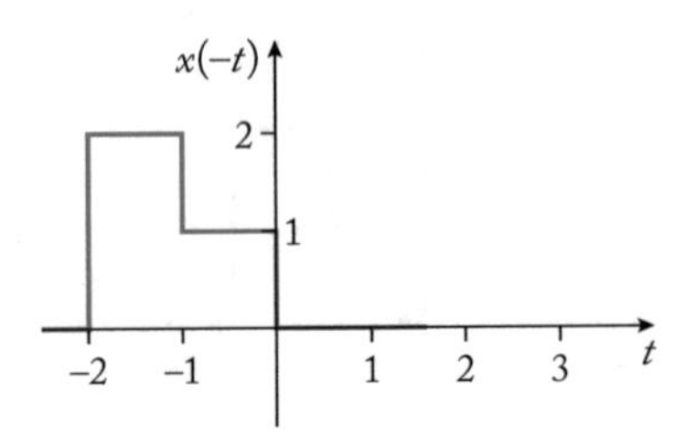

Figure 2.6: Flipping the function.

Multiplying through by -1 for the inequalities on the right hand side (and in the process reversing the inequalities) we get

$$x(-t) = \begin{cases} 1 & 0 \geq t > -1 \\ 2 & -1 \geq t > -2 \\ 0 & \text{otherwise} \end{cases}$$

or we can write this as

$$x(-t) = \begin{cases} 1 & -1 < t \leq 0 \\ 2 & -2 < t \leq -1 \\ 0 & \text{otherwise} \end{cases}$$

which we have plotted in Figure 2.6. The effect of replacing t with $-t$ has been to flip the function around the vertical axis. We can also think of this as mirroring the function around the vertical axis.

We can do many things to the argument and get many different effects. The important thing to note is that regardless of what we do to the argument of the function we can figure out the effect if we treat the argument as the input to a machine whose input/output relationship is given by the expression defining the function. As long as we do this we will always be able to figure out what kind of transformation we are working with.

2.1 TYPES OF FUNCTIONS

Functions can be classified in many different ways; the classification used depending on the application. Let's look at a couple of ways of classifying functions that will be useful to us.

2.1.1 CONTINUOUS AND DISCRETE TIME FUNCTIONS

Depending on the domain of the function we can define two type of functions, discrete functions, and continuous functions. Discrete functions are defined over a countable domain. That is, the domain can at least in theory be listed. Here is a function with a domain consisting of integers.

$$x[n] = \begin{cases} 1 & n = 0 \\ -1 & n = 1 \\ 3 & n = 2 \\ 2 & n = 4 \\ 0 & \text{all other integers} \end{cases}$$

Here, in theory, I could write down all values of the domain for which the function x generates an answer. Outside of these values the function is not defined. In this example we know the value of $x[1]$ or $x[10]$; but there is no value for $x[1.5]$. It is not that $x[1.5]$ is zero, the expression $x[1.5]$ simply does not make sense in this context.

A continuous function is defined over a continuous connected range of values. For example

$$x(t) = \begin{cases} t & \text{for} - 1 \leq t \leq 1 \\ 0 & \text{otherwise} \end{cases}$$

In this case the function is defined for all values of t. The values for which it is defined are continuous. If the function tells us the value at -1 and at -0.5 it will also tell us the value of the function at any point between -1 and -0.5.

The reason we are interested in continuous and discrete functions is because the signals we are interested in will often be one type or another. When I say the word "testing" into a microphone the output is a continuous function of time. At each instant of time, no matter how close it is to another instant of time the speech waveform has a value. It is a continuous function that maps instants of time into voltage values. There are an infinite number of time instances in each second and the function can (at least in theory) map each of those instances of time into a value. If we now sample this speech signal at 8000 samples per second—as is in your phone—we have a discrete function. This function only has values for 8000 points in time during each second. It is not defined for any other values of time. If we save this sampled speech we can think of this signal as a function of integer indices independent of time. The first sample would be $x[1]$, the second sample would be $x[2]$, etc. We can process this speech sample without any reference to the "actual" time. Only when we need to replay this signal do we have to be concerned with the "actual" time again. In this class we will examine and analyze both types of signals and the systems that operate on them.

2.1.2 EVEN AND ODD FUNCTIONS

Analysis means to break things up into its constituent parts; to decompose a complex problem into simpler components. In this class we will look at different ways of taking a signal and breaking it up. Most of these approaches will require breaking things up into an infinite number of constituents. Before we head off to infinity lets take a look at a very simple decomposition which will be useful to us: the decomposition of a signal into an even function and an odd function.

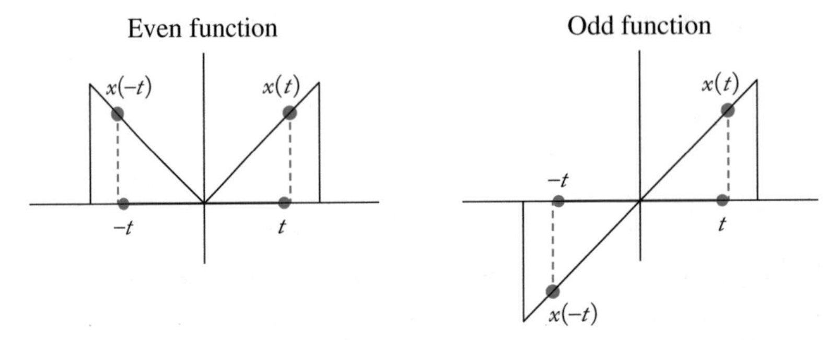

Figure 2.7: An even function and an odd function.

A function $x(t)$ is said to be even if

$$x(-t) = x(t)$$

It is said to be odd if

$$x(-t) = -x(t)$$

As we can see from Figure 2.7 an even function is one which the function on both sides of the vertical axis are mirror images of each other. This should have been evident from the example we did earlier with $x(-t)$. We can also think of an odd function in terms of symmetry. But this time the mirror would be placed at a $-45°$ angle to the vertical axis. Thus, what is positive on one side of the axis becomes negative on the other side and vice versa.

There are some properties of even and odd functions that at times come in handy. One such property is that the integral of a continuous odd function over a symmetric interval is zero while the integral of a continuous (discrete) even function is equal to twice its integral over the positive or negative half of the integral. This means that if we could decompose a signal into an even and an odd part its integral over a symmetric interval would be equal to twice the integral of just the even part over half the interval. Of course in order to do this we would have to decompose our signal into an even and an odd function. It turns out that we can always decompose any signal into an even and odd part. Let's call the even component of $x(t)$, $x_e(t)$, and the odd part of $x(t)$, $x_o(t)$. Then we can obtain $x_e(t)$ and $x_o(t)$ by

$$\begin{aligned} x_e(t) &= \frac{x(t) + x(-t)}{2} \\ x_o(t) &= \frac{x(t) - x(-t)}{2} \end{aligned}$$

We can check that $x_e(t)$ is really even and $x_o(t)$ is really odd by seeing if they satisfy the definition of an even function and an odd function.

$$\begin{aligned} x_e(-t) &= \frac{x(-t)+x(+t)}{2} \\ &= \frac{x(t)+x(-t)}{2} \\ &= x_e(t) \end{aligned}$$

$$\begin{aligned} x_o(-t) &= \frac{x(-t)-x(+t)}{2} \\ &= -\frac{x(t)-x(-t)}{2} \\ &= -x_o(t) \end{aligned}$$

If we add the even and odd parts back we should get back the original function.

$$\begin{aligned} x_e(t)+x_o(t) &= \frac{x(t)+x(-t)}{2}+\frac{x(t)-x(-t)}{2} \\ &= \frac{x(t)+x(-t)+x(t)-x(-t)}{2} \\ &= \frac{2x(t)}{2} \\ &= x(t) \end{aligned}$$

Let's find the even and odd parts of

$$x(t) = 3t + 2$$

$$\begin{aligned} x_e(t) &= \frac{3t+2-3t+2}{2} = 2 \\ x_o(t) &= \frac{3t+2+3t-2}{2} = 3t \end{aligned}$$

Clearly adding $x_e(t) = 2$ and $x_o(t) = 3t$ will give us back $x(t) = 3t + 2$.

2.2 SUMMARY

In this module we looked at functions as rules for generating an output given an input. Keeping this view of functions in mind will help you in what follows.

2.3 EXERCISES

(Answers on the following page)

1. Given the function $x(t)$

$$x(t) = t^2 + 2$$

   (a) What is $x(3)$?

   (b) What is $x(-100)$?

   (c) What is $x(2t + 1)$?

2. Given the signal

$$x(t) = \begin{cases} t & 0 \leq t \leq 1 \\ 0 & \text{otherwise} \end{cases}$$

   Find $x(2 - t)$

3. Given the signal

$$x[n] = \begin{cases} \frac{1}{n} & n = 1, 2, 3 \\ 0 & \text{otherwise} \end{cases}$$

   Find $x[2 - n]$

4. Find the even and odd components of $x(t) = e^{j3t}$?

5. Which of these are even, which of these are odd, and which of these are neither even nor odd?

   (a) $x(t) = t^2$

   (b) $x(t) = t$

   (c) $x(t) = \cos(t)$

   (d) $x(t) = \sin(t)$

   (e) $x(t) = e^{j3t}$

   (f) $x(t) = 3t + 2$

   (g) $x[n] = 4n$

   (h) $x[n] = 2n - 2$

6. Given

   (a) Plot $x(t)$

   (b) Plot $x(t - 1)$

(c) Plot $x(1-t)$

7. Find the even and odd parts of the following functions

   (a) $x(t) = 1 - e^{-2t}$

   (b) $x(t) = 1 - e^{-2t^2}$

   (c) $x(t) = 3t^3$

   (d) $x(t) = e^{j2\pi t}$

   (e) $x(t) = \cos(2\pi t)u(t)$

   (f) $x(t) = \sin(2\pi t)u(t)$

2.4 ANSWERS

1. (a) 11
   (b) 10,002
   (c) $4t^2 + 4t + 3$

2. $$x(2-t) = \begin{cases} 2-t & 1 \leq t \leq 2 \\ 0 & \text{otherwise} \end{cases}$$

3. Given the signal
$$x[2-n] = \begin{cases} \dfrac{1}{2-n} & n = -1, 0, 1 \\ 0 & \text{otherwise} \end{cases}$$
(If you plot both $x[n]$ and $x[2-n]$ you might find it instructive.)

4. $\cos(3t)$, $j\sin(3t)$

5. (a) even
   (b) odd
   (c) even
   (d) odd
   (e) neither even nor odd
   (f) neither even nor odd
   (g) odd
   (h) neither even nor odd

6. See Figure 2.8.

7. (a)
$$\begin{aligned} x_e(t) &= 1 - \frac{1}{2}\left(e^{2t} + e^{-2t}\right) \\ x_o(t) &= \frac{1}{2}\left(e^{2t} - e^{-2t}\right) \end{aligned}$$
   (b)
$$\begin{aligned} x_e(t) &= 1 - e^{-2t^2} \\ x_o(t) &= 0 \end{aligned}$$

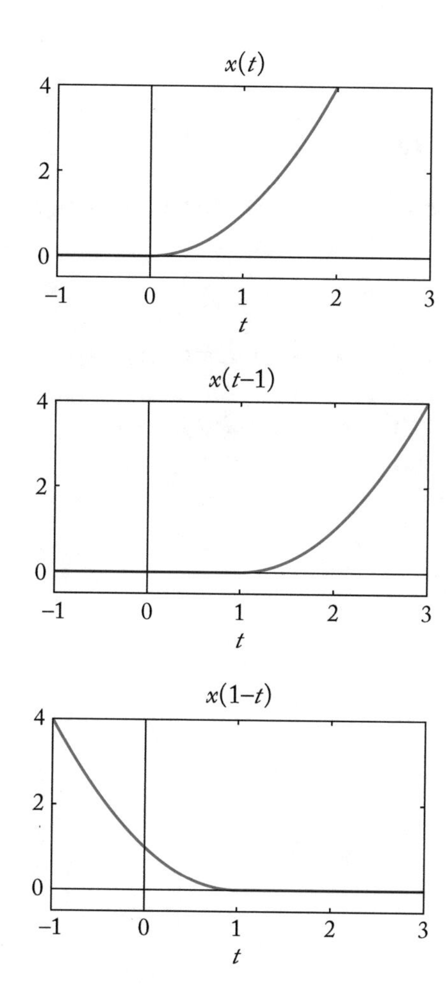

Figure 2.8: Plots of $x(t)$, $x(t-1)$, and $x(1-t)$.

(c)

$$
\begin{aligned}
x_e(t) &= 0 \\
x_o(t) &= 3t^3
\end{aligned}
$$

(d)

$$
\begin{aligned}
x_e(t) &= \cos(2\pi t) \\
x_o(t) &= j\sin(2\pi t)
\end{aligned}
$$

(e)

$$
\begin{aligned}
x_e(t) &= \frac{1}{2}\cos(2\pi t) \\
x_o(t) &= \frac{1}{2}(\cos(2\pi t)u(t) - \cos(2\pi t)u(-t))
\end{aligned}
$$

(f)

$$
\begin{aligned}
x_e(t) &= \frac{1}{2}(\sin(2\pi t)u(t) - \sin(2\pi t)u(-t)) \\
x_o(t) &= \frac{1}{2}\sin(2\pi t)
\end{aligned}
$$

MODULE 3

Special Functions

There are some special functions that are going to be particularly useful to us when we develop tools for the analysis of signals. The two most important special functions we will be using are the step function and the delta function (also known as the impulse function). There is both a discrete and a continuous version of each.

3.1 STEP FUNCTION

The discrete unit step function has a value of one when its argument is greater than or equal to zero, and it has a value of zero when its argument is less than zero.

$$u[n] = \begin{cases} 1 & n \geq 0 \\ 0 & n < 0 \end{cases}$$

The continuous time unit step function is defined in a similar manner

$$u(t) = \begin{cases} 1 & t > 0 \\ 0 & t < 0 \end{cases}$$

The only difference is that we will be agnostic about the value of $u(t)$ at $t = 0$ (Figure 3.1).

Remember that these are functions and therefore, they have an input (whatever is in the parentheses) and an output (the value of the function). So given that

$$u(t) = \begin{cases} 1 & t > 0 \\ 0 & t < 0 \end{cases}$$

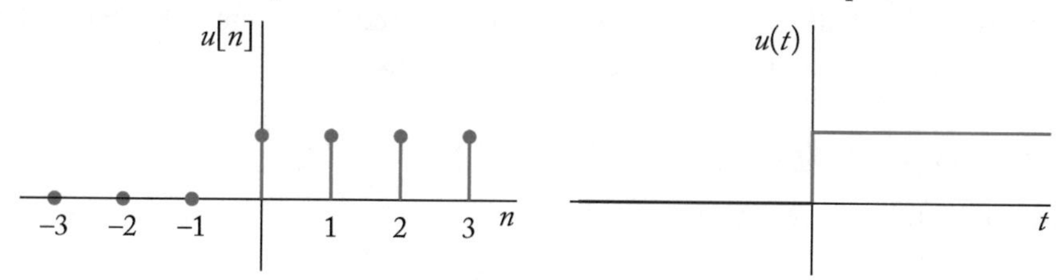

Figure 3.1: Discrete and continuous time step functions.

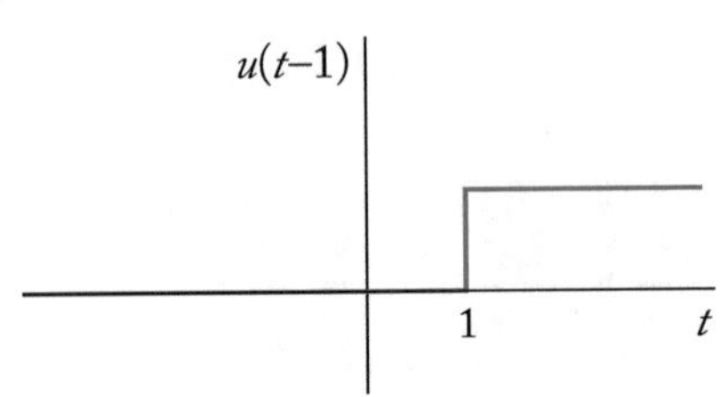

Figure 3.2: A step function shifted to the right by 1.

what is $u(t-1)$? In order to find this we substitute $t-1$ wherever we find t in the original definition.

$$u(t-1) = \begin{cases} 1 & t-1 > 0 \\ 0 & t-1 < 0 \end{cases}$$

or

$$u(t-1) = \begin{cases} 1 & t > 1 \\ 0 & t < 1 \end{cases}$$

We can plot this as shown in Figure 3.2.

Many of the signals we work with start at a particular time, prior to which they are zero. The unit step function allows us to write these functions using a shorthand notation. So, instead of

$$x(t) = \begin{cases} t & t > 0 \\ 0 & t < 0 \end{cases}$$

we can write

$$x(t) = tu(t)$$

Notice we are multiplying two functions. We do this by multiplying them point by point as shown in Figure 3.3. So for all values of t less than zero the function t is multiplied by zero and for values of t greater than 0 we multiply the function t by 1.

3.2 DELTA FUNCTION

The delta function or the impulse function is a function that has a nonzero value at only one point; the point where the argument of the function is zero.

3.2.1 DISCRETE TIME UNIT IMPULSE FUNCTION

In the discrete case this is a very simple function.

$$\delta[n] = \begin{cases} 1 & n = 0 \\ 0 & n \neq 0 \end{cases}$$

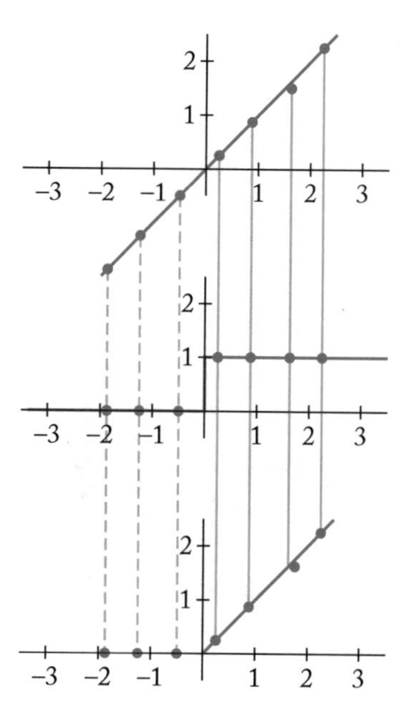

Figure 3.3: Multiplying t with the unit step function $u(t)$.

For all of its simplicity this is a function which can be used in a lot of interesting ways. We can obtain the unit step function from the delta function as

$$u[n] = \sum_{k=-\infty}^{n} \delta[k]$$

Think a bit about why this is true. The process will help you understand both functions and the ways we will use them.

We can also obtain the delta function from the unit step function by

$$\delta[n] = u[n] - u[n-1]$$

as shown in Figure 3.4.

We can also use the delta function and its shifted versions to write discrete functions in a compact way. In the previous module we had an example of a discrete function

$$x[n] = \begin{cases} 1 & n = 0 \\ -1 & n = 1 \\ 3 & n = 2 \\ 2 & n = 4 \\ 0 & \text{all other integers} \end{cases}$$

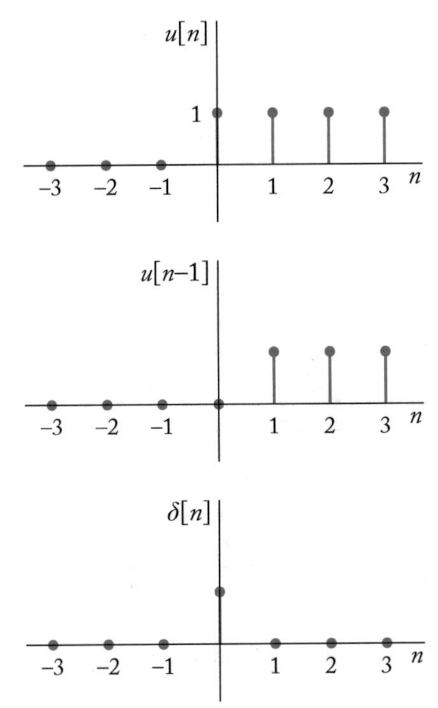

Figure 3.4: $\delta[n] = u[n] - u[n-1]$.

If we plot this function we see that it consists of a delta function at $n = 0$, -1 times a delta function at $n = 1$, 3 times a delta function at $n = 2$, and 2 times a delta function at $n = 4$. We can write this in a much more compact manner as

$$x[n] = \delta[n] - \delta[n-1] + 3\delta[n-2] + 2\delta[n-4]$$

We could also have written this as

$$x[n] = x[0]\delta[n] + x[1]\delta[n-1] + x[2]\delta[n-2] + x[4]\delta[n-4]$$

In general rather than write the value of $x[n]$ for each different value of n we can write

$$x[n] = \sum_{k=-\infty}^{\infty} x[k]\delta[n-k]$$

For this particular function most of the terms in the summation will be zero as $x[n]$ is zero for all n other than $n = 0, 1, 2$, and 4.

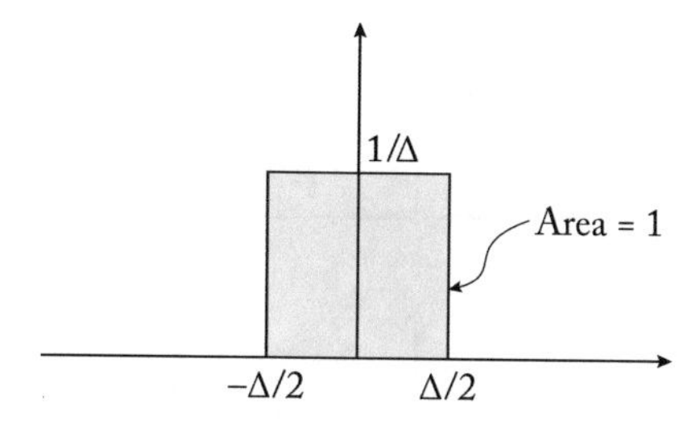

Figure 3.5: The function $\delta_\Delta(t)$.

3.2.2 CONTINUOUS TIME DELTA FUNCTION OR THE DIRAC DELTA FUNCTION

While the discrete time delta function is really simple the continuous time delta function is not. In fact it is not even clear that it is a function. It is better known by its properties than by a functional relationship. In fact, we can really only define it as a limit. Let δ_Δ be a rectangular function

$$\delta_\Delta = \begin{cases} \dfrac{1}{\Delta} & -\dfrac{\Delta}{2} < t < \dfrac{\Delta}{2} \\ 0 & \text{otherwise} \end{cases}$$

We plot this in Figure 3.5. Notice that regardless of the value of Δ the area under the curve $\delta_\Delta(t)$ is always 1. The continuous time delta function, or the impulse function is defined as the limit as Δ goes to zero of $\delta_\Delta(t)$.

$$\delta(t) = \lim_{\Delta \to 0} \delta_\Delta(t)$$

As Δ becomes smaller and smaller the function $\delta_\Delta(t)$ will become narrower and narrower. At the same time as the height of $\delta_\Delta(t)$ is $1/\Delta$ the function will become taller and taller. In the limit as Δ approaches zero the width of this function will go to zero and the height will go to infinity. This rather strange function is the continuous time delta function.

This function is zero everywhere except at 0. And at zero it has a value approaching infinity. You can see why we have difficulty talking about the continuous time delta function as a function. Instead we generally focus on its properties which are, after all, what makes the function useful for us. First, as we noted before, regardless of how small Δ becomes the area under the rectangle and hence the integral of this function is $\Delta \times \frac{1}{\Delta}$ or 1. Therefore, the integral of the continuous time delta function is 1. Of course, in order for this to be true, the integral limits have to include the one point where the delta function is nonzero. We can make use of this property in a number of different ways. Just as in the case of the discrete delta function we can obtain the unit step

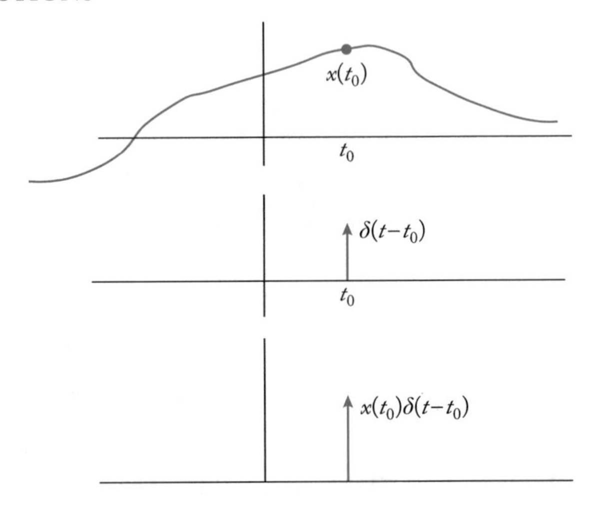

Figure 3.6: Sifting.

function from the delta function by using this property.

$$u(t) = \int_{-\infty}^{t} \delta(\tau) d\tau$$

As long as t is less than zero the region of integration does not include the one nonzero point of the delta function and the integral is zero. When t is greater than zero the region of integration includes the nonzero point of the function and the integral is one.

We can use the delta function to pick out particular values of other functions. By integrating any function with the delta function shifted to some point t_0, we can find, or sift out, the value of the function at t_0 (Figure 3.6).

$$\int_{-\infty}^{\infty} x(t)\delta(t - t_0)dt = x(t_0)$$

This is called the *sifting property* of the delta function. To see how this works consider the product $x(t)\delta(t - t_0)$. The value of $\delta(t - t_0)$ is zero everywhere except at $t = t_0$ so the product $x(t)\delta(t - t_0)$ is zero everywhere except at $t = t_0$. At $t = t_0$ the value of $x(t)$ is $x(t_0)$ so the product at $t = t_0$ of $x(t)$ and $\delta(t - t_0)$ is $x(t_0)\delta(t - t_0)$. Or, $x(t)\delta(t - t_0) = x(t_0)\delta(t - t_0)$ and

$$\int_{-\infty}^{\infty} x(t)\delta(t - t_0)dt = \int_{-\infty}^{\infty} x(t_0)\delta(t - t_0)dt = x(t_0) \int_{-\infty}^{\infty} \delta(t - t_0)dt = x(t_0)$$

where we have used the fact that

$$\int_{-\infty}^{\infty} \delta(t - t_0)dt = 1$$

If we change the order of t_0 and t we would get the same result. This is because the delta function is zero everywhere except when its argument is zero and the arguments $t - t_0$ and $t_0 - t$ are both zero when $t = t_0$. We can use this sifting property in a way that seems redundant but will be very helpful later on. Just like we used the discrete delta function to provide a compact representation of $x[n]$ we can use the continuous delta function to provide a representation of $x(t)$.

$$x(t) = \int_{-\infty}^{\infty} x(\tau)\delta(t-\tau)d\tau$$

The only difference between this equation and the previous equation are the variable symbols. Instead of t_0 in the previous equation we have used t and the variable of integration has changed from t to τ. So just as

$$\int_{-\infty}^{\infty} x(t)\delta(t_0-t)dt = x(t_0)$$

we have

$$\int_{-\infty}^{\infty} x(\tau)\delta(t-\tau)d\tau = x(t)$$

3.3 SUMMARY

In this module we introduced two very special functions, the unit step function, and the delta function. We also saw how we can use these special functions to obtain compact representations of many functions. In particular using the delta function we can write a discrete time function $x[n]$ as

$$x[n] = \sum_{k=-\infty}^{\infty} x[k]\delta[n-k]$$

and a continuous time function as

$$x(t) = \int_{-\infty}^{\infty} x(\tau)\delta(t-\tau)d\tau$$

3.4 EXERCISES

(Answers on the following page)

1. Plot the following:

   (a) $x(t) = u(t) - u(t-2)$

   (b) $x(t) = u(t+2) - u(t+1)$

   (c) $x[n] = u[n] - u[n-2]$

   (d) $x[n] = u[n+2] - u[n+1]$

   (e) $x(t) = u(t) + u(t-1) - 2u(t-2)$

2. Evaluate the integral

$$\int_{-\infty}^{\infty} e^{-t} u(t) dt$$

3. Evaluate the integral

$$\int_{-\infty}^{t} e^{-\tau} u(\tau) d\tau$$

4. Evaluate the sum

$$\sum_{n=-\infty}^{\infty} (0.5)^n u[n]$$

5. Evaluate the sum

$$\sum_{n=-\infty}^{\infty} (0.5)^n \delta[n-1]$$

6. Evaluate the integral

$$\int_{-\infty}^{\infty} t e^{-t} \delta(t-1) dt$$

7. Evaluate the integral

$$\int_{-\infty}^{\infty} t \delta(t-3) dt$$

8. Evaluate the integral

$$\int_{0}^{4} t \delta(t-3) dt$$

9. Evaluate the integral

$$\int_{0}^{2} t \delta(t-3) dt$$

10. Evaluate the integral

$$\int_{-\infty}^{\infty} [u(t) - u(t-1)]dt$$

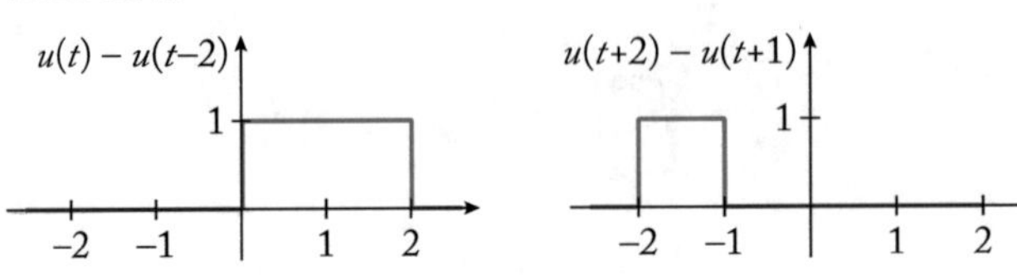

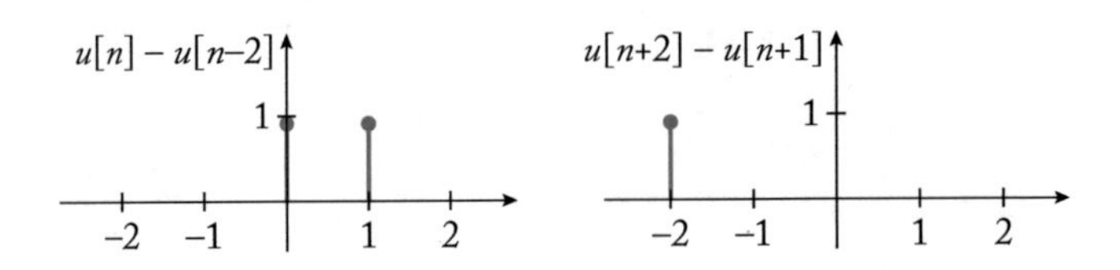

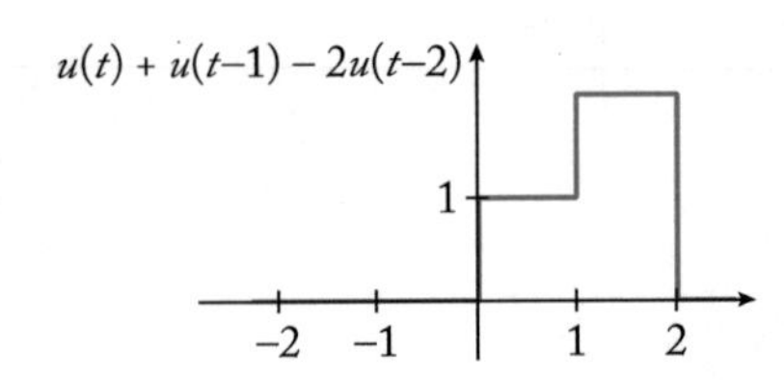

Figure 3.7

3.5 ANSWERS

1. Plot Figure 3.7.
2. 1
3. $1 - e^{-t}u(t)$
4. 2.0
5. 0.5
6. $1/e$
7. 3
8. 3
9. 0
10. 1

MODULE 4

Classification of Systems

Shakespeare did not say, all the world's a system and the signals merely inputs and outputs but, while it sounds silly the statement is not entirely bonkers. All around us we see systems, both natural and manufactured. Closest to us, our body is an incredibly complex system of systems with the different systems communicating with each other through chemical and electrical signals. Clearly, putting together the mathematical machinery required to handle all these systems would be close to impossible and ultimately self defeating. We want tools that are practical, easy to use, and still give us an insight into systems we are trying to analyze or design. The tools that would be able to handle the huge diversity of systems all around us would themselves need to be so complex as to be impractical. In order to develop practical tools for the analysis of systems what we need to do is to see if we can identify a group of systems which is broad enough to be useful and for the analysis of which simple tools can be developed. In order to define this set of systems we need to develop some terminology to describe systems in general. That is what we will do in this module. Our general framework is that of a system which has an input—which we will denote by $x(t)$ or $x[n]$ and an output which we will denote by $y(t)$ or $y[n]$. With this framework let us look at some properties of systems

4.1 MEMORY

We say a system has memory if the output depends not only on the input at the current time but also on the input that occurred in the past or will occur in the future. Otherwise the system is said to be memoryless. The current and past times are defined in terms of the index of the output. So if the output is $y(t)$ then the input $x(t)$ is the current input, the input $x(t-1)$ is the value of the input one unit of time in the past and $x(t+1)$ is the value of the input one unit of time in the future. Consider the system

$$y(t) = 10x(t)$$

If I wanted to know the value of the output $y(t)$ at time $t = 3$, or $y(3)$ we would only need the value of the input $x(t)$ at time $t = 3$, or $x(3)$. We wouldn't need $x(3.0001)$ or $x(2.9999)$. We wouldn't need to "remember" any value of the input. We would only need the current value of the input. This lack of the requirement for remembering is why we call a system memoryless. Other examples of memoryless systems include

$$y[n] = 3x[n]^2$$

$$y(t) = \cos(x(t))$$

In each case notice that in order to know the output at a particular time we only need the input at that particular time. The system does not have to "remember" the past or (and this sounds a bit weird) the future.

So, what does a system with memory look like. Here is an example. The system with an input/output relationship

$$y[n] = x[n] - x[n-1]$$

is not a memoryless system because it uses the value of the input one time instant in the past. The output at time $n = 3$ is given by

$$y[3] = x[3] - x[3-1] = x[3] - x[2]$$

So in order to find the output at time $n = 3$ we need the value of the input at both time $n = 3$ and at time $n = 2$. The system has to remember the value of the input one time instant prior to the current time instant, therefore, it has to have memory. Another system with memory is the system with the input/output relationship

$$y(t) = \frac{1}{2}(x(t+1) - x(t-1))$$

Here the system has to remember not only an input from one time unit in the past, it also has to "remember" the input from one time unit in the future.

Another example is an operation that you use a lot—the derivative operation. The derivative is not a memoryless operation. Think of the definition of the derivative:

$$\frac{dx}{dt} = \lim_{\Delta t \to 0} \frac{x(t+\Delta t) - x(t)}{\Delta t}$$

To differentiate a function $x(t)$ we need the current value of the function and the value of the function just after the current moment. Therefore, systems that implement a derivative—think of a differentiator—are not memoryless systems. Neither is another system we use a lot—an integrator. Integration is not a memoryless operation.

$$y(t) = \int_{-\infty}^{t} x(\tau)d\tau$$

In order to evaluate this integral we not only need the value of $x(\ )$ at time t, we need its values at all times prior to t. Therefore, integrators are not memoryless systems.

In terms of simple circuits, purely resistive circuits are memoryless systems. Circuits that contain capacitors or inductors are systems with memory.

4.2 INVERTIBILITY

If we can recover the input exactly from the output the system is said to be invertible. The system described by the input/output relationship

$$y(t) = 10x(t)$$

is invertible as we can obtain the input $x(t)$ given the output $y(t)$.

$$x(t) = \frac{1}{10}y(t)$$

The system described by the input/output relationship

$$y[n] = x(n)^2$$

is not invertible. If we are told that $y[n] = 4$ we cannot say if $x[n] = 2$ or $x[n] = -2$. Neither is the system described by the input/output relationship

$$y(t) = \cos(x(t))$$

If we know $y(t)$ we only know $x(t)$ within 2π. For example, if we know that $y(t) = 1/\sqrt{2}$, $x(t)$ could be $\pi/4$ or 2.25π, or 4.25π, and so on. Therefore, the this system is not invertible without making some assumptions.

For this case the invertibility or the lack of it was fairly obvious. In other cases it may not be that easy to see whether a system is invertible. Consider the summer

$$y[n] = \sum_{k=-\infty}^{n} x[k]$$

This looks fairly complicated and at first glance you may be tempted to say that this system is not invertible even though it is. One reason for this confusion might be that in the examples we have been talking about knowledge of the output seems to be the knowledge of the output at only the current time. Actually, when we say we have knowledge of the output we mean we have knowledge of the output for all time. Therefore, in this particular case, we not only know $y[n]$ we also know $y[n-1]$, and given these two we can find the value of $x[n]$. Consider,

$$\begin{aligned} y[n] = \sum_{k=-\infty}^{n} x[k] &= \cdots + x[-1] + x[0] + \cdots + x[n-1] + x[n] \\ y[n-1] = \sum_{k=-\infty}^{n-1} x[k] &= \cdots + x[-1] + x[0] + \cdots + x[n-1] \end{aligned}$$

Subtracting the lower sum from the top one we get

$$y[n] - y[n-1] = x[n]$$

In other words, given the output values we can obtain the input values.

The idea of invertibility becomes particularly important when we are trying to understand the behavior of systems we haven't designed. For example, we can understand that somebody has atrial fibrillation by observing the ECG—which in some sense reflects the output of the heart, but what we may be really interested in is what was causing the atrial fibrillation—in other words what inputs to the system which is the heart were causing the atrial fibrillation. Similar problems occur when we are trying to investigate geophysical systems, or even economic systems.

4.3 CAUSALITY

A system is said to be causal if the output is not based on the future values of the input. So,

$$y[n] = x[n] + y[n-1]$$

describes a causal system, while

$$y(t) = \frac{1}{2}(x(t+1) - x(t-1))$$

does not.

When we discussed memoryless systems the past and future were interchangeable. A system which needed to remember the past to generate an output needed memory as did a system which used a future value to generate an output. When we speak of causality we differentiate between past and future values. A system which does not use future values of the input to generate the current output is a causal system. The two systems described above are simple examples of causal and non-causal systems.

The term causality is a somewhat loaded term here. In our day to day life we generally always have the cause of something precede its effect. You hit the snooze button and the alarm stops. You open your hand and the cup you were holding drops to the floor. You turn the key in the ignition and the car starts. However, in terms of engineering, causality is not a necessary requirement for systems. Think of an image processing system in which you smooth out the image by replacing each pixel by the average of all the neighboring pixels. If we think in terms of a raster scan the neighboring pixels include pixels that occur prior to the current pixel as well as the pixel that occurs after the current pixel. So the smoothing system uses information from the "future" to process the current pixel.

Somewhat more complex are some of the systems you use everyday. When you are using your cell phone you are using a non-causal system. The cellphone does not transmit your voice samples. Instead it breaks up the voice samples into *frames* or sets of samples which are used to generate a model for the speech. This model, along with some other information is used to regenerate the speech segment at the receiver. In order to regenerate each sample of a frame the receiver is using the model which was created using all the samples in the frame—past, current, and future. The cellphone thus is a non-causal system. When you view digital videos you are generally using a non-causal system. Generally the video you see consists of three different kinds of frame, I frames P frames and B frames. The I frames can be thought as simply images. You could print one out and you would have a picture. A P frame is actually just the difference between pixel values of a a previous frame and the current frame and a B frame is the difference in pixel values between the current frame and *past and future* values. Because you need to use future values to reconstruct the current frame the system is non-causal.

Note that we are only talking about future values of the *input*. An input/output relationship given by

$$y(t) = (t+1)x(t)$$

is a causal system. The $t+1$ in the equation has nothing to do with the input. At time $t = 2$ you can figure out the value of $t+1$ without any need for precognition. Remember to keep this in mind when you are evaluating a system to figure out whether it is causal.

4.4 STABILITY

There are a number of ways one can define stability. The definition we use here is that a system is stable if for a bounded input the output is always bounded—this is also known as Bounded Input Bounded Output (BIBO) stability. A more formal way of saying this is if you are guaranteed that there is a finite constant B such that

$$|x(t)| < B \quad \text{for all} \quad t$$

then you can find a finite value C such that

$$|y(t)| < C \quad \text{for all} \quad t$$

So which of these is BIBO stable?

$$y(t) = 10x(t)$$

$$y[n] = e^{x[n]}$$

$$y(t) = \frac{1}{x(t)}$$

$$y(t) = \int_{-\infty}^{t} x(\tau)d\tau$$

Let's take each in turn. In the case of the first system

$$y(t) = 10x(t)$$

if we are guaranteed that $|x(t)| < B$ for some finite B, we can pick $C = 10B$ and we are guaranteed that $|y(t)|$ will be less than C for all t. You tell me that $|x(t)|$ will be less than 15 for all values of t and I can give you a guarantee that $|y(t)|$ will be less than 150 for all values of t.

For the system in which

$$y[n] = e^{x[n]}$$

If $x[n]$ is guaranteed to be less than B, we can pick $C = e^B$ and guarantee that $|y[n]|$ is less than C. So this system is also BIBO stable.

This is not the case for the third system defined by

$$y(t) = \frac{1}{x(t)}$$

No matter what value of C you pick we can find a value of $x(t)$ small enough (and hence less than any bound B) such that $1/x(t)$ is greater than any given value of C.

Our last system is also not BIBO stable. Let's say you give me a guarantee that $|x(t)|$ is less than some number B. Then

$$\begin{aligned} |y(t)| &< \int_{-\infty}^{t} B d\tau \\ &= B \int_{-\infty}^{t} d\tau \\ &\rightarrow \infty \end{aligned}$$

The infinite range of the equation makes it impossible to create any guarantee on the upper bound of $|y(t)|$. You can say that this is not reasonable because a practical integrator will never be integrating from $-\infty$. Also infinity is a weird concept, it can lead to strange result. As George Gamow well knew when he wrote:

> There was a young fellow from Trinity,
> Who took the square root of infinity.
> But the number of digits, Gave him the fidgets;
> He dropped Math and took up Divinity.

So let's modify the input/output relationship to

$$y(t) = \int_{0}^{t} x(\tau) d\tau$$

Again, let's assume there is a number B such that $|x(t)| < B$ for all t then

$$
\begin{aligned}
|y(t)| &< \int_0^t B d\tau \\
&= B \int_0^t d\tau \\
&= tB
\end{aligned}
$$

At first sight that doesn't look too bad. To see why this does not rescue this particular system from instability let's use some numbers. Let's say $B = 10$. Let's suppose we pick $C =$ 1,000,000. That is we are saying that as long as $|x(t)|$ is less than 10, we guarantee that $|y(t)|$ will be less than 1,000,000. All we have to do to break this guarantee is to pick $t = 100{,}001$! So even without the lower limit of infinity the integrator is not BIBO stable.

A system not being BIBO stable does not make it unusable. Integrators, after all, are an integral (no pun intended—well, maybe just a little bit) part of many many systems. But, when you are designing a large system with many subsystems it is useful to know when you are using something that can go unstable.

4.5 SUMMARY

In this module we defined four properties of systems:

1. **Memory:** A system has memory if it does not use values of future and past inputs to generate the current output.
2. **Invertible:** A system is invertible if we can recover the input given the output.
3. **Causality:** A system is causal if it does not use future values of the input to generate the current output.
4. **Stability:** A system is BIBO stable if when the input is guaranteed to be bounded by a finite value we can give a guarantee of a finite bound for the output.

4.6 EXERCISES

(Answers on the following page)

1. Are the following systems memoryless?

   (a) $y[n] = x[n] + y[n-1]$

   (b) $y(t) = e^{x(t)}$

   (c) $y(t) = \log(x(t))$

   (d) $y(t) = \cos(x(t)) + \sin(x(t-1))$

2. Is the system described by the following input-output relationship invertible

$$y[n] = x[n] + y[n-1] \quad y[n] = 0 \text{ for } n < 0?$$

3. Are the systems described by the following input output relationships causal?

   (a) $y[n] = x[-n]$

   (b) $y(t) = x(t)\cos(t+1)$

4. Are the following systems BIBO stable?

   (a) $y(t) = e^{x(t)}$

   (b) $y[n] = x[n+4] - x[n-4]$

   (c) $y(t) = t^2 x(t)$

   (d) $y[n] = nx[n]$

5. Determine whether the following systems are (i) causal, (ii) BIBO stable, (iii) memoryless, and (iv) invertible. In each case show your reasoning. In each case $x(t)$ or $x[n]$ is the input and $y(t)$ or $y[n]$ is the output.

   (a) $y(t) = ax(t)$ where a is a constant

   (b) $y(t) = x(t)\sin(t)$

   (c) $y(t) = \frac{1}{x(t)}$

   (d) $y(t) = \int_{-\infty}^{t} x(\tau)d\tau$

   (e) $y[n] = 2x[n] - x[n-1]$

   (f) $y[n] = \sum_{k=-\infty}^{n} x[k+1]$

   (g) $y(t) = tx(t-2)$

   (h) $y(t) = \cos(x(t))$

Table 4.1: Causal, BIBO statle, memoryless, and invertible systems

Input/Output Relationship	Causal	BIBO Stable	Memoryless	Invertible
$y(t) = 3x(t) + 2$				
$y[n] = nx[n]$				
$y(t) = \int_{-\infty}^{2t} x(\tau)d\tau$				
$y[n] = x[n-2] - 2x[n-8]$				

6. Determine whether the following systems are causal, BIBO stable, memoryless, and invertible. In each case $x(t)$ or $x[n]$ is the input and $y(t)$ or $y[n]$ is the output (see Table 4.1).

Table 4.2: Causal, BIBO statle, memoryless, and invertible systems

Input/Output Relationship	Causal	BIBO Stable	Memoryless	Invertible
$y(t) = 3x(t) + 2$	Yes	Yes	Yes	Yes
$y[n] = nx[n]$	Yes	No	Yes	Yes
$y(t) = \int_{-\infty}^{2t} x(\tau)d\tau$	No	No	No	Yes
$y[n] = x[n-2] - 2x[n-8]$	Yes	Yes	No	Yes

4.7 ANSWERS

1. (a) No
   (b) Yes
   (c) Yes
   (d) No
2. Yes
3. (a) No
   (b) Yes
4. (a) Yes
   (b) Yes
   (c) No
   (d) No
5. (a) Causal, BIBO Stable, Memoryless, Invertible
   (b) Causal, BIBO Stable, Memoryless, Invertible
   (c) Causal, Not BIBO Stable, Memoryless, Invertible
   (d) Causal, Not BIBO Stable, Not Memoryless, Invertible
   (e) Causal, BIBO Stable, Not Memoryless, Invertible
   (f) Not Causal, Not BIBO Stable, Not Memoryless, Invertible
   (g) Causal, Not BIBO Stable, Not Memoryless, Invertible
   (h) Causal, BIBO Stable, Memoryless, Not Invertible (see Table 4.2)

MODULE 5

Linearity and Time Invariance

We have left the two most important properties of systems—the properties which we will use to define the set of systems which will be our focus throughout this course for last. These properties are Linearity and Time Invariance.

5.1 LINEARITY

The linearity property is one of the most important properties we will discuss. It is this property that allows you to use superposition when you solve circuits. It is this property together with time invariance which will allow us to develop a mathematical characterization of many systems of interest. And it is very straightforward. The linearity property actually consists of two properties scaling and additivity.

5.1.1 SCALING

The scaling property (also known as the homogeneity property) states that in systems that have this property if we scale an input by a certain amount the output gets scaled by the same amount. Symbolically we could write this property as; if

$$x(t) \Rightarrow y(t)$$

then

$$\alpha x(t) \Rightarrow \alpha y(t)$$

So, if we double the input, pick $\alpha = 2$, the output gets doubled. Most importantly if we pick α to be zero, that is we zero out the input, the output gets zeroed out as well. So the system given by the input/output equation

$$y(t) = 2x(t) + 3$$

does not satisfy the scaling property. In order to see this let's represent this system as shown in Figure 5.1.

Whatever we put into the system the system multiplies it by 2 and then adds 3 to the product. So if $x(t)$ is the input the output is

$$y(t) = 2 \times [x(t)] + 3 = 2x(t) + 3$$

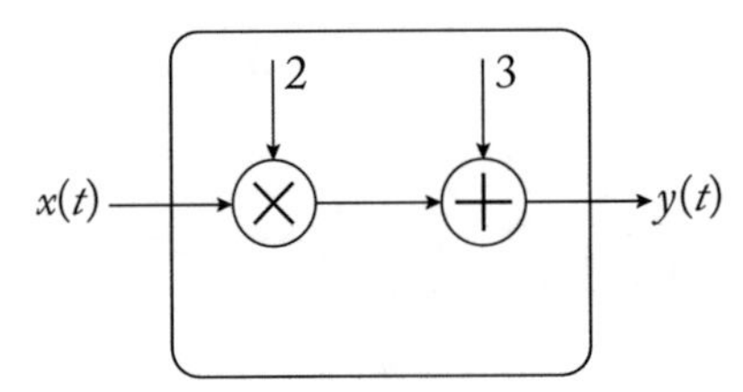

Figure 5.1: A system with input/output relationship: $y(t) = 2x(t) + 3$.

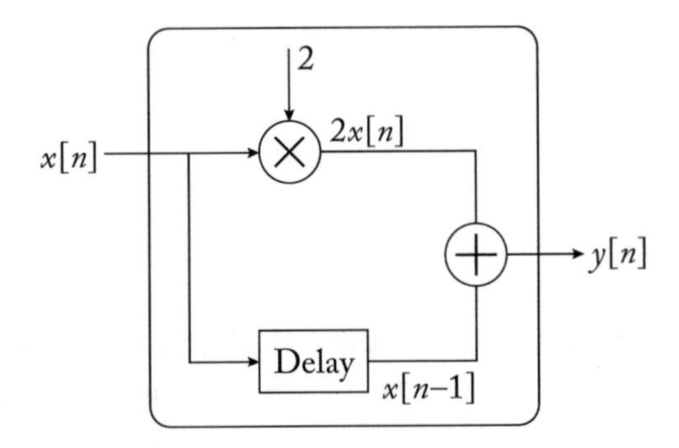

Figure 5.2: A system with input/output relationship: $y[n] = 2x[n] + x[n-1]$.

If we now multiply the input with α to get $\alpha x(t)$ as the new input to the system, the system operates in the same way. The output is given by

$$2 \times [\alpha x(t)] + 3 = 2\alpha x(t) + 3$$

Notice that this is not what we would have obtained if we had multiplied $y(t)$ with α. If we did that we would have obtained

$$\alpha y(t) = \alpha\,[2x(t) + 3] = 2\alpha x(t) + 3\alpha$$

Clearly

$$\alpha x(t) \not\Rightarrow \alpha y(t)$$

So lets take a look at a system that does satisfy the scaling property. Let the input output relationship of a discrete time system be

$$y[n] = 2x[n] + x[n-1]$$

shown schematically in Figure 5.2

If $x[n]$ is the input to the system, from the upper arm we will get $2 \times [x[n]]$ and from the lower arm we will get $x[n]$ delayed by one unit, or $x[n-1]$. The adder adds these two terms and we get $y[n]$ as shown above.

Now let's replace $x[n]$ with $\alpha x[n]$ in the upper arm we will get $2\alpha x[n]$ and from the lower arm we will get $\alpha x[n]$ delayed by one unit or $\alpha x[n-1]$. Adding the two terms we get

$$2\alpha x[n] + \alpha x[n-1]$$

which is exactly $\alpha y[n]$. So this system satisfies the scaling property.

Let's look at a couple of more examples. Let's suppose we have a squarer.

$$y[n] = (x[n])^2$$

If we multiplied $y[n]$ by α we would get

$$\alpha y[n] = \alpha\,(x[n])^2$$

What would happen if we put $\alpha x[n]$ into the squarer? We would get

$$(\alpha x[n])^2 = \alpha^2\,(x[n])^2$$

which is not the same as $\alpha y[n]$.

Let's consider a hard limiter—a useful system in many applications. We can model this with the input output relationship

$$y(t) = 2u(x(t)) - 1$$

The output of this system is 1 if $x(t)$ is positive and -1 if $x(t)$ is negative. Clearly multiplying $x(t)$ with α is not going to change the output from being either $+1$ or -1. So replacing $x(t)$ with $\alpha x(t)$ will not change $y(t)$ to $\alpha y(t)$.

A more familiar system is the diode. Let's assume the diode is ideal and the input/output relationship is given by

$$y(t) = \begin{cases} x(t) & x(t) > 0 \\ 0 & x(t) < 0 \end{cases}$$

At first sight this looks like it satisfies the scaling property but if we take alpha to be negative it clearly does not.

5.1.2 ADDITIVITY

In a system which satisfies the additivity property, the response to a sum of inputs is the sum of the responses to the individual inputs. Or symbolically, if

$$x_1(t) \Rightarrow y_1(t)$$

and

$$x_2(t) \Rightarrow y_2(t)$$

then

$$x(t) = x_1(t) + x_2(t) \Rightarrow y(t) = y_1(t) + y_2(t)$$

What are some of the systems that satisfy additivity. Clearly resistive circuits satisfy additivity otherwise you would not be able to use superposition. We can easily show that the system shown in Figure 5.2 also satisfies additivity. Let's apply the definition to it. If we give the system $x_1[n]$ as the input the output will be

$$y_1[n] = 2x_1[n] + x_1[n-1]$$

If the input is $x_2[n]$ the output will be

$$y_2[n] = 2x_2[n] + x_2[n-1]$$

Now, let's suppose the input is $x_1[n] + x_2[n]$. In the upper branch the system will multiply whatever the input is by 2, so in this branch we will get $2 \times [x_1[n] + x_2[n]]$. In the lower branch we will get the input delayed by one, $[x_1[n-1] + x_2[n-1]]$. Adding the two we get $2x_1[n] + 2x_2[n] + x_1[n-1] + x_2[n-1]$ which is exactly $y_1[n] + y_2[n]$. Therefore, the system possesses the additivity property.

Another system that satisfies the additivity property is an integrator.

$$y(t) = \int_{-\infty}^{t} x(\tau)d\tau$$

We can validate this easily by noting that

$$y_1(t) = \int_{-\infty}^{t} x_1(\tau)d\tau$$

$$y_2(t) = \int_{-\infty}^{t} x_2(\tau)d\tau$$

And if we give the system the input $x_1(t) + x_2(t)$ the output will be

$$\begin{aligned} y(t) &= \int_{-\infty}^{t} [x_1(\tau) + x_2(\tau)]d\tau \\ &= \int_{-\infty}^{t} x_1(\tau)d\tau + \int_{-\infty}^{t} x_2(\tau)d\tau \\ &= y_1(t) + y_2(t) \end{aligned}$$

(Note that for all this to be true all these integrals have to exist. In practice this is generally not a problem.)

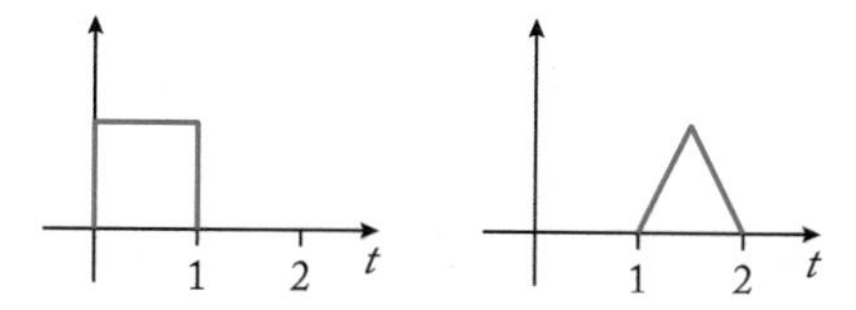

Figure 5.3: Inputs for which outputs are known.

What are examples of systems that do not have the additivity property. All the examples described above that didn't have the scaling property also do not have the additivity property. Another one is the system with the input output relationship

$$y(t) = e^{x(t)}$$

We can easily see that

$$y(t) = e^{x_1(t)+x_2(t)} \neq e^{x_1(t)} + e^{x_2(t)} = y_1(t) + y_2(t)$$

So if a system possesses the scaling property does it always also satisfy additivity? Not necessarily. Consider

$$y[n] = \frac{x[n]x[n-1]}{x[n-2]}; \quad x[n] \neq 0 \text{ for all } n$$

This will satisfy the scaling property but not the additivity property.

Together the two properties of scaling and additivity give us the linearity property which says that if a particular input is the weighted sum of other known inputs, the output will be the weighted sum of the outputs to those known inputs.

$$\alpha x_1(t) + \beta x_2(t) \Rightarrow \alpha y_1(t) + \beta y_2(t)$$

One way in which this property can be very helpful is if we can express our inputs as a weighted sums of known functions. We could record the outputs of the linear system to the known functions and then we could find the output of any weighted combinations of these inputs simply by taking the weighted combinations of the known outputs. Suppose our known functions are the ones shown in Figure 5.3.

If we have a linear system for which we know the outputs for these inputs. We can find the outputs of the infinite linear combinations of these inputs. A few of the possible combinations of the inputs are shown in Figure 5.4. Just knowing the outputs for two inputs, we can immediately obtain the corresponding outputs for any of these infinite inputs if we can find the weights of the linear combinations of these two inputs.

5.2 TIME INVARIANCE

Time invariance is easy to define and hard to demonstrate. A system is said to be time invariant if its behavior does not change with time. If you give it an input today and get a particular output

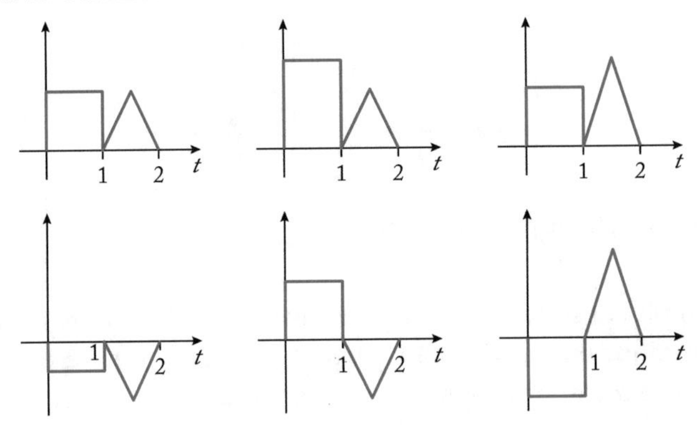

Figure 5.4: A few signals obtained using a linear combination of the signals in the previous figure.

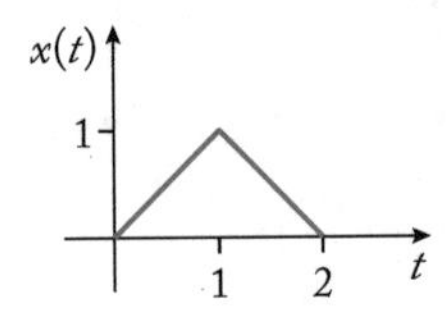

Figure 5.5: Triangular function.

then you will also get the same output if you give it the same input tomorrow. Mathematically we can say that if $x(t)$ results in an output $y(t)$ then for a time-invariant system $x(t - t_0)$ results in the output $y(t - t_0)$—shifting the input simply shifts the output. To see the issues involved when trying to demonstrate wether a system is time invariant or time varying let's pick a particular input $x(t)$—the triangular function (Figure 5.5)

$$x(t) = \begin{cases} t & 0 \leq t \leq 1 \\ -t + 2 & 1 \leq t \leq 2 \\ 0 & \text{otherwise} \end{cases}$$

If we delay this function by t_0 we will simply move the entire function to the left by t_0. To see this simply replace t with $t - t_0$ in the equation above (Figure 5.6).

$$x(t - t_0) = \begin{cases} t - t_0 & 0 \leq t - t_0 \leq 1 \\ -(t - t_0) + 2 & 1 \leq t - t_0 \leq 2 \\ 0 & \text{otherwise} \end{cases}$$

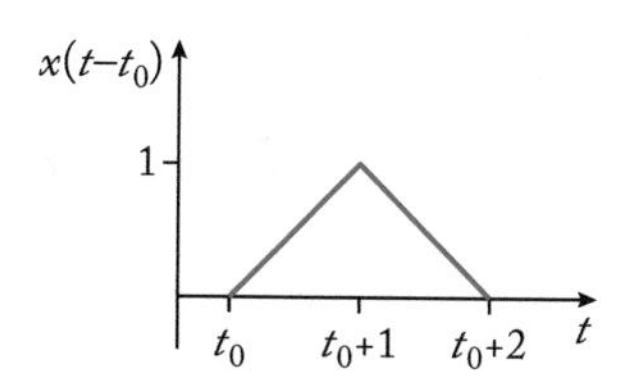

Figure 5.6: Replacing t with $t - t_0$.

from which we get

$$x(t - t_0) = \begin{cases} t - t_0 & t_0 \le t \le t_0 + 1 \\ -(t - t_0) + 2 & t_0 + 1 \le t \le t_0 + 2 \\ 0 & \text{otherwise} \end{cases}$$

Now let's see what happens to the output given two different systems with the following input output relationships.

$$y(t) = 2x(t)$$

$$y(t) = x\left(\frac{t}{2}\right)$$

Lets name the output $y(t)$ when the input is $x(t)$ and let's name it $y_d(t)$ when the input is $x(t - t_0)$—the original input delayed by t_0. The system is time invariant if when we shift the output $y(t)$ by t_0 to get $y(t - t_0)$, $y(t - t_0) = y_d(t)$.

In the first system if the input is $x(t)$, the output will be

$$y(t) = 2x(t) = \begin{cases} 2t & 0 \le t \le 1 \\ -t + 2 & 1 \le t \le 2 \\ 0 & \text{otherwise} \end{cases}$$

If we shift the input by t_0 so that the input is $x(t - t_0)$

$$y_d(t) = 2x(t - t_0) = \begin{cases} 2(t - t_0) & t_0 \le t \le t_0 + 1 \\ -2(t - t_0) + 4 & t_0 + 1 \le t \le t_0 + 2 \\ 0 & \text{otherwise} \end{cases}$$

If we plot these two functions we see that $y_d(t)$ is simply $y(t)$ shifted by t_0 (Figure 5.7).

So, clearly the system given by the input-output relationship $y(t) = 2x(t)$ is time invariant. Now let's consider the second system. In this case we replace t with $t/2$.

$$y(t) = x\left(\frac{t}{2}\right) = \begin{cases} \frac{t}{2} & 0 \le \frac{t}{2} \le 1 \\ -\frac{t}{2} + 2 & 1 \le \frac{t}{2} \le 2 \\ 0 & \text{otherwise} \end{cases}$$

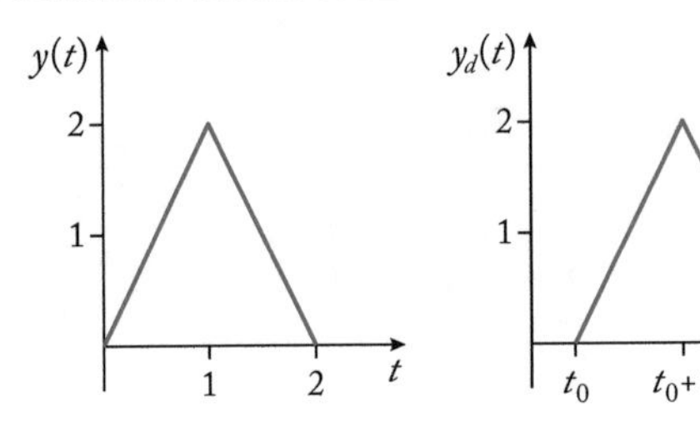

Figure 5.7: $y_d(t)$ is simply $y(t)$ delayed by t_0.

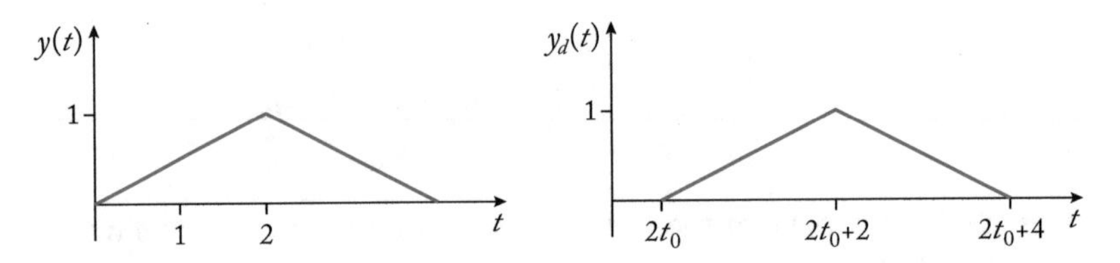

Figure 5.8: The system is time varying.

or

$$y(t) = \begin{cases} \frac{t}{2} & 0 \leq t \leq 2 \\ -\frac{t}{2} + 2 & 2 \leq t \leq 4 \\ 0 & \text{otherwise} \end{cases}$$

Now let's do the same for $x(t - t_0)$, i.e., replace t with $t/2$ in the expression for $x(t - t_0)$

$$y_{t_0}(t) = x\left(\frac{t}{2} - t_0\right) = \begin{cases} \frac{t}{2} - t_0 & t_0 \leq \frac{t}{2} \leq t_0 + 1 \\ -\left(\frac{t}{2} - t_0\right) + 2 & t_0 + 1 \leq \frac{t}{2} \leq t_0 + 2 \\ 0 & \text{otherwise} \end{cases}$$

or

$$y_{t_0}(t) = x\left(\frac{t}{2} - t_0\right) = \begin{cases} \frac{t}{2} - t_0 & 2t_0 \leq t \leq 2t_0 + 2 \\ -\left(\frac{t}{2} - t_0\right) + 2 & 2t_0 + 21 \leq t \leq 2t_0 + 4 \\ 0 & \text{otherwise} \end{cases}$$

Clearly y_{t_0} is $y(t)$ shifted by $2t_0$, not t_0 (Figure 5.8). In other words $y(t - t_0) \neq y_{t_0}(t)$ and the system described by the input output relationship $y = x(t/2)$ is not time invariant. Another way of saying the same thing is to say that this system is time varying.

In general you will find that whenever in the input/output relationship the argument of the input is scaled we will get a system that is not time invariant. Another type of system which is

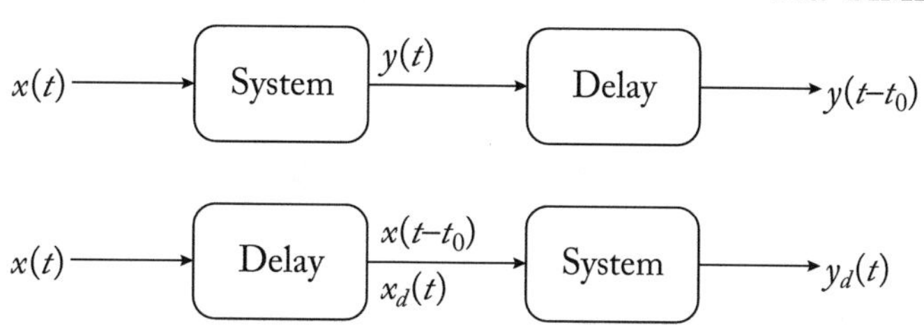

Figure 5.9: Setup for testing for time-invariance.

always time-varying is one in which the input gets multiplied by a function of time. For example the system

$$y(t) = tx(t)$$

is a time varying system. As is it's discrete time counterpart

$$y[n] = nx[n]$$

If we replace t or n with functions of t in the continuous time and functions of n in the discrete time we again get systems that are time varying. So

$$y(t) = \cos(t)x(t)$$

is the input output relationship of a time varying system.

We can show that these systems are time varying using a simple approach. In this approach we will first pass the input $x(t)$ through the system to get $y(t)$ and then pass $y(t)$ through a delay to get $y(t - t_0)$. In parallel we will pass $x(t)$ through a delay to get $x(t - t_0)$ which we will call $x_d(t)$. We will then pass $x_d(t)$ through the system to get $y_d(t)$. If $y_d(t) = y(t - t_0)$ the system is time invariant. We show the setup in Figure 5.9. Let's test the system with the input output relationship

$$y(t) = tx(t)$$

We can picture this system as shown in Figure 5.10. The system takes the input multiplies it with t and spits it out. In the top configuration

$$y(t) = tx(t)$$

is the system output. After the delay we get

$$y(t - t_0) = (t - t_0)x(t - t_0)$$

In the bottom configuration, after the delay we get

$$x_d(t) = x(t - t_0)$$

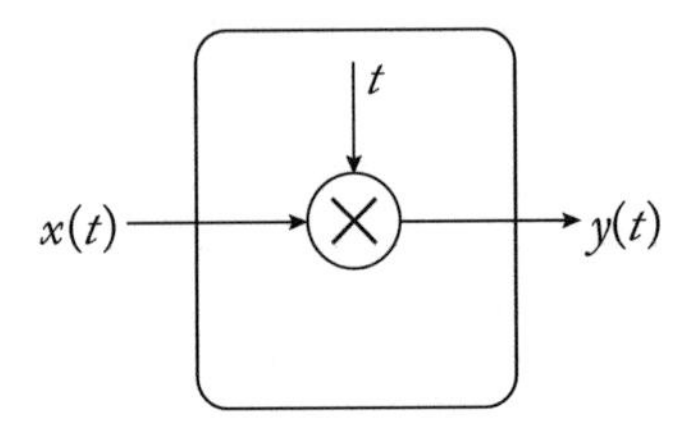

Figure 5.10: System with input/output relationship $y(t) = tx(t)$.

When we now put $x_d(t)$ into the system, the system multiplies it with t and we get

$$y_d(t) = tx_d(t)$$

Replacing $x_d(t)$ with $x(t - t_0)$ we get

$$y_d(t) = tx(t - t_0)$$

which is not the same as $y(t - t_0)$. Hence this system is not time-invariant.

Why is this property important? When we design a system we would like it to give us the same output for the same input regardless of when the input was provided. Whether we turn on the switch at noon or at midnight we would like the light, or the coffee maker, or the car to behave in the same way. In practice, the car will behave a bit differently if we start it at noon during the summer or at midnight during the winter so it is not strictly time invariant. But in practice what we find is that sometimes "good enough for government work" is good enough.

5.3 LINEAR TIME-INVARIANT (LTI) SYSTEMS

If our system was not simply linear but also time invariant then knowing that the input $x(t)$ results in the output $y(t)$, we can find the output for a shifted input $x(t - t_0)$ for any value of t_0. This means that knowing the output for the two inputs of Figure 5.3 we would also know the output of the system for any linear combination of the shifted inputs.

$$x(t) = \alpha x_1(t - t_1) + \beta x_2(t - t_2)$$

In Figure 5.11 we have a few of the infinitely many combinations of the weighted and shifted signals shown in Figure 5.3. Because of this property, which allows us to obtain the outputs of an infinite number of possible inputs just by knowing the outputs for a few inputs we can build some powerful analysis tools for these systems—systems that are both linear and time invariant. This is why for the rest of the course we will focus almost exclusively on linear time-invariant (LTI) systems.

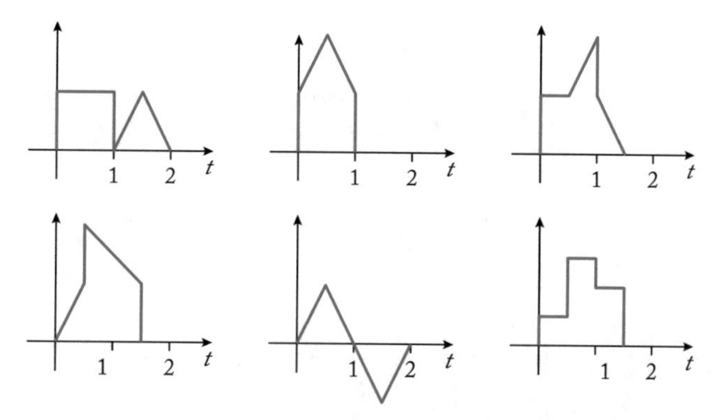

Figure 5.11: A few signals obtained using a linear combination of shifts of the signals in Figure 5.3.

5.4 SUMMARY

In this module we looked at two of the most important properties (for us) of systems - linearity and time invariance. If we know the output of a linear time-invariant system for a particular input we also know the output for any linear combination of time shifted versions of that input. This property will become enormously useful as we will see in the next module.

Table 5.1: Time invariant and linear systems

Input/Output Relationship	Linear	Time-invariant
$y(t) = 3x(t) + 2$		
$y[n] = nx[n]$		
$y(t) = \int_{-\infty}^{2t} x(\tau)d\tau$		
$y[n] = x[n-2] - 2x[n-8]$		
$y[n] = x[\sin[n]]$		
$y(t) = e^{x(t)}$		
$y(t) = \frac{d}{dt}x(t)$		

5.5 EXERCISES

(Answers on the following page)

1. Determine whether the following systems are linear and/or time invariant. In each case $x(t)$ or $x[n]$ is the input and $y(t)$ or $y[n]$ is the output.
   (a) $y(t) = ax(t)$ where a is a constant
   (b) $y(t) = x(t)\sin(t)$
   (c) $y(t) = \frac{1}{x(t)}$
   (d) $y(t) = \int_{-\infty}^{t} x(\tau)d\tau$
   (e) $y[n] = 2x[n] - x[n-1]$
   (f) $y[n] = \sum_{k=-\infty}^{n} x[k+1]$
   (g) $y(t) = tx(t-2)$
   (h) $y(t) = \cos(x(t))$
2. Determine whether the following systems are time invariant, and linear. In each case $x(t)$ or $x[n]$ is the input and $y(t)$ or $y[n]$ is the output (see Table 5.1).

Table 5.2: Time invariant and linear systems

Input/Output Relationship	Linear	Time-invariant
$y(t) = 3x(t) + 2$	No	Yes
$y[n] = nx[n]$	Yes	No
$y(t) = \int_{-\infty}^{2t} x(\tau)d\tau$	Yes	No
$y[n] = x[n-2] - 2x[n-8]$	Yes	Yes
$y[n] = x[\sin[n]]]$	Yes	No
$y(t) = e^{x(t)}$	No	Yes
$y(t) = \frac{d}{dt}x(t)$	Yes	Yes

5.6 ANSWERS

1. (a) Linear, time-invariant
   (b) Linear, not time-invariant
   (c) Not linear, time-invariant
   (d) Linear, time invariant
   (e) Linear, time-invariant
   (f) Linear, time-invariant
   (g) Linear, not time-invariant
   (h) Not linear, time-invariant (see Table 5.2)

MODULE 6

Linearity, Time-Invariance, and the Role of the Impulse Response

Let's take a particular signal and see how it plays with LTI systems. That particular signal is the unit impulse. We will see how the response to the unit impulse of a linear time-invariant system completely characterizes the system for both continuous time systems and discrete time systems. We will begin with discrete systems and then go through the process again with continuous systems.

6.1 THE RESPONSE TO AN IMPULSE FOR A DISCRETE-TIME SYSTEM

For shorthand we will adopt the notation

$$x[n] \Rightarrow y[n]$$

to denote a system with input $x[n]$ and output $y[n]$. Let's suppose the input in Figure 6.1 is a discrete impulse function. In other words the input $x[n] = \delta[n]$. Let's give the output a name. Given that this output is the response of the system to an impulse we will call it the *impulse response* of the system and denote it by $h[n]$. In other words, for $x[n] = \delta[n]$, $y[n] = h[n]$. In terms of our notation

$$\delta[n] \Rightarrow h[n]$$

Because this is a time invariant system if we now use a shifted impulse as the input $x[n] = \delta[n - n_o]$ the output will be a shifted version of the impulse response $y[n] = h[n - n_o]$.

$$\delta[n - n_o] \Rightarrow h[n - n_o]$$

If we also scaled the input, because of linearity the output would be scaled as well. Let's scale the input by the value α. Then

$$\alpha\delta[n - n_o] \Rightarrow \alpha h[n - n_o]$$

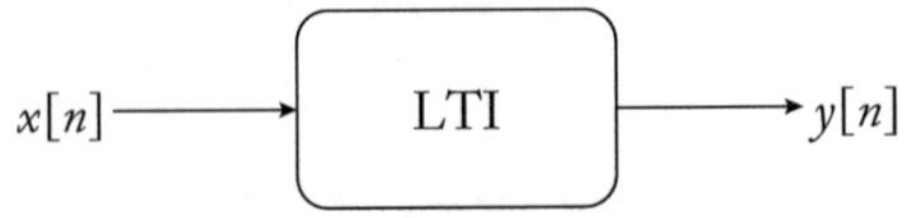

$x[n] \longrightarrow$ LTI $\longrightarrow y[n]$

Figure 6.1: A discrete linear time invariant system.

Let's make the input a little more complicated and add another delta function, so $x[n] = \alpha\delta[n - n_0] + \beta\delta[n - n_1]$ then the additivity part of the linearity property kicks in and the output is now $y[n] = \alpha h[n - n_0] + \beta h[n - n_1]$ or

$$\alpha\delta[n - n_0] + \beta\delta[n - n_1] \Rightarrow \alpha h[n - n_0] + \beta h[n - n_1]$$

So if I can write an input solely in terms of weighted and shifted delta functions the output can be written as the sum of weighted and shifted impulse responses. However, we know we can write any discrete function as the weighted sum of shifted delta functions. If

$$x[n] = \begin{cases} 1 & n = 0 \\ 1.5 & n = 1 \\ .5 & n = 2 \\ 1 & n = 3 \\ 0 & \text{otherwise} \end{cases}$$

we can write this as

$$x[n] = \delta[n] + 1.5\delta[n - 1] + .5\delta[n - 2] + \delta[n - 3]$$

Using the fact that the response to $\delta[n]$ is $h[n]$, using the linearity and time-invariance property of the system the response to this input would be

$$y[n] = h[n] + 1.5h[n - 1] + .5h[n - 2] + h[n - 3]$$

or

$$\delta[n] + 1.5\delta[n - 1] + .5\delta[n - 2] + \delta[n - 3] \Rightarrow h[n] + 1.5h[n - 1] + .5h[n - 2] + h[n - 3]$$

Looking at the expression for $x[n]$ above we can see that the value of $x[n]$ for $n = 0$ is 1, for $n = 1$ is 1.5, for $n = 2$ is .5 and for $n = 3$ is 1. In other words $x[0] = 1,\ x[1] = 1.5,\ x[2] = .5$, and $x[3] = 1$. So we could have written $x[n]$ as

$$x[n] = x[0]\delta[n] + x[1]\delta[n - 1] + x[2]\delta[n - 2] + x[3]\delta[n - 3]$$

and $y[n]$ as

$$y[n] = x[0]h[n] + x[1]h[n - 1] + x[2]h[n - 2] + x[3]h[n - 3]$$

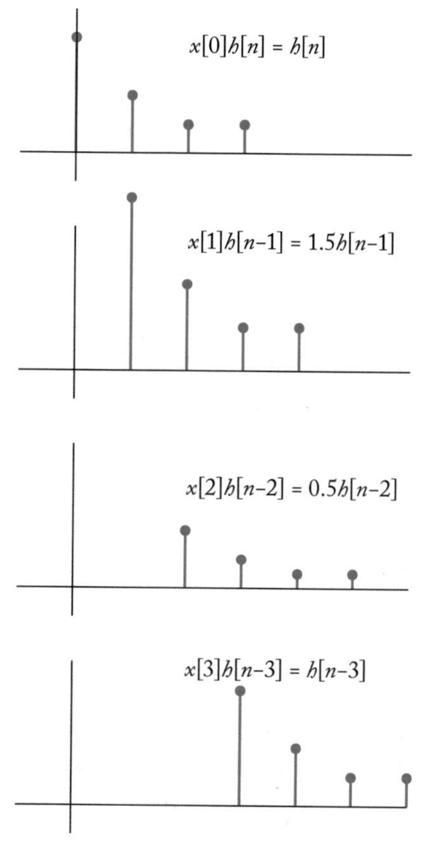

Figure 6.2: The terms in the sum which result in $y[n]$.

In fact, in general we can write any $x[n]$ as

$$x[n] = \ldots x[-2]\delta[n+2] + x[-1]\delta[n+1] + x[0]\delta[n] + x[1]\delta[n-1] + x[2]\delta[n-2] + \cdots$$

So, in general (for a linear time invariant system) we can write the output of the system as

$$y[n] = \cdots + x[-2]h[n+2] + x[-1]h[n+1] + x[0]h[n] + x[1]h[n-1] + x[2]h[n-2] + \cdots$$

or

$$y[n] = \sum_{k=-\infty}^{\infty} x[k]h[n-k]$$

We can shorten the description above as follows

$$
\begin{aligned}
\delta[n] \quad & \Rightarrow h[n] \quad \text{By definition} \\
\delta[n-k] \quad & \Rightarrow h[n-k] \quad \text{By time-invariance} \\
x[k]\delta[n-k] \quad & \Rightarrow x[k]h[n-k] \quad \text{By Linearity} \\
\sum_{k=-\infty}^{\infty} x[k]\delta[n-k] \quad & \Rightarrow \textstyle\sum_{k=-\infty}^{\infty} x[k]h[n-k] \quad \text{By Linearity}
\end{aligned}
$$

But the left hand side is simply $x[n]$. Therefore, the right hand side is $y[n]$.

We can show through variable substitution that this sum can also be written as

$$y[n] = \sum_{k=-\infty}^{\infty} x[n-k]h[k]$$

This sum is called the *convolution sum* and can be viewed as an operation between $x[n]$ and $h[n]$. We denote this operation by $\circledast$. So,

$$y[n] = x[n] \circledast h[n] = \sum_{k=-\infty}^{\infty} x[k]h[n-k] = \sum_{k=-\infty}^{\infty} x[n-k]h[k]$$

We will have more to say about the convolution operation later in this module. For now let's consider what this equation says. What it says is that in order to obtain the output of an LTI system for a given input all we need is the impulse response of the system. Think how cool that is. Because knowing the impulse response allows us to compute the output of the LTI system for every input we can say that if the system is a linear time invariant system it can be completely characterized by the impulse response. As we will see we can tell whether the system has memory, whether it's causal, whether it's stable; all this by knowing how it responds to an input which is unity at $n = 0$ and zero everywhere else. While the theory is lovely the actual computation can sometimes (OK often) be a pain. Let's delay the pain for a while and see how we can get a similar characterization for continuous time systems.

6.2 THE RESPONSE TO AN IMPULSE FOR CONTINUOUS TIME SYSTEMS

We saw how the impulse response can be used with the convolution sum to give the output of an LTI discrete time system for any input. We can do the same for continuous time systems. In this case we use the response of the system to a continuous time delta function and a convolution integral. Recall the sifting property of the continuous time impulse

$$\int_{-\infty}^{\infty} x(t)\delta(t-t_o) = x(t_o)$$

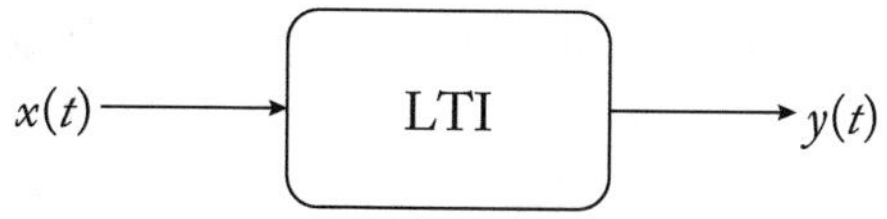

$x(t)$ → LTI → $y(t)$

Figure 6.3: A continuous time linear-time-invariant system.

From which we can derive

$$x(t) = \int_{-\infty}^{\infty} x(\tau)\delta(t-\tau)d\tau$$

Now, consider the continuous time linear time-invariant system shown in Figure 6.3. Let's consider the case where the input is the delta function $x(t) = \delta(t)$. Let's call the response of the system to an impulse the *impulse response* and denote it by $h(t)$. In other words when the input $x(t) = \delta(t)$, the output $y(t) = h(t)$. We can denote this symbolically like this:

$$\delta(t) \Rightarrow h(t)$$

Because the system is time invariant if we now shift the input by τ so that if the input is $\delta(t-\tau)$ the output will be a shift of the impulse response by the same amount—$h(t-\tau)$.

$$\delta(t-\tau) \Rightarrow h(t-\tau)$$

Now let's scale the input by $x(\tau)$. Because the system is linear scaling the input by $x(\tau)$ will scale the output by $x(\tau)$ as well. So, when the input is $x(\tau)\delta(t-\tau)$ the output will be $x(\tau)h(t-\tau)$.

$$x(\tau)\delta(t-\tau) \Rightarrow x(\tau)h(t-\tau)$$

In the final step we will use the summability part of the linearity property. Recall that for a linear system the response of a sum of inputs is the sum of the corresponding outputs. Instead of simply taking a sum we will take the integral of the input. You might remember from calculus that the integral can be defined as the limit of a sum, so the same rules apply. If we integrate the input over a certain range the system will respond with the integral of the corresponding output over the same range. So

$$\int_{-\infty}^{\infty} x(\tau)\delta(t-\tau)d\tau \Rightarrow \int_{-\infty}^{\infty} x(\tau)h(t-\tau)d\tau$$

But $\int_{-\infty}^{\infty} x(\tau)\delta(t-\tau)$ is simply $x(t)$, or

$$x(t) \Rightarrow \int_{-\infty}^{\infty} x(\tau)h(t-\tau)d\tau$$

Therefore, the output of a linear time invariant system with impulse response $h(t)$ to an input $x(t)$ is given by the *convolution integral.*

$$y(t) = \int_{-\infty}^{\infty} x(\tau)h(t-\tau)d\tau$$

As in the case of the digital systems the convolution operation is denoted by $\circledast$.

$$y(t) = x(t) \circledast h(t) = \int_{-\infty}^{\infty} x(\tau)h(t-\tau)d\tau$$

6.3 THE CONVOLUTION OPERATION

Convolution can be thought of as a binary operation like addition multiplication etc. and like those operations it has some useful properties. One of the most useful is that the convolution operation is commutative. That is

$$x(t) \circledast h(t) = h(t) \circledast x(t) = \int_{-\infty}^{\infty} x(t-\tau)h(\tau)d\tau$$

We show this using variable substitution. Consider the integral

$$\int_{-\infty}^{\infty} x(\tau)h(t-\tau)d\tau$$

Define

$$\sigma = t - \tau$$

which we will use as the variable of integration in the integral. This means that everywhere we see τ we need to replace it with a function of σ. If we write τ in terms of σ and t we get

$$\tau = t - \sigma$$

and, therefore, taking the differential of both sides we get

$$d\tau = -d\sigma$$

Notice there is no dt in here. This is because as far as the integral is concerned t is a constant. Finally, we need to see how this substitution effects the limits of integration. The lower limit in the original integral was $\tau = -\infty$. If $\tau = -\infty$ then after the variable substitution $t - \sigma = -\infty$ or, taking t over to the other side and multiplying by -1, $\sigma = +\infty$. Similarly where the upper limit in the original integral was $+\infty$, after the variable substitution it becomes $-\infty$. Plugging all this in we get

$$y(t) = \int_{-\infty}^{\infty} x(\tau)h(t-\tau)d\tau = \int_{+\infty}^{-\infty} x(t-\sigma)h(\sigma)(-d\sigma)$$

We can swap the limits of integration if we multiply the integral with -1. This -1 will cancel out the minus in front of $d\sigma$ and we get

$$y(t) = \int_{-\infty}^{\infty} x(t-\sigma)h(\sigma)d\sigma$$

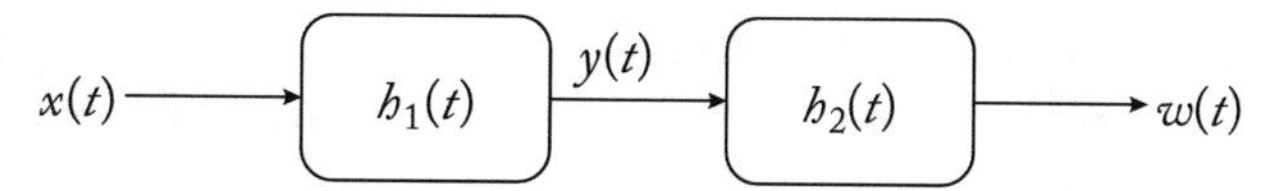

Figure 6.4: A series connection of linear-time-invariant systems.

σ is simply a dummy variable. We can replace it with any variable we want. We could replace it with *Tom* in which case we would get

$$y(t) = \int_{-\infty}^{\infty} x(t - Tom)h(Tom)d\,Tom$$

The *Toms* of the world may not like this so we replace σ with τ to get

$$y(t) = \int_{-\infty}^{\infty} x(t - \tau)h(\tau)d\tau$$

But this is simply $h(t) \circledast x(t)$. So the convolution operation is commutative.

The convolution operation, like addition and multiplication, is also associative, that is

$$(x(t) \circledast h_1(t)) \circledast h_2(t) = x(t) \circledast (h_1(t) \circledast h_2(t))$$

This property is especially useful when dealing with systems that are connected in series. Consider the setup in Figure 6.4 consisting of a series connection of two linear time-invariant systems. The input to the first system with impulse response $h_1(t)$ is $x(t)$ therefore the output $y(t)$ is given by

$$y(t) = x(t) \circledast h_1(t)$$

For the second system with impulse response $h_2(t)$ the input is $y(t)$ and the output is $w(t)$. As this is a linear time-invariant system

$$w(t) = y(t) \circledast h_2(t)$$

Substituting for $y(t)$ from the previous equation

$$w(t) = (x(t) \circledast h_1(t)) \circledast h_2(t)$$

If we use the associative property we can write this as

$$w(t) = x(t) \circledast (h_1(t) \circledast h_2(t))$$

which means that we could replace the two systems with a single system whose impulse response would be $h_1(t) \circledast h_2(t)$. A more useful application of this property is that we can take a more complex system with impulse response $h_1(t) \circledast h_2(t)$ and replace it with a series connection of two simpler systems. The question of how you would take the more complicated impulse response and extract the simpler impulse responses from them we leave aside for the moment. Turns out it is not as difficult as it seems but we need one more tool before we can tackle that problem.

Finally, just as multiplication is distributive over addition

$$\alpha \times (a + b) = \alpha \times a + \alpha \times b$$

the convolution operation is also distributive over addition

$$x(t) \circledast (h_1(t) + h_2(t)) = x(t) \circledast h_1(t) + x(t) \circledast h_2(t)$$

Again we can see how we could take a more complex system with impulse response $h_1(t) + h_2(t)$ and break the impulse response into simpler components.

6.4 SUMMARY

In this module we showed that a linear time-invariant system can be completely characterized by its response to an impulse. If we denote the response of the system to an impulse - the impulse response - as $h[n]$ or $h(t)$ the output of the system for an arbitrary input $x[n]$ or $x(t)$ is given by the convolution sum

$$y[n] = x[n] \circledast h[n] = \sum_{k=-\infty}^{\infty} x[k]h[n-k] = \sum_{k=-\infty}^{\infty} x[n-k]h[k]$$

or the convolution integral

$$y(t) = x(t) \circledast h(t) = \int_{-\infty}^{\infty} x(\tau)h(t-\tau)d\tau = \int_{-\infty}^{\infty} x(t-\tau)h(\tau)d\tau$$

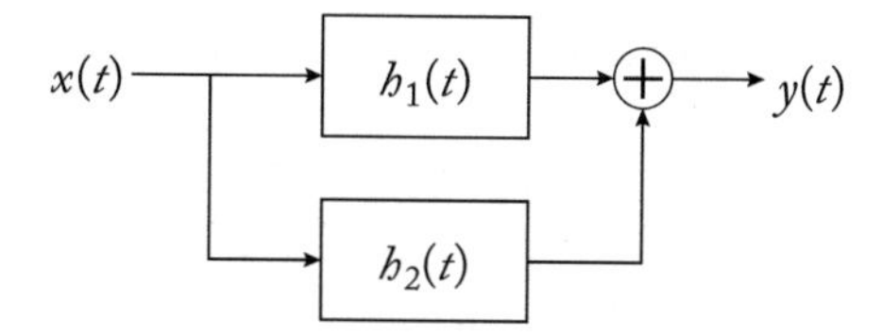

Figure 6.5: A parallel connection of linear-time-invariant systems.

6.5 EXERCISES

(Answers on the following page)

1. The input-output relationship of a linear time-invariant system is given by

$$y[n] = x[n] - x[n-1]$$

   What is the impulse response of this system?

2. The input-output relationship of a linear time-invariant system is given by

$$y[n] = x[n+1] - 2x[n] + x[n-1]$$

   What is the impulse response of this system?

3. The input-output relationship of a linear time-invariant system is given by

$$y[n] = x[n] + 0.9y[n-1]$$

   What is the impulse response of this system?

4. For the system shown in Figure 6.5.
   Write the output $y(t)$ in terms of $x(t)$, $h_1(t)$, and $h_2(t)$.

5. The input-output relationship of a linear time-invariant system is given by

$$y(t) = -\frac{1}{\alpha}\int_{-\infty}^{t} x(\tau)d\tau$$

   What is the impulse response of this system? Can you come up with a circuit that would have this impulse response?

6. The input-output relationship of a linear time-invariant system is given by

$$y(t) = \int_{-\infty}^{t} e^{-(t-\tau)}x(\tau-1)d\tau$$

   What is the impulse response of this system?

7. The input-output relationship of a linear time-invariant system is given by

$$y(t) = e^{-t} \int_{-\infty}^{t} x(\tau) e^{\tau} d\tau$$

What is the impulse response of this system?

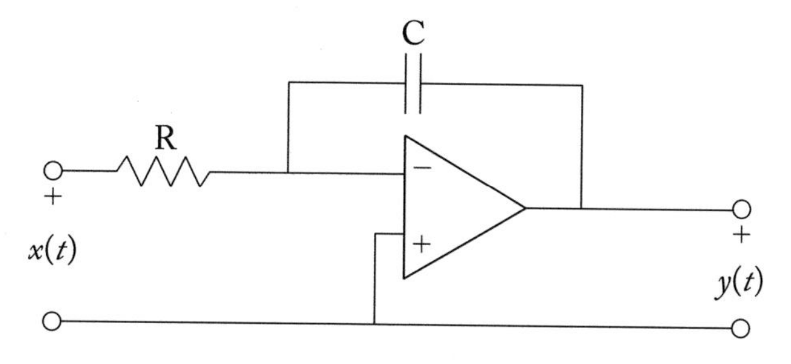

Figure 6.6: $\alpha = RC$.

6.6 ANSWERS

1. $h[n] = \delta[n] - \delta[n-1]$
2. $h[n] = \delta[n+1] - 2\delta[n] + \delta[n-1]$
3. $h[n] = (0.9)^n u[n]$
4. $y(t) = x(t) \circledast h_1(t) + x(t) \circledast h_2(t)$
5. $h(t) = -\frac{1}{\alpha} u(t)$ (Figure 6.6)
6. $h(t) = e^{-(t-1)} u(t-1)$
7. $h(t) = e^{-t} u(t)$

MODULE 7

Properties of LTI Systems

We have made the claim that the impulse response of a linear time-invariant system completely determines the behavior the system. If that is the case we should be able to determine the various system properties we discussed earlier from the impulse response. Let's take each in turn and demonstrate that we can.

7.1 MEMORY IN LTI SYSTEMS

A system is said to be memoryless when the output depends on only the current value of the input. How is that reflected in the impulse response? Let's consider a discrete system first. For a discrete system the input and output are related by the convolution sum

$$y[n] = \sum_{k=-\infty}^{\infty} x[k]h[n-k]$$

For the system to be memoryless we want only terms containing $x[n]$ on the right hand side. The only term in this summation that contains $x[n]$ is the term with $k = n$, or $x[n]h[n-n]$. Given that the input can be anything, for only this term to be nonzero we need the only nonzero term in the impulse response to be $h[n-n]$ or $h[0]$. The discrete function that is only nonzero for $n = 0$ is $\delta[n]$. So, for a memoryless system the impulse response has to be of the form

$$h[n] = K\delta[n]$$

A system with impulse response

$$h[n] = u[n] - u[n-4]$$

is a system with memory. To see this let's write the input output relationship for this and see if the output depends on more than the current input. Recall that given the impulse response the output is given by

$$y[n] = \sum_{k=-\infty}^{\infty} x[k]h[n-k] = \sum_{k=-\infty}^{\infty} x[n-k]h[k]$$

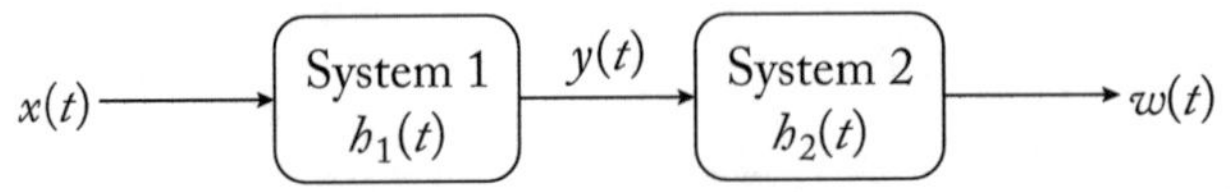

Figure 7.1: A series of LTI systems.

If we use the second form of the summation we can expand this summation as

$$\begin{aligned} y[n] &= h[0]x[n] + h[1]x[n-1] + h[2]x[n-2] + h[3]x[n-3] \\ &= x[n] + x[n-1] + x[n-2] + x[n-3] \end{aligned}$$

because $h[0], h[1], h[2]$, and $h[3]$ are the only nonzero values of the impulse response $h[n]$. Therefore, the output at time n depends not only on the input at time n but also on the input at time $n-1$ and $n-2$ and $n-3$. Hence the system is not memoryless.

What about continuous systems? Consider the convolution integral.

$$y(t) = \int_{-\infty}^{\infty} x(\tau)h(t-\tau)d\tau$$

We want the right hand side to consist of terms containing only $x(t)$. The only way that can happen is if

$$h(t-\tau) = K\delta(t-\tau)$$

Therefore, a system is memoryless if the impulse response is given by

$$h(t) = K\delta(t)$$

So, for both discrete and continuous time systems if the impulse response is zero for all values of the argument except 0 the system is memoryless, otherwise it has memory.

7.2 INVERTIBILITY OF LTI SYSTEMS

Given the series connection of linear time invariant systems in Figure 7.1 we can say that the system labeled *System 1* is invertible if we can find a *System 2* such that $w(t) = x(t)$. The output of *System 1* is given by

$$y(t) = x(t) \circledast h_1(t)$$

and the output of *System 2* is given by

$$w(t) = y(t) \circledast h_2(t)$$

Substituting for $y(t)$ in this equation we get

$$w(t) = (x(t) \circledast h_1(t)) \circledast h_2(t)$$

At this point we will use the *associative property* of convolution. Using this property we can rewrite the expression for the output as

$$(x(t) \circledast h_1(t)) \circledast h_2(t) = x(t) \circledast (h_1(t) \circledast h_2(t))$$

For an invertible system in the strict sense we want $w(t) = x(t)$, which means that

$$x(t) \circledast (h_1(t) \circledast h_2(t)) = x(t)$$

We know that

$$x(t) \circledast \delta(t) = x(t)$$

So we want

$$h_1(t) \circledast h_2(t) = \delta(t)$$

If we allow some delay in the system so that $w(t) = x(t - t_o)$ then we can relax our requirement to

$$h_1(t) \circledast h_2(t) = \delta(t - t_0)$$

We can go through the same procedure and find that for a system with impulse response $h_1[n]$ to be invertible there has to exist a system with impulse response $h_2[n]$ such that $h_1[n] \circledast h_2[n] = \delta[n]$.

Remember we said that the inverse system to

$$y[n] = \sum_{k=-\infty}^{n} x[k]$$

is

$$y[n] = x[n] - x[n-1]$$

Let's figure out the impulse response of each of these systems and show that their convolution results in a delta function. The easiest way to find the impulse response of a linear time invariant system is to replace the input with an impulse function. In other words substitute $x[k]$ with $\delta[k]$ in the input output relationship. Therefore

$$h_1[n] = \sum_{k=-\infty}^{n} \delta[k]$$

In this sum there is only one term—$\delta[0]$—which is nonzero. Therefore, as long as this term is included in the summation the sum will be one. If this term is not included in the summation the sum is 0. This term will be included in the summation as long as $n \geq 0$. Therefore,

$$h_1[n] = \begin{cases} 1 & n \geq 0 \\ 0 & \text{otherwise} \end{cases}$$

or

$$h_1[n] = u[n]$$

A somewhat roundabout way of finding the same impulse response is to note that we have two expressions for the output—the convolution sum and the input output relationship. The two must by necessity be equal.

$$y[n] = \sum_{k=-\infty}^{\infty} x[k]h_1[n-k] = \sum_{k=-\infty}^{n} x[k]$$

This means

$$h_1[n-k] = \begin{cases} 1 & k \leq n \\ 0 & \text{otherwise} \end{cases}$$

Which in turn implies

$$h_1[-k] = \begin{cases} 1 & k \leq 0 \\ 0 & \text{otherwise} \end{cases}$$

and

$$h_1[k] = \begin{cases} 1 & -k \leq 0 \\ 0 & \text{otherwise} \end{cases}$$

or

$$h_1[k] = \begin{cases} 1 & k \geq 0 \\ 0 & \text{otherwise} \end{cases}$$

or $h_1[n] = u[n]$. Clearly the first approach is preferable wherever it is available.

For the second system using the same argument

$$h_2[n] = \delta[n] - \delta[n-1]$$

and

$$h_1[n] \circledast h_2[n] = u[n] \circledast (\delta[n] - \delta[n-1]) = u[n] - u[n-1] = \delta[n]$$

Thus, the two systems—the summer and the differencer are inverse of each other.

7.3 CAUSALITY OF LTI SYSTEMS

We want the output at time n not to depend on the future values of the input, namely $x[k]$ for $k > n$. In other words

$$y[n] = \sum_{k=-\infty}^{\infty} x[k]h[n-k] = \sum_{k=-\infty}^{n} x[k]h[n-k]$$

In order for this to be true

$$\begin{aligned} & h[n-k] = 0 \quad \text{for} \;\; k > n \\ \Rightarrow \; & h[n-k] = 0 \quad \text{for} \;\; k - n > 0 \\ \Rightarrow \; & h[n-k] = 0 \quad \text{for} \;\; n - k < 0 \end{aligned}$$

or

$$h[k] = 0 \;\; \text{for} \;\; k < 0$$

So the impulse response of a causal system is zero for all values of its argument less than zero. For a discrete system this means that the impulse response is zero for $n < 0$ and for continuous systems the impulse response is zero for $t < 0$.

We can also see that this has to be true by considering what an impulse response is. The impulse response is the response of the system to an input at time $t = 0$ or $n = 0$. If the impulse response is nonzero prior to this time this would mean that the system responded before the input occurred which would mean that the system was non-causal.

Let's take a look at a few examples of causal and non-causal systems. Consider the system with impulse response

$$h(t) = e^{-t}u(t)$$

The output from this system is given by

$$y(t) = \int_{-\infty}^{\infty} x(\tau)h(t-\tau)d\tau = \int_{-\infty}^{\infty} x(\tau)e^{-(t-\tau)}u(t-\tau)d\tau$$

Let's look for a moment at the behavior of $u(t-\tau)$. Remember that

$$u(t) = \begin{cases} 1 & t > 0 \\ 0 & t < 0 \end{cases}$$

Therefore,

$$u(t-\tau) = \begin{cases} 1 & t-\tau > 0 \\ 0 & t-\tau < 0 \end{cases} = \begin{cases} 1 & -\tau > -t \\ 0 & -\tau < -t \end{cases} = \begin{cases} 1 & \tau < t \\ 0 & \tau > t \end{cases}$$

Because $u(t-\tau)$ is equal to zero for $t < \tau$, the integrand $x(\tau)e^{-(t-\tau)}u(t-\tau)$ is zero for $t < \tau$, therefore,

$$y(t) = \int_{-\infty}^{t} x(\tau)e^{-(t-\tau)}d\tau$$

or the only values of the input that effect the output at time t, are the values at t and prior to t. Hence, as expected this system is causal.

What about the system with an impulse response which is nonzero for $t < 0$.

$$h(t) = e^{-a|t|} \quad a > 0$$

We can write this impulse response as

$$h(t) = \begin{cases} e^{-at} & t > 0 \\ e^{at} & t < 0 \end{cases} = e^{-at}u(t) + e^{at}u(-t)$$

So we can write the convolution integral as

$$\begin{aligned} y(t) &= \int_{-\infty}^{\infty} x(\tau)\left[e^{-a(t-\tau)}u(t-\tau) + e^{a(t-\tau)}u(-(t-\tau))\right]d\tau \\ &= \int_{-\infty}^{\infty} x(\tau)e^{-a(t-\tau)}u(t-\tau)d\tau + \int_{-\infty}^{\infty} x(\tau)e^{a(t-\tau)}u(\tau-t)d\tau \\ &= \int_{-\infty}^{t} x(\tau)e^{-a(t-\tau)}d\tau + \int_{t}^{\infty} x(\tau)e^{a(t-\tau)}d\tau \end{aligned}$$

where, in the second integral, we have used the fact that $u(\tau - t)$ is zero for $\tau < t$. The second integral also means that we need to use values of the input beyond the time t in order to obtain the output at time t. Hence the system is noncausal.

Turning to discrete systems, consider the system with impulse response

$$h[n] = u[n]$$

which is zero for $n < 0$. If we examine the output of this system we find that

$$y[n] = \sum_{k=-\infty}^{\infty} x[k]h[n-k] = \sum_{k=-\infty}^{\infty} x[k]u[n-k]$$

The function $u[n-k]$ is zero for $k > n$, therefore,

$$y[n] = \sum_{k=-\infty}^{n} x[k]$$

and the output at time n depends only on the input at time n and times prior to n. Hence the system with this impulse response is causal.

7.3.1 STABILITY FOR LTI SYSTEMS

Suppose you are guaranteed that $|x(t)| < B$ then

$$\begin{aligned} |y(t)| &= \left| \int_{-\infty}^{\infty} x(t-\tau)h(\tau)d\tau \right| \\ &\leq \int_{-\infty}^{\infty} |x(t-\tau)h(\tau)|\, d\tau \\ &< \int_{-\infty}^{\infty} B\, |h(\tau)|\, d\tau \\ &= B \int_{-\infty}^{\infty} |h(\tau)|\, d\tau \end{aligned}$$

So if we want $|y(t)| < \infty$ then we need

$$\int_{-\infty}^{\infty} |h(\tau)|\, d\tau < \infty$$

Another way of stating this is to say that we want $h(t)$ to be absolutely integrable.

Similarly for the discrete case we want $h[n]$ to be absolutely summable

$$\sum_{k=-\infty}^{\infty} |h[k]| < \infty$$

Consider for example the integrator with impulse response

$$h(t) = u(t)$$

Clearly

$$\int_{-\infty}^{\infty} h(\tau)d\tau = \int_{0}^{\infty} d\tau \rightarrow \infty$$

Therefore, this system is unstable in the bounded input bounded output sense.

7.4 SUMMARY

In this module we examined our assertion that a linear time-invariant system is completely characterized by its impulse response. We showed that the impulse response can tell us whether a system has memory, is stable, or is causal.

- A memoryless LTI system has an impulse response of the form

$$h(t) = K\delta(t) \qquad h[n] = K\delta[n]$$

- The impulse response of a causal system is zero for negative values of its argument.

$$h(t) = 0 \text{ for } t < 0 \qquad h[k] = 0 \text{ for } k < 0$$

- An LTI system is stable if the impulse response if absolutely integrable or absolutely summable.

$$\int_{-\infty}^{\infty} |h(\tau)|\, d\tau < \infty \qquad \sum_{k=-\infty}^{\infty} |h[k]| < \infty$$

- An LTI system with impulse response $h_1(t)$ or $h_1[n]$ is invertible if we can find another system with impulse response $h_2(t)$ or $h_2[n]$ such that

$$h_1(t) \circledast h_2(t) = \delta(t) \qquad h_1[n] \circledast h_2[n] = \delta[n]$$

7.5 EXERCISES

(Answers on the following page)

Given the following impulse responses determine if the system is memoryless, causal, or stable.

1. $h[n] = u[n]$
2. $h[n] = \delta[n] - \delta[n-1]$
3. $h(t) = u(t+1) - u(t-1)$
4. $h(t) = u(t+1)$
5. $h[n] = \alpha^n u[n] |\alpha| < 1$
6. $h(t) = e^{-3t} u(t)$
7. $h(t) = e^{-3t} u(-t)$
8. $h(t) = e^{3t} u(t)$
9. $h(t) = e^{3t} u(-t)$
10. $h(t) = e^{3t} u(t+1)$
11. $h[n] = \alpha^n u[n] |\alpha| > 1$
12. $h[n] = u[n] - u[n-1]$

7.6 ANSWERS

1. $h[n] = u[n]$; Not memoryless, Causal, Not stable
2. $h[n] = \delta[n] - \delta[n-1]$; Not memoryless, Causal, Stable
3. $h(t) = u(t+1) - u(t-1)$; Not memoryless, Not Causal, Stable
4. $h(t) = u(t+1)$; Not memoryless, Not Causal, Not Stable
5. $h[n] = \alpha^n u[n] |\alpha| < 1$; Not memoryless, Causal, Stable
6. $h(t) = e^{-3t} u(t)$; Not memoryless, Causal, Stable
7. $h(t) = e^{-3t} u(-t)$; Not memoryless, Not Causal, Not Stable
8. $h(t) = e^{3t} u(t)$; Not memoryless, Causal, Not Stable
9. $h(t) = e^{3t} u(-t)$; Not memoryless, Not Causal, Stable
10. $h(t) = e^{3t} u(t+1)$; Not memoryless, Not Causal, Not Stable
11. $h[n] = \alpha^n u[n] |\alpha| > 1$; Not memoryless, Causal, Not Stable
12. $h[n] = u[n] - u[n-1]$; Memoryless, Causal, Stable

MODULE 8
Discrete Time Convolution

In this module we will look in some detail at discrete time convolution—mostly through examples. Discrete time convolution is not simply a mathematical construct, it is a roadmap for how a discrete system works. This becomes especially useful when designing or implementing systems in discrete time such as digital filters and others which you may need to implement in embedded systems.

To reiterate, the input and output of a linear time invariant system are related through convolution with the impulse response.

$$y[n] = \sum_{k=-\infty}^{\infty} x[k]h[n-k]$$

We saw how this came about mathematically. Let's see how this equation comes about using a very simple example. Let's suppose we have a linear time-invariant system with impulse response

$$h[n] = 1.5\delta[n] + \delta[n-1] + 0.5\delta[n-2] + 0.5\delta[n-3]$$

Suppose we excite it with an input

$$x[n] = \delta[n] + 1.5\delta[n-1] + 1.5\delta[n-2] + \delta[n-3]$$

Let's work through this one input sample at a time and see how the system responds to each input value (Figure 8.1). We will then use the additive property of linearity to add the responses to the individual samples and reconstruct the input and output.

We start out with the sample at $n = 0$. This being an impulse the output is simply the impulse response (Figure 8.2).

The next input value is a delta function of weight 1.5. As the input occurs at time $n = 1$, because of time invariance the output is the impulse response shifted by one and weighted by the value of the input (Figure 8.3).

The next input occurs at time $n = 2$ and is an impulse with a weight of 1.5. The output then is the impulse response shifted by 2 and weighted by 1.5 (Figure 8.4). Finally, the last input is an impulse at $n = 3$ with a weight of 1. The response to this is the impulse response shifted by 3 (Figure 8.5).

To obtain the response to the complete input $x[n]$ we simply add the response to each individual sample. In Figure 8.6 we show the input and the corresponding response at time 0,

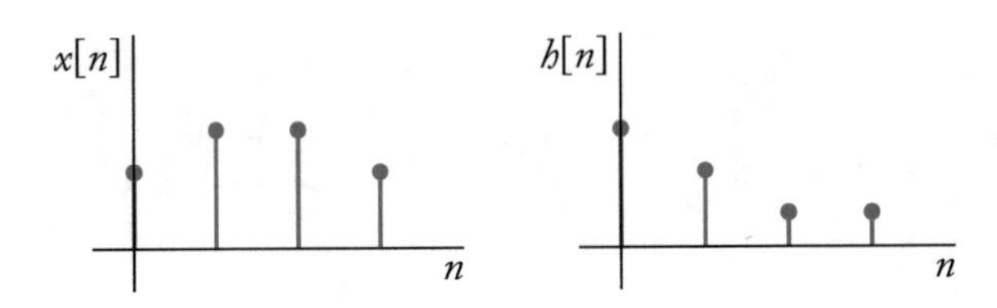

Figure 8.1: The input $x[n]$ and the impulse response $h[n]$.

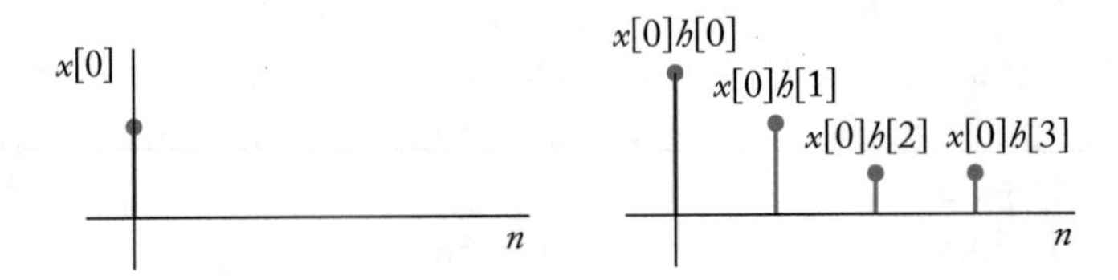

Figure 8.2: The response of the system to the input $x[0]$.

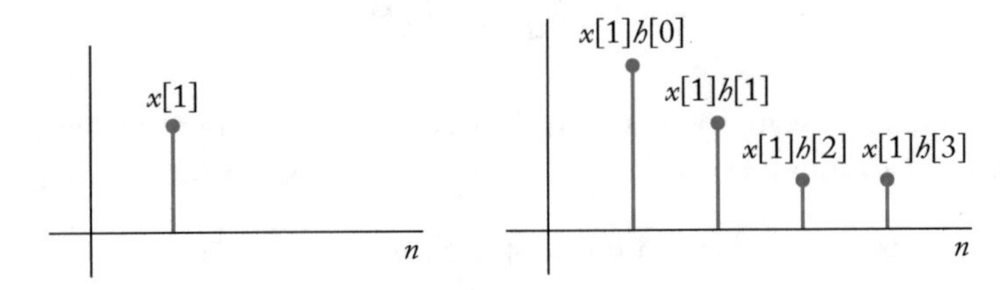

Figure 8.3: The response of the system to the input $x[1]$.

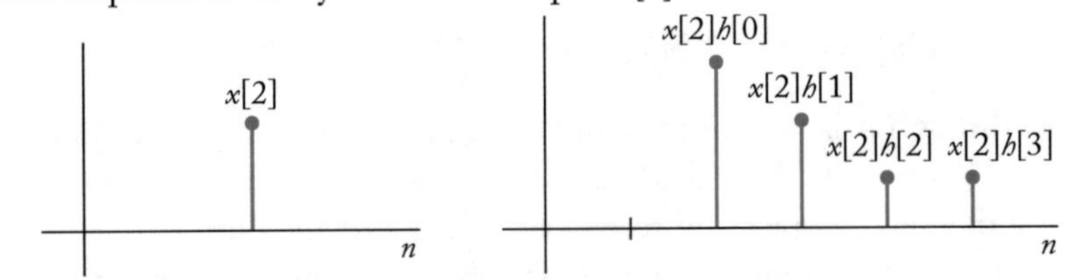

Figure 8.4: The response of the system to the input $x[2]$.

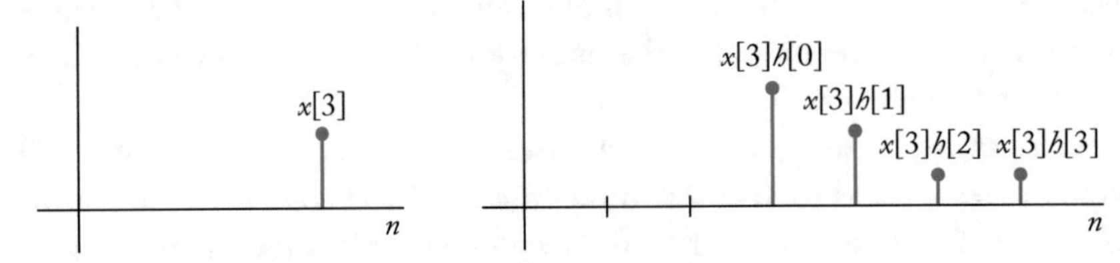

Figure 8.5: The input $x[n]$ and the impulse response $h[n]$.

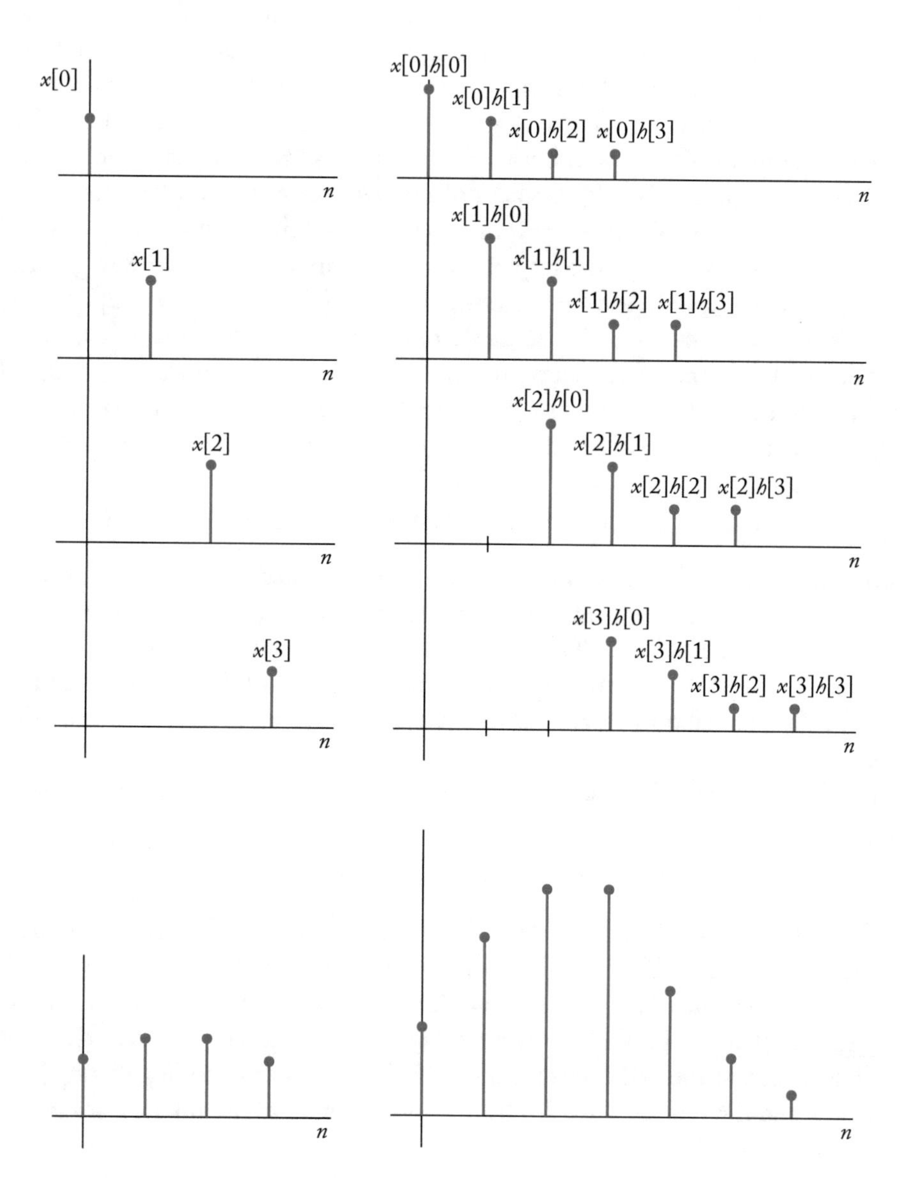

Figure 8.6: The input $x[n]$ and the output $y[n]$.

1, 2, and 3. In order to get the complete output all we need to do is add these outputs.

At time $n = 0$ the only non-zero term is the response to the input at time 0. Therefore, the output is simply $x[0]h[0]$. At time $n = 1$ there is nonzero output from the input at time 0 and the input at time 1. The output at time 1 resulting from the input at time 0 is $x[0]h[1]$. Recall that in this example an impulse at time $n = 0$ results in an impulse response which continues until $n = 3$. So the effect of the input $x[0]$ is continuing. But now we also have the system responding to the input at time $n = 1$. Because at $n = 1$ the system has just "seen" the input $x[1]$ it responds with the output $x[1]h[0]$. Because of linearity these two responses add up and the output is $x[0]h[1] + x[1]h[0]$. At time $n = 2$ the response from the input at time $n = 0$ and $n = 1$ are still continuing and we add in the response from the input at time $n = 2$ to get the output $x[0]h[2] + x[1]h[1] + x[2]h[0]$. At $n = 3$ the output is $x[0]h[3] + x[1]h[2] + x[2]h[1] + x[3]h[0]$. Notice how in each sum we have the index for one factor go up and the index for the other factor goes down. This is the effect that we see in the terms of the summation $x[k]h[n-k]$, as k increases the argument of $x[\ ]$ increases and the argument of $h[\ ]$ decreases. To really emphasize this point, let's rewrite the output at time $n = 3$ to match the convolution expression

$$y[3] = x[0]h[3-0] + x[1]h[3-1] + x[2]h[3-2] + x[3]h[3-3]$$

At time $n = 4$ the effect of the input at time $n = 0$ has died out and there is no new input so all we have are the responses from the input at times $n = 1, 2,$ and 3 continuing to generate the output $x[1]h[3] + x[2]h[2] + x[2]h[1]$. At time $n = 5$ the only responses left are from the input at time $n = 2$ and $n = 3$ which generate the output $x[2]h[3] + x[3]h[2]$. And finally at time $n = 6$ the only contribution is from the response to the input at time $n = 3$ and the output is $x[3]h[3]$.

Hopefully it is clear why the output for a linear time invariant system is given by the convolution equation

$$y[n] = \sum x[k]h[n-k]$$

Having hopefully convinced ourselves of the physical validity of the convolution sum, let's turn our attention to the problem of computing the convolution sum. In the convolution sum notice that k is the variable in the summation and n is fixed. So, for *each* n we need to find the sum $\sum_{k=-\infty}^{\infty} x[k]h[n-k]$. Each time we change n, nothing happens to $x[k]$, but $h[n-k]$ changes. So let's look at the behavior of $h[n-k]$. We could do this for a generic $h[n]$ but it is easier to understand the process if we do this for specific impulse responses. Let's take the easy way out.

Example 8.1 Let's suppose

$$\begin{aligned} h[n] &= (0.9)^n u[n] \\ x[n] &= u[n] \end{aligned}$$

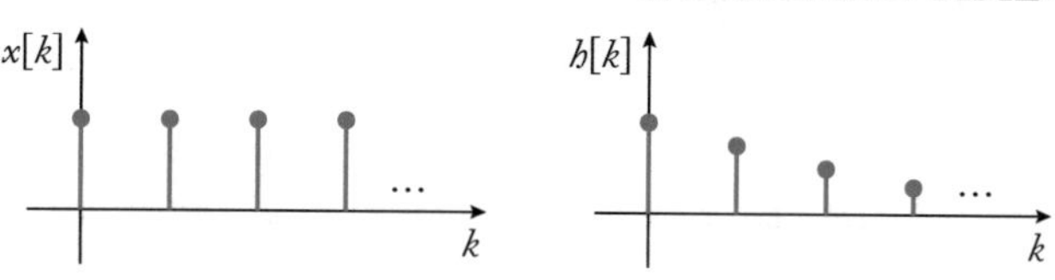

Figure 8.7: The input $x[k]$ and the impulse response $h[k]$.

These signals are plotted in Figure 8.7. We will take the transformation of $h[k]$ to $h[n-k]$ in two steps. First lets look at $h[-k]$. Recall that replacing k with $-k$ in the argument involves a mirroring across the vertical axis. We plot $h[k]$ and $h[-k]$ in Figure 8.8.

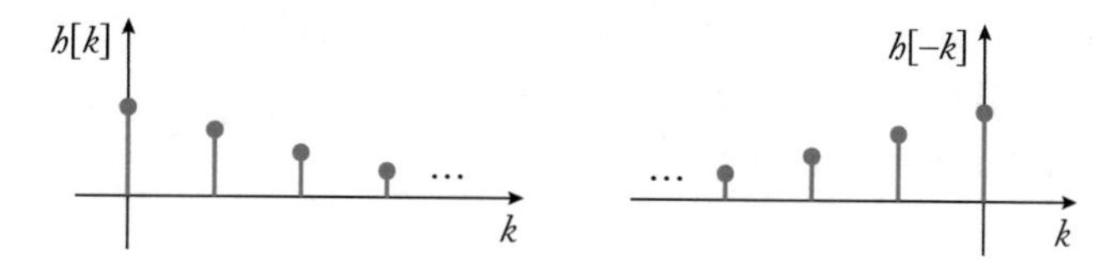

Figure 8.8: The impulse response $h[k]$ and it's mirrored version $h[-k]$.

In Figure 8.9 we plot $h[n-k]$. Notice that we have drawn two versions of $h[n-k]$, one for $n > 0$ and one for $n < 0$. We have also drawn $x[n]$ to provide context. In the case $n < 0$ wherever $h[n-k] \neq 0$ $x[k] = 0$, and wherever $x[k] \neq 0$ $h[n-k] = 0$, therefore, when we take the product $x[k]h[n-k]$, regardless of the value of k the product will be zero. Hence, for $n < 0$, $y[n] = 0$. To see what the output is when $n > 0$ let's turn to the convolution sum.

$$y[n] = \sum_{k=-\infty}^{\infty} x[k]h[n-k] = \sum_{k=-\infty}^{\infty} u[k](0.9)^{n-k}u[n-k]$$

Before we compute this lets plot $x[k]$ and $h[n-k]$. As we mentioned before while we can plot $x[k]$ directly as a function of k, the plot of $h[n-k]$ will change depending on the value of n.

Let's see if we can simplify this. Before we start let's assume that $n \geq 0$. We have a pretty good idea of what happens when $n < 0$. Let's split the summation into the three regions.

$$\sum_{k=-\infty}^{-1} u[k](0.9)^{n-k}u[n-k] + \sum_{k=0}^{n} u[k](0.9)^{n-k}u[n-k] + \sum_{k=n+1}^{\infty} u[k](0.9)^{n-k}u[n-k]$$

The first sum is zero because $u[k] = 0$ for $k < 0$. The third sum is zero because $u[n-k] = 0$ for $k > n$. Therefore we can rewrite the summation as

$$y[n] = \sum_{k=0}^{n} u[k](0.9)^{n-k}u[n-k]$$

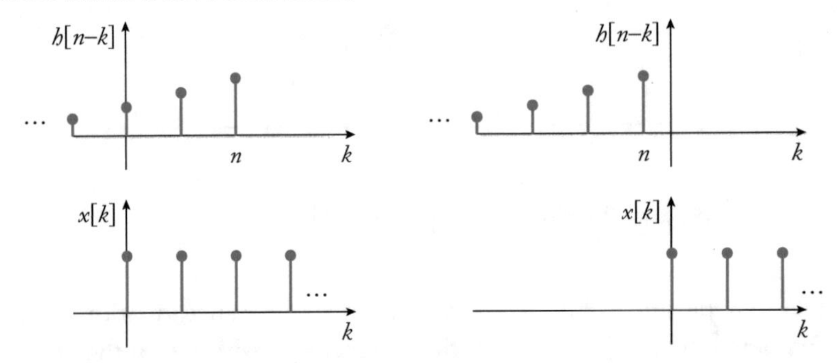

Figure 8.9: $h[n-k]$ for positive and negative values of n with the input $x[k]$ for context.

In this range (from $k = 0$ to $k = n$) both $u[k]$ and $u[n-k]$ are equal to 1 so we can drop them from the summation altogether and we get

$$y[n] = \sum_{k=0}^{n} (0.9)^{n-k}$$

Or, given that the summation is over k and $(0.9)^{n-k} = (0.9)^n (0.9)^{-k}$ we can write this as

$$y[n] = (0.9)^n \sum_{k=0}^{n} (0.9)^{-k} = (0.9)^n \sum_{k=0}^{n} (0.9^{-1})^k$$

This final summation is a form that we will encounter quite often. So, let's take a brief detour and explain how to evaluate sums of this form.

Geometric Sum This kind of sum is called a geometric sum. Let's define the sum as

$$S_{m,n} = \sum_{k=m}^{k=n} \alpha^k$$

We can derive the value of $S_{m,n}$ pretty easily.

$$\begin{aligned} S_{m,n} &= \alpha^m + \alpha^{m+1} + \alpha^{m+2} + \cdots + \alpha^n \\ \alpha S_{m,n} &= \alpha^{m+1} + \alpha^{m+2} + \cdots + \alpha^n + \alpha^{n+1} \end{aligned}$$

Subtracting the bottom equation from the top equation we get

$$S_{m,n} - \alpha S_{m,n} = \alpha^m - \alpha^{n+1}$$

or

$$(1-\alpha) S_{m,n} = \alpha^m - \alpha^{n+1}$$

from which we get

$$S_{m,n} = \frac{\alpha^m - \alpha^{n+1}}{1-\alpha}$$

Applying the geometric sum formula we get

$$y[n] = (0.9)^n \left(\frac{(0.9^{-1})^0 - (0.9^{-1})^{n+1}}{1 - 0.9^{-1}} \right)$$

Anything to the power 0 is 1 and 0.9^{-1} is $1/0.9$ so we can simplify this to

$$y[n] = \frac{(0.9)^n \left(1 - \frac{1}{(0.9)^{n+1}}\right)}{1 - \frac{1}{0.9}}$$

which after simplification gives us

$$y[n] = 10\left(1 - (0.9)^{n+1}\right)$$

But remember this was all for the case where $n \geq 0$. What happens when $n < 0$? If $n < 0$ then all values of k from $-\infty$ to n are negative and in this range $u[k] = 0$ and, therefore, $y[n] = 0$. Therefore, to account for all n the output $y[n]$ is given by

$$y[n] = 10\left(1 - (0.9)^{n+1}\right) u[n]$$

If you plot this response you can see that the $y[n]$ starts at 1 and reaches an asymptote at 10.

We used

$$y[n] = \sum_{k=-\infty}^{\infty} x[k]h[n-k]$$

to find $y[n]$. What would have happened if we used

$$y[n] = \sum_{k=-\infty}^{\infty} x[n-k]h[k]$$

Let's find out.

$$y[n] = \sum_{k=-\infty}^{\infty} x[n-k]h[k] = \sum_{k=-\infty}^{\infty} u[n-k](0.9)^k u[k]$$

Using the same arguments as before we note that for $n < 0$ the sum is over all zero values and thus is zero. For $n > 0$ because of $u[k]$ the lower limit becomes zero and because of $u[n-k]$ the upper limit becomes n and we have

$$y[n] = \sum_{k=0}^{n} (0.9)^k$$

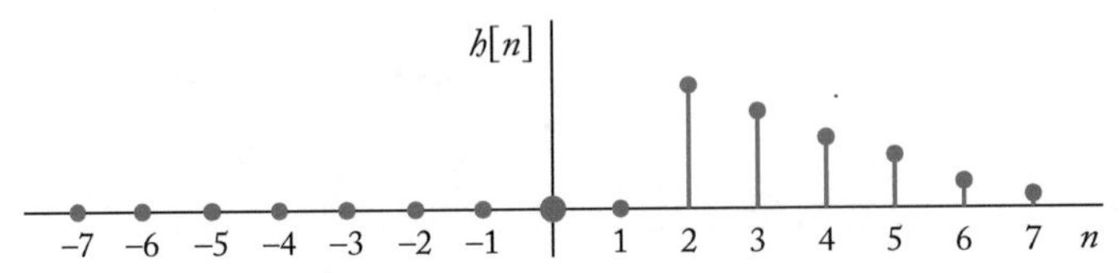

Figure 8.10: The impulse response $h[n]$.

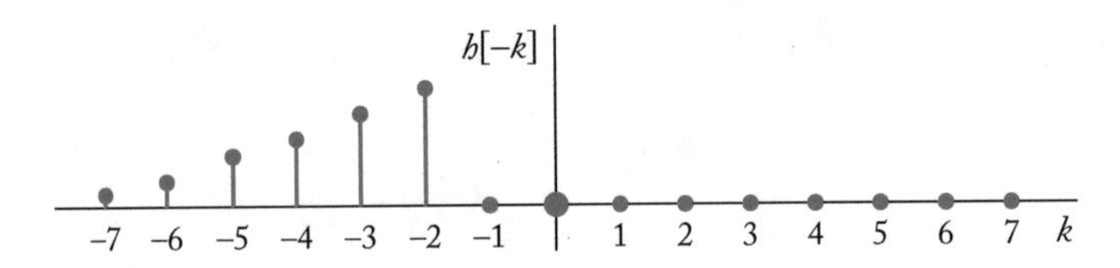

Figure 8.11: The impulse response flipped around the y-axis $h[-k]$.

Applying the geometric sum formula we get

$$y[n] = \frac{(0.9)^0 - (0.9)^{n+1}}{1 - 0.9} = 10\left(1 - (0.9)^{n+1}\right)$$

which is the same result as we had before. And arguably it is also a much simpler path. When faced with a convolution problem it is always a good idea to check which configuration will give you an easier path to the solution.

Example 8.2 Let's try a different example with the same input but a slightly different impulse response.

$$h[n] = 0.9^{n-2}u[n-2]$$

This is almost the same as the previous impulse response—just with a slight shift. The impulse response is plotted in Figure 8.10. But remember the variable is k so we should plot $h[k]$. Of course $h[k]$ looks exactly the same with the x-axis labeled differently. What about $h[-k]$? As we pointed out earlier, it will be $h[k]$ swiveled around the y-axis. What was at 1 will now be at -1 and what was at -1 will be at 1, and so on. We have plotted $h[-k]$ in Figure 8.11.

Notice that we have exaggerated the point marking the value of $h[0]$ to indicate that we flipped the impulse response around that point—this will become important in just a moment. Let's now plot $h[n-k]$ for different values of n.

Look at Figure 8.12 where we have drawn $h[n-k]$ for different values of n. Can you see the pattern? As expected, we can see that as n gets larger we shift the response to the right. But more specifically we can see that for each value of n we shift the origin of the flipped impulse response (which we have drawn as a larger black disk) to that value of n. The point around which you flip either $h[k]$ (to get $h[n-k]$) or $x[k]$ (to get $x[n-k]$) is important.

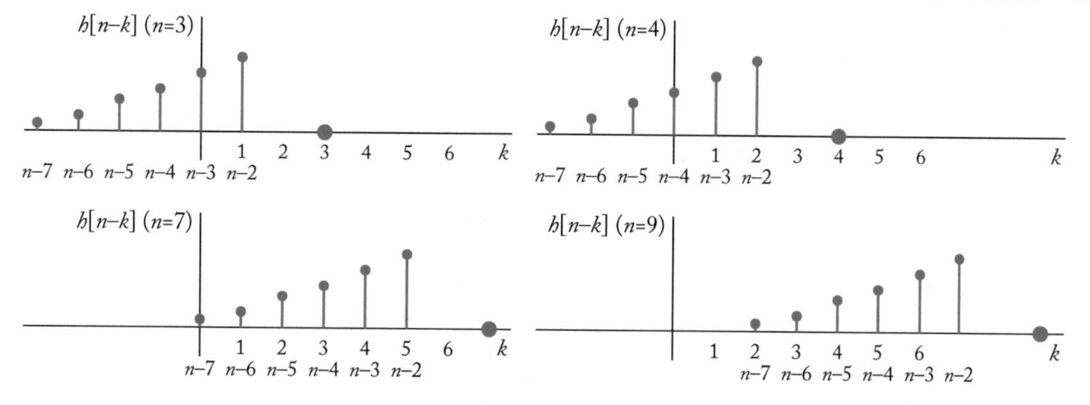

Figure 8.12: The impulse response flipped around the y-axis and moved by n, $h[n-k]$.

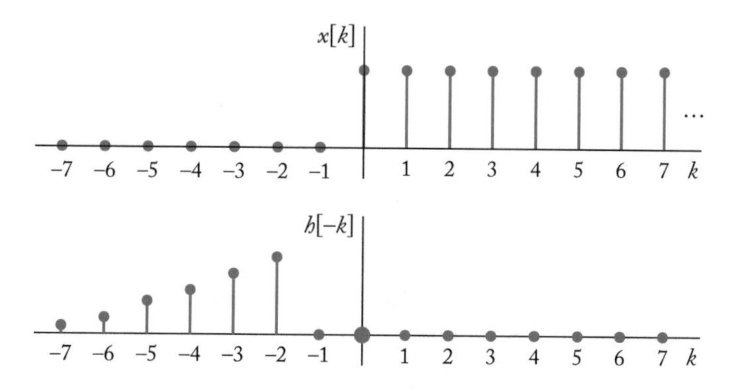

Figure 8.13: The input and impulse response plotted for $n = 0$.

If we look at the convolution sum now we can see that we need to then multiply the corresponding values of $x[k]$ and $h[n-k]$ and take the sum of the products.

$$y[n] = \sum_{k=-\infty}^{\infty} x[k]h[n-k]$$

For $n = 0$ the plots of $h[n-k]$ and $x[k]$ are shown in Figure 8.13.

If we take the point-by-point product of these two functions the result will be zero; wherever $x[k]$ is nonzero, $h[-k]$ is zero and wherever $h[-k]$ is nonzero, $x[k]$ is zero. If we pull the flipped impulse response to the right by one by picking $n = 1$ (the larger black dot will be at $n = 1$), nothing about the product will change. The same is true if we pull the flipped impulse response to the left by picking values of n less than zero. So we can say that for $n < 2$ for each k, the product $x[k]h[n-k]$ is zero. As the sum of zeros (even an infinite number of them) is zero, this means that for $n < 2$, $y[n] = 0$.

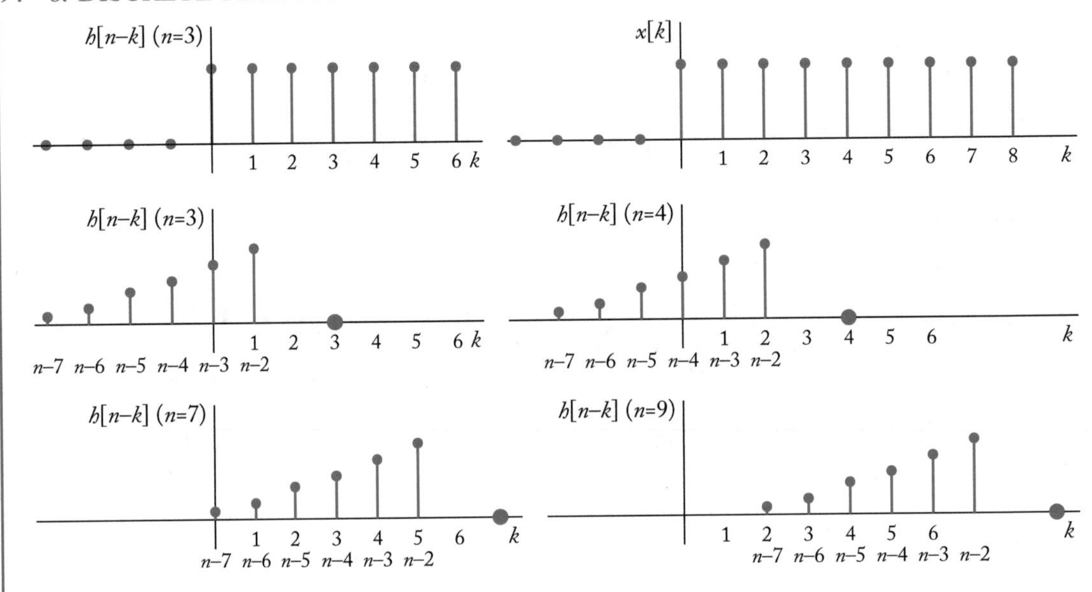

Figure 8.14: The input and impulse response plotted for $n = 3, 4, 7, 9$.

So, what happens for $n \geq 2$? Lets take a look at Figure 8.14. This is simply Figure 8.12 with $x[k]$ plotted above it. We can see that the values of k for which the product between $x[k]$ and $h[n-k]$ is nonzero are the values $0 \leq k \leq n-2$. We can therefore change our limits in our convolution sum to

$$y[n] = \sum_{k=-\infty}^{\infty} x[k]h[n-k] = \sum_{k=0}^{n-2} x[k]h[n-k]$$

In this interval $(0 \leq k \leq n-2)$ $x[k] = 1$ and $h[n-k] = 0.9^{n-k-2}$, therefore,

$$y[n] = \sum_{k=0}^{n-2} 1 \cdot 0.9^{n-k-2} = \sum_{k=0}^{n-2} 0.9^{n-k-2} = 0.9^{n-2} \sum_{k=0}^{n-2} 0.9^{-k} = 0.9^{n-2} \sum_{k=0}^{n-2} \left(\frac{1}{0.9}\right)^{k}$$

Plugging the values of the upper and lower limits into the geometric sum formula we obtain the value of $y[n]$ for $n \geq 2$.

$$y[n] = 0.9^{n-2} \frac{\left(\frac{1}{0.9}\right)^{0} - \left(\frac{1}{0.9}\right)^{n-2+1}}{1 - \frac{1}{0.9}}$$

Table 8.1: Output for different values of n

n	$y[n]$
2	1
3	1.9
4	2.71
⋮	⋮
11	6.51
⋮	⋮
21	8.78
⋮	⋮
31	9.58

or

$$y[n] = 0.9^{n-2} \frac{1 - \left(\frac{1}{0.9}\right)^{n-1}}{1 - \frac{1}{0.9}} = \frac{0.9^{n-2} - \frac{1}{0.9}}{1 - \frac{1}{0.9}}$$

Multiplying top and bottom by 0.9

$$y[n] = \frac{0.9^{n-1} - 1}{0.9 - 1} = \frac{1 - 0.9^{n-1}}{1 - 0.9} = 10(1 - 0.9^{n-1})$$

We can write the value of $y[n]$ for both ranges as

$$y[n] = \begin{cases} 0 & n < 2 \\ 10(1 - 0.9^{n-1}) & n \geq 2 \end{cases}$$

Or we can make use of the discrete unit step function and write this as

$$y[n] = 10(1 - 0.9^{n-1})u[n-2]$$

What does this output look like for different values of n? (see Table 8.1).

Does this make sense?

Example 8.3 If it does, let's do a more difficult example (Figure 8.15). In this case we will keep the impulse response the same as our first example but change the input from the unit step to a rectangular pulse, So,

$$x[n] = u[n] - u[n-4]$$

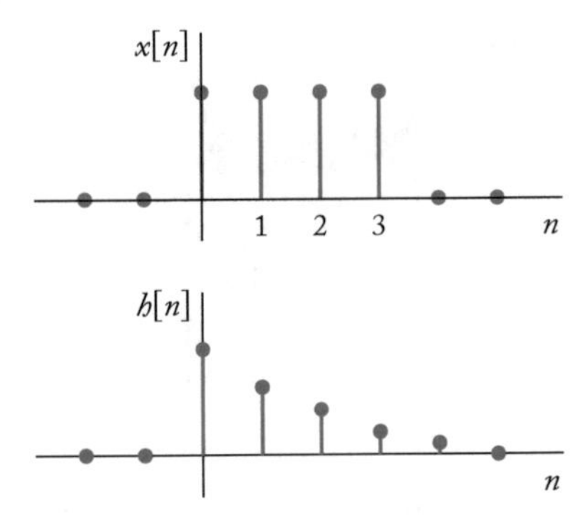

Figure 8.15: The input and impulse response.

and the impulse response is

$$h[n] = 0.9^n u[n]$$

We will use the convolutional sum

$$y[n] = \sum_{k=-\infty}^{\infty} x[k]h[n-k]$$

Again we have to compute the output $y[n]$ for all n. In the previous example we ended up computing $y[n]$ in two regions, $n < 2$ and $n \geq 2$. In this case we will end up finding $y[n]$ for three separate regions. Take a look at Figure 8.16.

Notice that for $n < 0$ there is no overlap between the nonzero range for $x[k]$ and $h[n-k]$. For n between 0 and 3 the overlap of the nonzero portion of $x[k]$ and $h[n-k]$ is between 0 and n, and finally when $n > 3$ the overlap of the nonzero portion is between 0 and 3. Let's compute $y[n]$ for each of these intervals.

For $n < 0$ this is easy. There is no overlap between the nonzero portions of $x[k]$ and $h[n-k]$ so the product $x[k]h[n-k]$ is always zero and hence $y[n] = 0$.

For $0 \leq n \leq 3$ the limits of the convolution sum become 0 and n.

$$y[n] = \sum_{k=0}^{n} x[k]h[n-k] = \sum_{k=0}^{n} 1 \cdot (0.9)^{n-k} = 0.9^n \sum_{k=0}^{n} (0.9)^{-k}$$

or

$$y[n] = 0.9^n \sum_{k=0}^{n} \left(\frac{1}{0.9}\right)^k$$

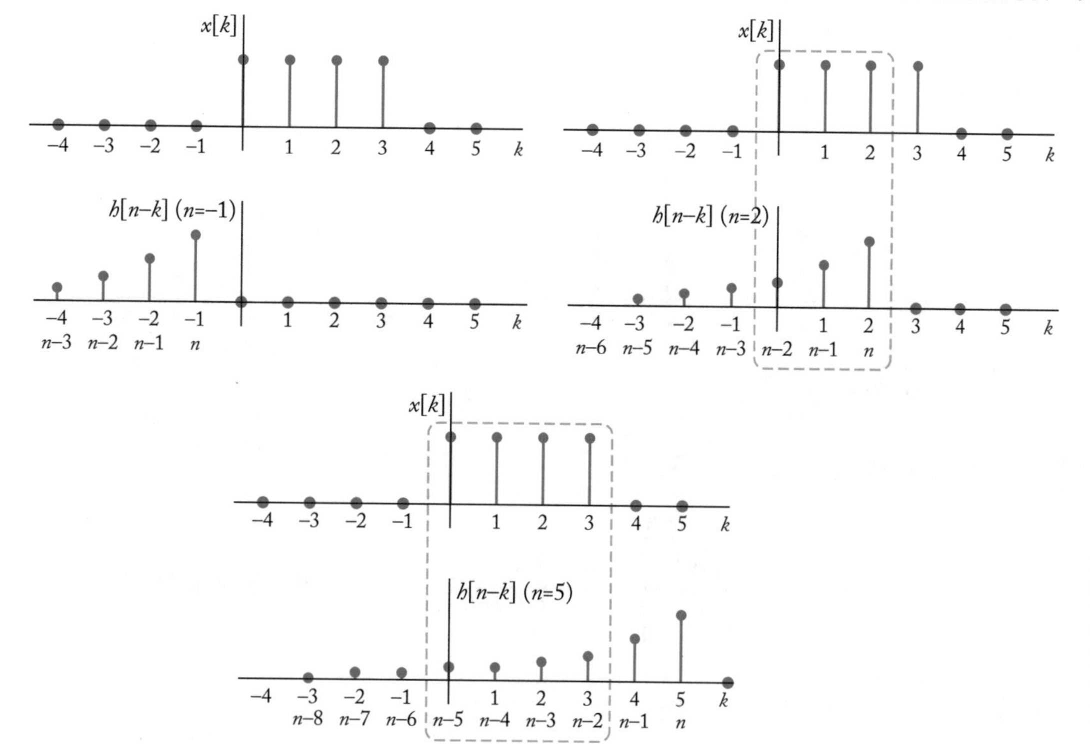

Figure 8.16: The input and impulse response.

Using the geometric sum formula with lower limit 0 and upper limit n we obtain

$$y[n] = 0.9^n \left(\frac{\left(\frac{1}{0.9}\right)^0 - \left(\frac{1}{0.9}\right)^{n+1}}{1 - \frac{1}{0.9}} \right)$$

Working out the fiddly bits and multiplying through by 0.9 we get

$$y[n] = \frac{0.9^n - \frac{1}{0.9}}{1 - \frac{1}{0.9}} = \frac{0.9^{n+1} - 1}{0.9 - 1} = 10\left(1 - 0.9^{n+1}\right)$$

For $n > 3$ the limits of the convolution sum become 0 and 3.

$$y[n] = \sum_{k=0}^{3} x[k]h[n-k] = \sum_{k=0}^{3} 1 \cdot (0.9)^{n-k} = 0.9^n \sum_{k=0}^{3} \left(\frac{1}{0.9}\right)^k$$

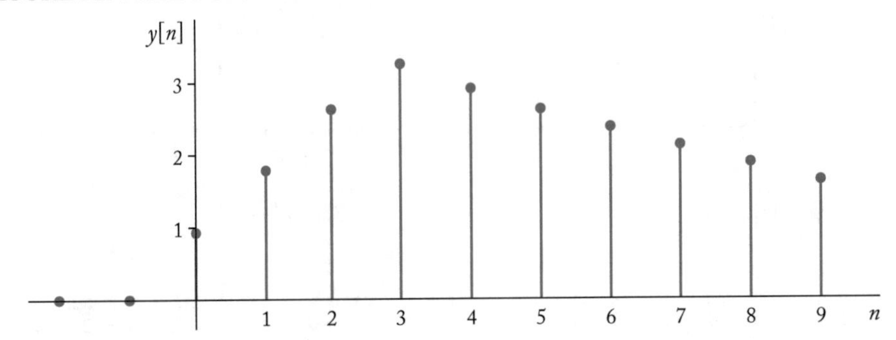

Figure 8.17: The output.

Applying the geometric sum formula

$$y[n] = 0.9^n \left(\frac{1 - \left(\frac{1}{0.9}\right)^4}{1 - \frac{1}{0.9}} \right) = \frac{0.9^{n+1} - 0.9^{n-3}}{0.9 - 1} = 10 \left(0.9^{n-3}(1 - 0.9^4)\right)$$

Put all three parts together and you get

$$y[n] = \begin{cases} 0 & n < 0 \\ 10\left(1 - 0.9^{n+1}\right) & 0 \le n \le 3 \\ 10\left(0.9^{n-3}(1 - 0.9^4)\right) & n > 3 \end{cases}$$

The output is plotted in Figure 8.17.

Example 8.4 Let's look at a final example to explore the effect of the changing regions in which the product $x[k]h[n-k]$ is non-zero. This time we will take both the input and the impulse response to be rectangular pulses.

$$\begin{aligned} x[n] &= u[n] - u[n-4] \\ h[n] &= u[n] - u[n-4] \end{aligned}$$

The input signal and the impulse response are shown in Figure 8.18.

Note one of the differences between this set of signals and the ones we have used in previous examples. In the previous examples at least one of the signals was non-zero over a semi-infinite interval. In this case both signals are non-zero over a finite interval. What this does is increases the number of boundaries between regions over which the signal is non-zero and the region over which the signal is zero. The practical effect of this is that the resulting

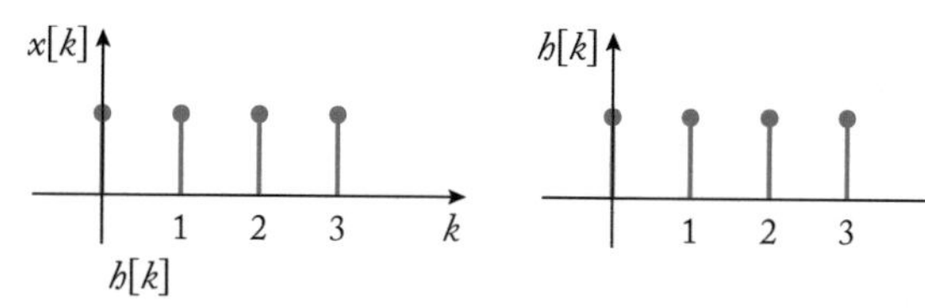

Figure 8.18: The input and impulse response.

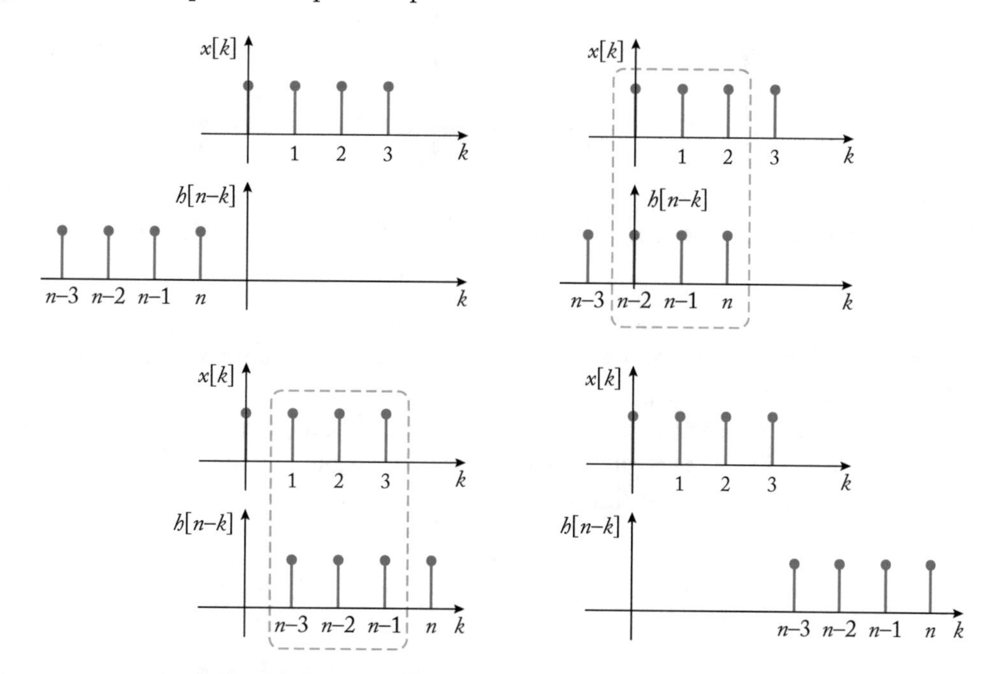

Figure 8.19: Four different situations for the input and impulse response; for $n < 0$, for $0 < n \leq 3$, for $3 \leq n \leq 6$, and $n > 6$.

output has to be computed based on the relative locations of the boundaries. Let's work through this example and try to clarify this.

In Figure 8.19 we show the input $x[k]$ and the flipped and shifted impulse response $h[n-k]$ for four different values of n.

Let's look at the output for each of these conditions. The first condition is easiest to handle. When $n < 0$ there is no range in which both signals are nonzero. So in this case $y[n] = 0$. When $0 \leq n \leq 3$ we have the condition in the top right panel of Figure 8.19. In this case the region of overlap of non-zero values is from 0 to n. Therefore, the output is given by

$$y[n] = \sum_{k=0}^{n} x[k]h[n-k] = \sum_{k=0}^{n} 1 = n+1$$

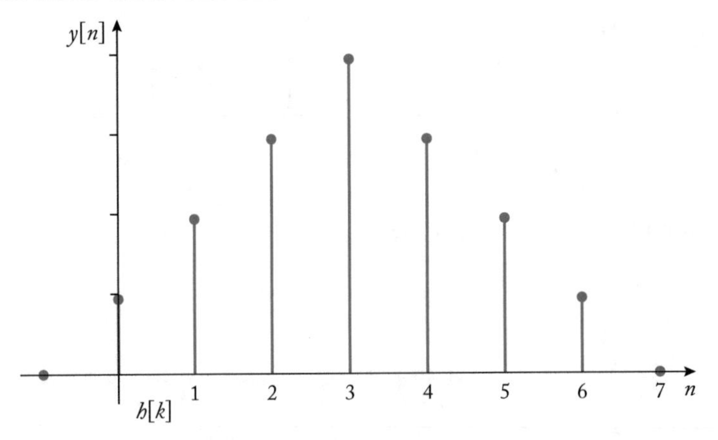

Figure 8.20: Output fot the case where the input and impulse response are identical rectangular pulses.

where we have used the fact that in the region $0 \leq k \leq n$ for $n \leq 3$ both $x[k]$ and $h[n-k]$ are one and therefore the product $x[k]h[n-k]$ will also be one.

The third situation depicted in the bottom left panel of Figure 8.19 is the case where $3 \leq n \leq 6$. In this case from the figure we can see that the region of overlap between nonzero values is between $n-3$ and 3. Therefore,

$$y[n] = \sum_{k=n-3}^{3} x[k]h[n-k] = \sum_{k=n-3}^{3} 1 = 3 - (n-3) + 1 = 7 - n$$

The final case is that depicted in the bottom right panel of Figure 8.19. This is the case where $n > 6$. In this case again there is no overlap between non-zero regions of $x[k]$ and $h[n-k]$ and therefore the output is zero. Putting all this together

$$y[n] = \begin{cases} 0 & n < 0 \\ n+1 & 0 \leq n \leq 3 \\ 7-n & 3 \leq n \leq 6 \\ 0 & n > 6 \end{cases}$$

which is plotted in Figure 8.20.

After all of these examples you can see that computing the discrete time convolution can be tedious but it is really not that difficult. To get reasonably competent at it requires only a modicum of practice.

8.1 SUMMARY

In this module we worked through several examples of discrete time convolution. I strongly recommend that once you have understood the examples and their solutions you close the book and see if you can solve the examples without looking at the solutions.

8.2 EXERCISES

(Answers on the following page)

Given the functions $a[n]$ and $b[n]$, find $c[n]$, where $c[n] = a[n] \circledast b[n]$ where $\circledast$ denotes convolution.

$$c[n] = a[n] \circledast b[n] = \sum_{k=-\infty}^{\infty} a[k]b[n-k] = \sum_{k=-\infty}^{\infty} a[n-k]b[k]$$

1. $a[n] = \delta[n] - \delta[n-4]$, $b[n] = u[n]$
2. $a[n] = (0.2)^n u[n]$, $b[n] = u[n]$
3. $a[n] = (0.2)^n u[n]$, $b[n] = \delta[n-3]$
4. $a[n] = (0.2)^n u[n]$, $b[n] = u[n] - u[n-5]$
5. $a[n] = u[n] - u[n-4]$, $b[n] = u[n-1] - u[n-7]$
6. $a[n] = u[n] - u[n-5]$, $b[n] = u[n] - u[n-5]$
7. $a[n] = u[n] - u[n-5]$, $b[n] = u[n+5] - u[n]$
8. $a[n] = u[n+3] - u[n-4]$, $b[n] = \delta[n-2]$
9. $a[n] = u[n] - u[n-5]$, $b[n] = u[n-1] - u[n-4]$

8.3 ANSWERS

1.

$$c[n] = u[n] - u[n-4]$$

2.

$$c[n] = 1.25\left(1 - (o.2)^{n+1}\right)$$

3.

$$c[n] = (0.2)^{n-3}u[n-3]$$

4.

$$c[n] = 1.25\left((0.2^{n-4}(1 - (0.2)^5)\right)$$

5.

$$c[n] = \begin{cases} 0 & n < 0 \\ n+1 & 0 \le n \le 3 \\ 4 & 3 \le n \le 6 \\ 10-n & 6 \le n \le 9 \\ 0 & n > 9 \end{cases}$$

6.

$$c[n] = \begin{cases} 0 & n < 0 \\ n+1 & 0 \le n \le 4 \\ 9 - n4 \le n \le 8 \\ 0 & n > 8 \end{cases}$$

7.

$$c[n] = \begin{cases} 0 & n < 5 \\ n+6 & -5 \le n \le -1 \\ 4-n & -1 \le n \le 3 \\ 0 & n > 3 \end{cases}$$

8.

$$c[n] = u[n+1] - u[n-6]$$

9.

$$c[n] = \begin{cases} 0 & n < 1 \\ n & 1 \le n \le 3 \\ 3 & 3 \le n \le 5 \\ 8-n & 5 \le n \le 7 \end{cases}$$

MODULE 9

Continuous Time Convolution

We turn our attention now to the response of continuous time linear time invariant systems. Given the impulse response $h(t)$ of a continuous time linear time invariant system (Figure 9.1) we can find the output $y(t)$ for any input $x(t)$ by solving the convolution integral.

$$y(t) = \int_{-\infty}^{\infty} x(\tau)h(t-\tau)d\tau = \int_{-\infty}^{\infty} x(t-\tau)h(tau)d\tau$$

As in the case of the discrete time convolution we will try and understand how to compute this integral through a number of examples. In order to connect these examples to a system that you are familiar with let's first derive the impulse response of a simple RC circuit. We can then look at the response of this circuit to inputs you are familiar with using the convolution integral and see how well the mathematical approach matches physical reality.

9.1 THE IMPULSE RESPONSE OF A SIMPLE RC CIRCUIT

We will later see an easy way to find the impulse response of this circuit. For now let's just use Kirchoff's laws and differential equations. From the circuit we can see that the input voltage $x(t)$ is equal to the voltage across the resistor plus the voltage across the capacitor. The voltage across the capacitor is simply the output $y(t)$ while the voltage across the resistor is the resistance times the current through the resistor. The current through the resistor is the same current that flows through the capacitor and can be obtained as $C\frac{dy}{dt}$. Putting all this together we get

$$RC\frac{dy}{dt} + y(t) = x(t)$$

or

$$\frac{dy}{dt} + \frac{1}{RC}y(t) = \frac{1}{RC}x(t)$$

Let's now multiply through by the integrating factor $e^{t/RC}$.

$$\frac{dy}{dt}e^{t/RC} + \frac{1}{RC}y(t)e^{t/RC} = \frac{1}{RC}x(t)e^{t/RC}$$

Using the product rule of differentiation we can write the left hand side of this equation as

$$\frac{d}{dt}\left(y(t)e^{t/RC}\right) = \frac{1}{RC}x(t)e^{t/RC}$$

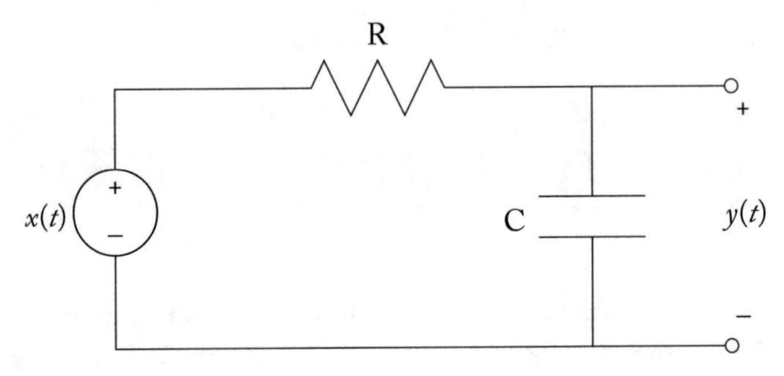

Figure 9.1: An example of a continuous time linear-time-invariant system.

To find the impulse response let's set $x(t) = \delta(t)$ which means $y(t) = h(t)$

$$\frac{d}{dt}\left(h(t)e^{t/RC}\right) = \frac{1}{RC}\delta(t)e^{t/RC}$$

Let's take a look at the product at the right hand side. Recall that $\delta(t)$ is zero everywhere except at $t = 0$, which means that

$$\frac{1}{RC}\delta(t)e^{t/RC} = \frac{1}{RC}\delta(t)e^{0} = \frac{1}{RC}\delta(t)$$

Plugging this back into the differential equation we get

$$\frac{d}{dt}\left(h(t)e^{t/RC}\right) = \frac{1}{RC}\delta(t)$$

We can replace t with τ without changing anything

$$\frac{d}{d\tau}\left(h(\tau)e^{\tau/RC}\right) = \frac{1}{RC}\delta(\tau)$$

Integrating both sides:

$$\int_{-\infty}^{t}\frac{d}{d\tau}\left(h(\tau)e^{\tau/RC}\right)d\tau = \int_{-\infty}^{t}\frac{1}{RC}\delta(\tau)d\tau$$

or

$$h(\tau)e^{\tau/RC}|_{-\infty}^{t} = \frac{1}{RC}u(t)$$

or

$$h(t)e^{t/RC} - 0 = \frac{1}{RC}u(t)$$

from which we can get the impulse response as

$$h(t) = \frac{1}{RC}e^{-t/RC}u(t)$$

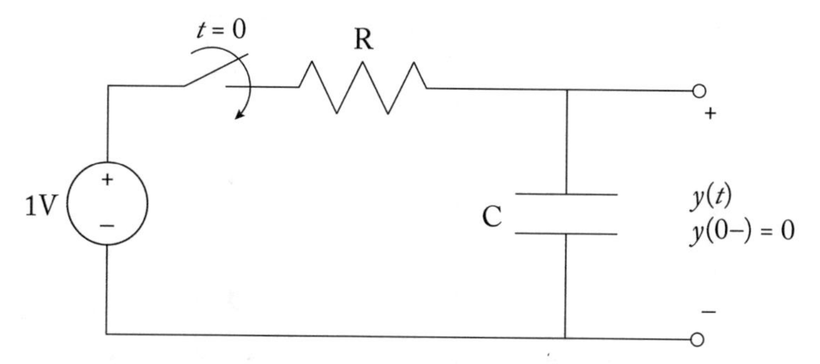

Figure 9.2: The RC filter with a unit step input.

For convenience let's define

$$\alpha = \frac{1}{RC}$$

which means

$$h(t) = \alpha e^{-\alpha t} u(t)$$

where $\alpha > 0$.

Example 9.1 For our first example let's examine the response of the familiar circuit shown in Figure 9.2. We have already derived the impulse response of this circuit.

$$h(t) = \alpha e^{-\alpha t} u(t)$$

The closing of the switch at $t = 0$ can be modeled with unit step input.

$$x(t) = u(t)$$

We know what happens from our previous knowledge; the capacitor will charge up to 1 V with the charging time depending on the RC time constant. Let's see if the convolution integral verifies this result.

$$y(t) = \int_{-\infty}^{\infty} x(\tau)h(t-\tau)d\tau = \int_{-\infty}^{\infty} u(\tau)\alpha e^{-\alpha(t-\tau)}u(t-\tau)d\tau$$

Before we try to solve this integral let's do the same thing with $h(\tau)$ that we did with $h[k]$ to see what happens when we replace τ with $t - \tau$ in the argument. Let's begin with the functions themselves in Figure 9.3 where we have picked $\alpha = 1$.

If we replace τ with $-\tau$ in the impulse response we will flip the response around the y axis. If we now replace it with $t - \tau$ we will drag it back and forth with t being the point around which we flipped the original function. In Figure 9.4 we show $x(\tau)$ and $h(t - \tau)$ for $t < 0$.

In Figure 9.5 we show $x(\tau)$ and $h(t - \tau)$ for $t > 0$.

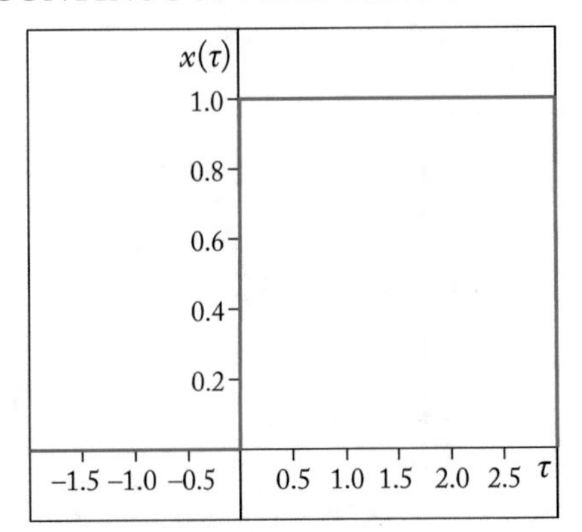

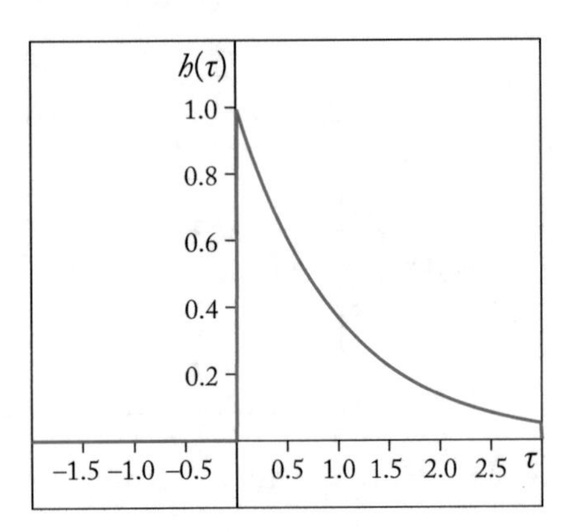

Figure 9.3: Input $x(t)$ and impulse response $h(t)$.

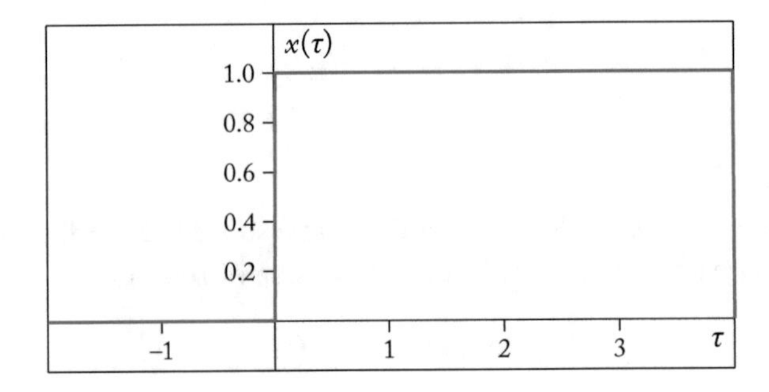

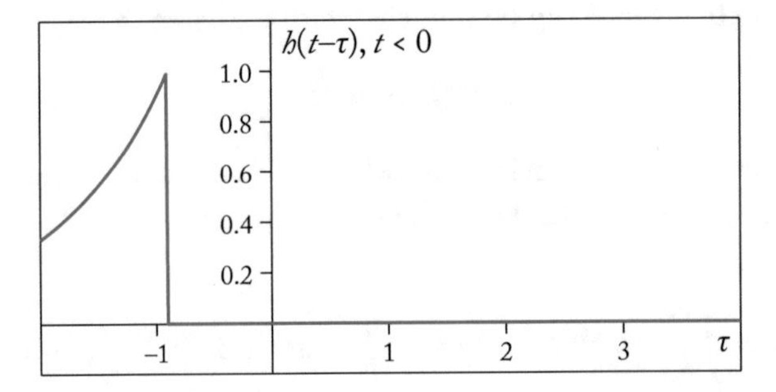

Figure 9.4: $x(\tau)$ and $h(t - \tau)$ for $t < 0$.

Notice that for $t < 0$ there is no overlap between the nonzero portions of $x(\tau)$ and $h(t - \tau)$, therefore, $x(\tau)h(t - \tau)$ will be zero for all values of τ and, therefore, the integral will also be zero.

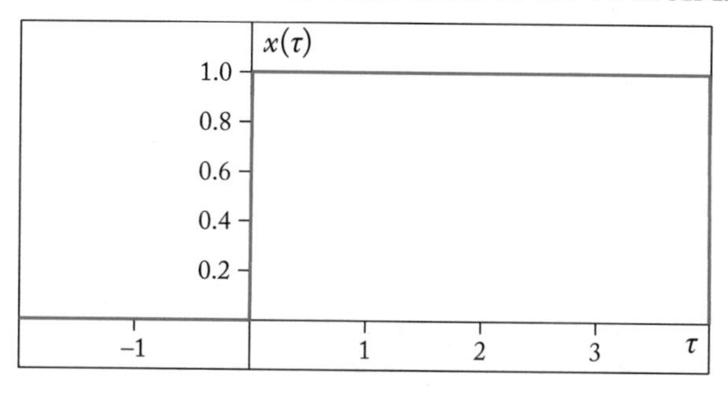

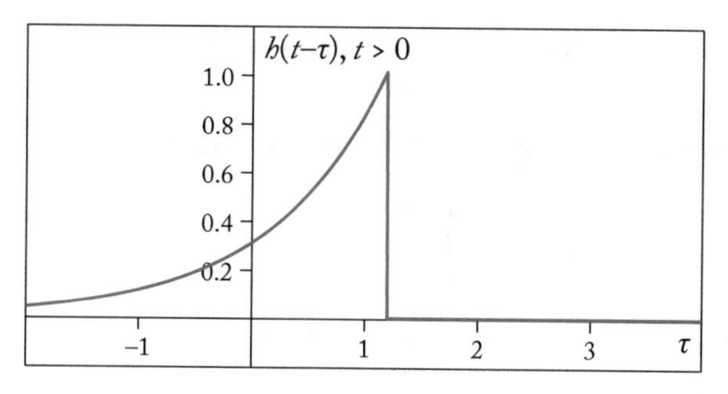

Figure 9.5: $x(\tau)$ and $h(t-\tau)$ for $t > 0$.

For $t > 0$ the overlap between the nonzero portions of $x(\tau)$ and $h(t-\tau)$ is the region between 0 and t, therefore,

$$y(t) = \int_0^t u(\tau)\alpha e^{-\alpha(t-\tau)} u(t-\tau) d\tau$$

In this region $u(\tau) = 1$ as $\tau > 0$. Because $\tau < t$, $t - \tau$ is greater than zero and $u(t-\tau) = 1$. Therefore,

$$\begin{aligned} y(t) &= \alpha \int_0^t e^{-\alpha(t-\tau)} d\tau \\ &= \alpha e^{-\alpha t} t \int_0^t e^{\alpha\tau} d\tau \\ &= \alpha \frac{1}{\alpha} e^t e^{\alpha\tau}|_0^t \\ &= e^{-\alpha t}\left[e^{\alpha t} - e^0\right] \\ &= e^{-\alpha t}\left[e^{\alpha t} - 1\right] \\ &= \left[1 - e^{-\alpha t}\right] \end{aligned}$$

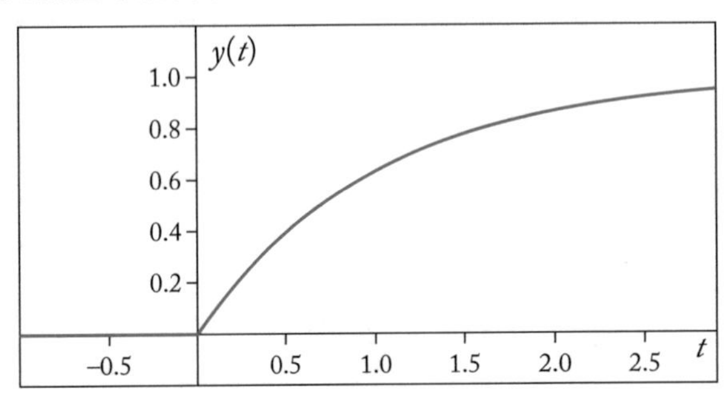

Figure 9.6: Output of the RC filter.

Putting the results for $t < 0$ and $t > 0$ together we have

$$\begin{aligned} y(t) &= \begin{cases} 0 & t < 0 \\ \alpha\left[1 - e^{-\alpha t}\right] & t > 0 \end{cases} \\ &= \alpha\left[1 - e^{-\alpha t}\right]u(t) \end{aligned} \tag{9.1}$$

Plotting this in Figure 9.6, for $\alpha = 1$, we see that indeed the voltage across the capacitor charges up to 1 V and the result of the convolution integral agrees with what we already know from circuit analysis.

Example 9.2 For our second example let's look at how the RC circuit responds to an input pulse with a pulsewidth of two time units. We can represent this pulse as

$$x(t) = u(t) - u(t - 2)$$

The impulse response of the RC system is as before. For the figures in this example we have used $\alpha = 3$.

We will use the same version of the convolution integral

$$y(t) = \int_{-\infty}^{\infty} x(\tau)h(t - \tau)d\tau$$

As in the previous case we will flip the impulse response and move it around. In the previous case we had two situations, $t < 0$ and $t > 0$ which resulted in different expressions. In this case we end up with three situations. The first case is where $t < 0$. As we can see from Figure 9.7 in this case wherever $x(\tau)$ is nonzero $h(t - \tau)$ is zero and vice versa. Therefore, the product $x(\tau)h(t - \tau)$ is zero for all values of τ resulting in $y(t)$ being zero for $t < 0$.

Let's move the impulse response to the right. For $0 < t < 2$ the region over which the product is nonzero is from 0 to t.

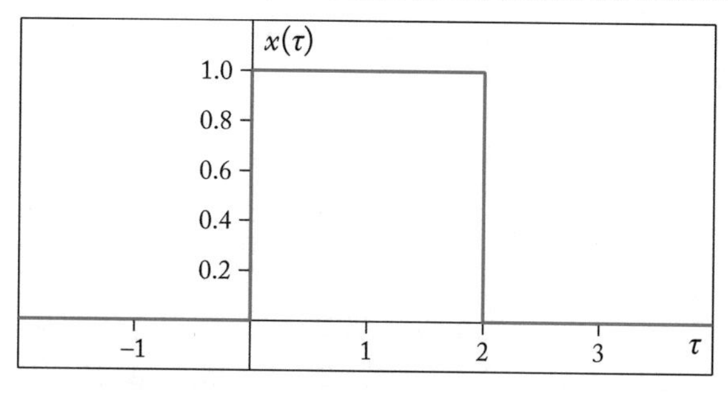

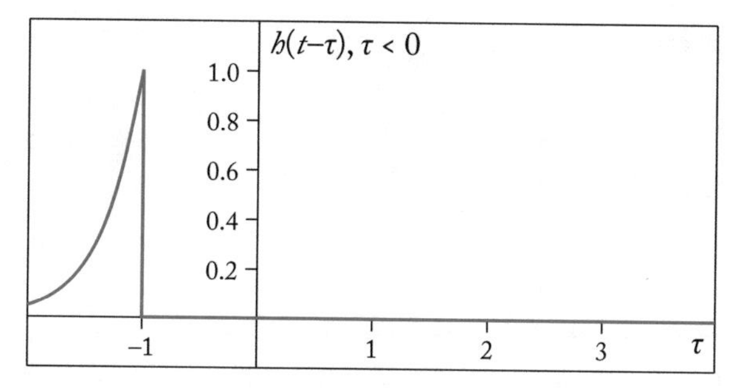

Figure 9.7: $x(\tau)$ and $h(t-\tau)$ for $t < 0$.

Therefore we can compute $y(t)$ as

$$
\begin{aligned}
y(t) &= \int_0^t x(\tau)h(t-\tau)d\tau \\
&= \int_0^t \alpha e^{-\alpha(t-\tau)}d\tau \\
&= \alpha e^{-\alpha t}\int_0^t e^{\alpha\tau}d\tau \\
&= \alpha e^{-\alpha t}\frac{1}{\alpha}e^{\alpha\tau}|_0^t \\
&= e^{-\alpha t}\left[e^{\alpha t}-e^0\right] \\
&= 1-e^{-\alpha t}
\end{aligned}
$$

If we move the impulse response further to the right so that the leading edge of the flipped impulse response t is beyond $\tau = 2$ we end up in the situation shown in Figure 9.9. In this configuration the interval over which the product is nonzero is the interval $[0, 2]$. The output

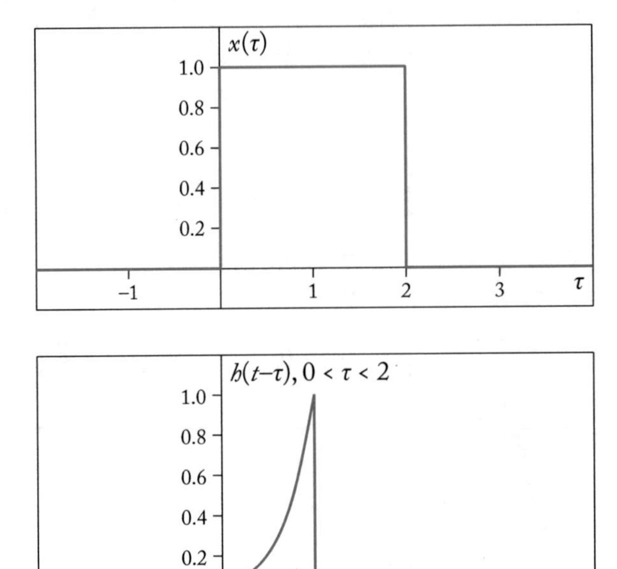

Figure 9.8: $x(\tau)$ and $h(t-\tau)$ for $0 < t < 2$.

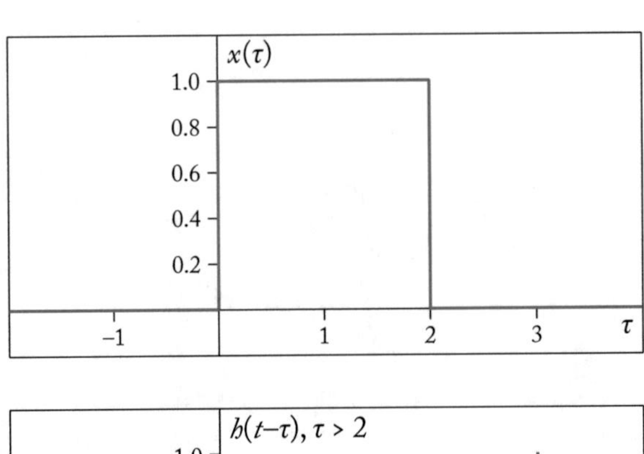

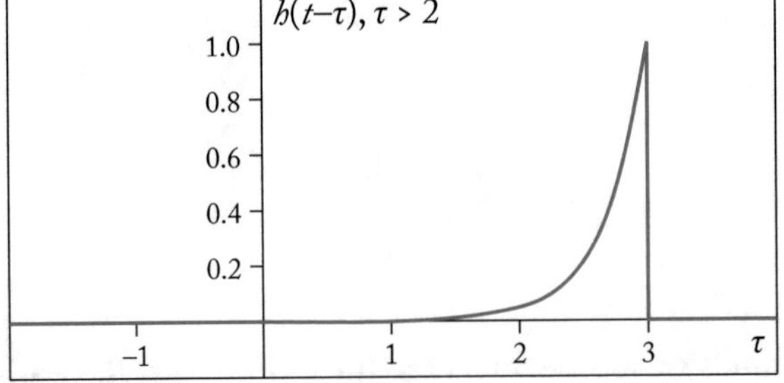

Figure 9.9: $x(\tau)$ and $h(t-\tau)$ for $t > 2$.

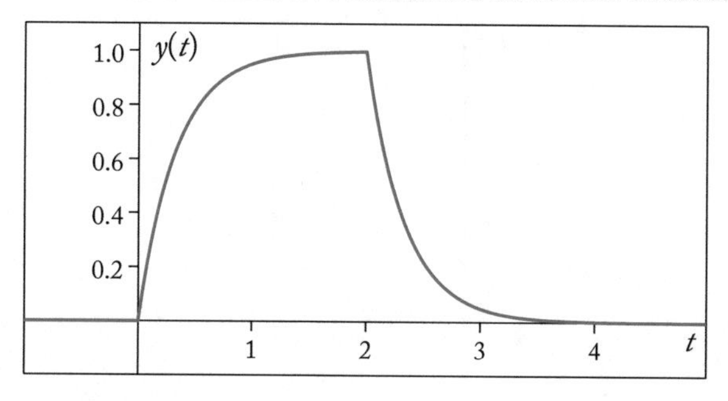

Figure 9.10: Response of the RC circuit to a pulse input.

then is given by

$$\begin{aligned} y(t) &= \int_0^2 x(\tau)h(t-\tau)d\tau \\ &= \int_0^2 \alpha e^{-\alpha(t-\tau)} d\tau \\ &= \alpha e^{-\alpha t} \int_0^2 e^{\alpha\tau} d\tau \\ &= \alpha e^{-\alpha t} \frac{1}{\alpha} e^{\alpha\tau} |_0^2 \\ &= e^{-\alpha t}\left[e^{2\alpha t} - e^0\right] \\ &= e^{-\alpha(t-2)} - e^{-\alpha t} \end{aligned}$$

Putting all of this together we get

$$y(t) = \begin{cases} 0 & t < 0 \\ 1 - e^{-\alpha t} & 0 < t < 2 \\ 1 - e^{-\alpha(t-2)} & t > 2 \end{cases}$$

We plot this in Figure 9.10 for $\alpha = 3$. As you would expect the response shows the capacitor charging up during the time the pulse is active and then discharging after the pulse terminates.

Example 9.3 Finally, let's look at the convolution of two rectangular pulses of different widths. Let's take $x(t)$ to be a pulse of width one and $h(t)$ to be a pulse of width two.

$$\begin{aligned} x(t) &= u(t) - u(t-1) \\ h(t) &= u(t) - u(t-2) \end{aligned}$$

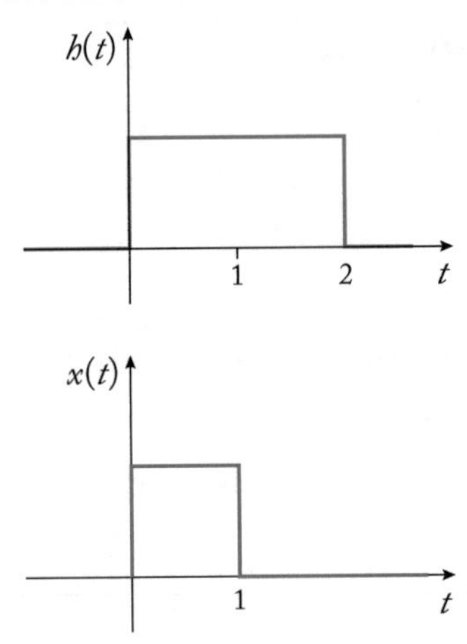

Figure 9.11: The impulse response $h(t)$ and the input $x(t)$.

This time we will use the integral

$$y(t) = \int_{-\infty}^{\infty} x(t-\tau)h(\tau)d\tau$$

The input and impulse response are plotted in Figure 9.11. Generally when convolving signals that are both nonzero over a finite interval it is more convenient to flip and move the narrower signals. However, this might be a matter of preference. You should use whichever formulation feels more convenient to you. Because of the commutative property of the convolution operation both will give exactly the same answer.

The signal $x(t-\tau)$ is obtained by flipping $x(t)$ around the vertical axis and then shifting it by t. When t is less than zero the impulse response $h(\tau)$ and the flipped and shifted signal $x(t-\tau)$ are shown in Figure 9.12. As can be seen from the figure there is no region in which both $h(\tau)$ and $x(t-\tau)$ are nonzero. The impulse response $h(\tau)$ is nonzero for $0 < \tau < 2$ and $x(t-\tau)$ is zero in this region. Therefore the product $x(t-\tau)h(\tau)$ is zero for all values of τ and the integral, and hence $y(t)$ is zero for $t < 0$.

Let's move $x(t-\tau)$ to the right by letting t be greater than zero while keeping $t-1$ to be less than zero. In other words $0 < t < 1$. This condition is shown in Figure 9.13. In this situation the region in which the product $x(t-\tau)h(\tau)$ is nonzero is the interval $0 < \tau < t$. Therefore,

$$y(t) = \int_{0}^{t} d\tau = t$$

where we have used the fact that in the region of integration the product $x(t-\tau)h(\tau)$ is 1.

Continuing to move $x(t-\tau)$ to the right let's increase t so that $t-1$ is greater than zero but keep t to be less than 2. That is $1 < t < 2$. This situation is shown in Figure 9.14. We have

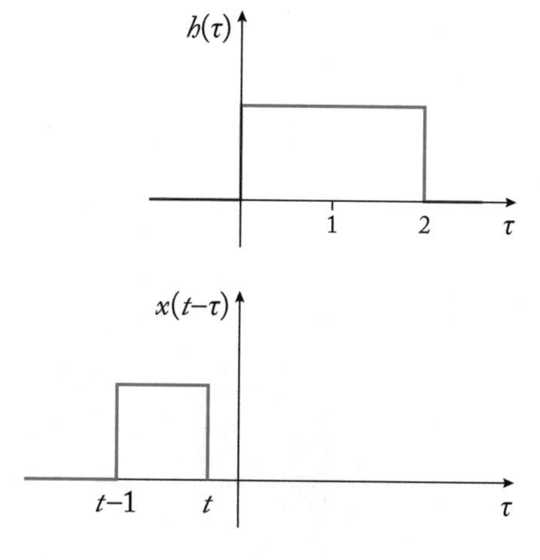

Figure 9.12: The impulse response $h(\tau)$ and the input $x(t-\tau)$ for $t<0$.

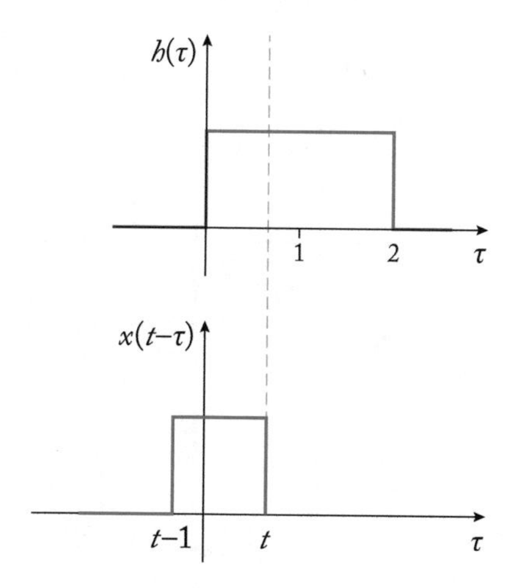

Figure 9.13: The impulse response $h(\tau)$ and the input $x(t-\tau)$ for $0<t<1$.

shown the region in which both $h(\tau)$ and $x(t-\tau)$ are nonzero with the dashed lines. In this condition the integral limits will be $t-1$ and t and

$$y(t)=\int_{t-1}^{t} d\tau = t-(t-1)=1$$

where we have again used the fact that the product $x(t-\tau)h(\tau)$ is one in this region. Notice that this is the region in which the nonzero range of the narrower pulse is completely enclosed within the range of the wider pulse. As both pulses have constant values the integral under this condition will remain constant.

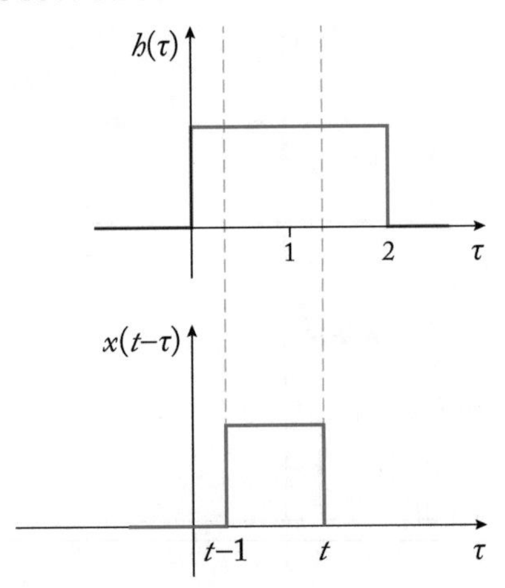

Figure 9.14: The impulse response $h(\tau)$ and the input $x(t-\tau)$ for $1 < t < 2$.

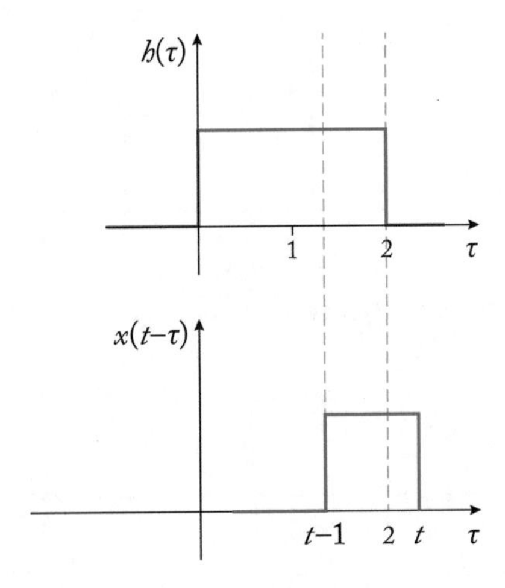

Figure 9.15: The impulse response $h(\tau)$ and the input $x(t-\tau)$ for $2 < t < 3$.

If we now increase t to be larger than 2 while keeping $t-1$ to be less than 2 we will get the situation shown in Figure 9.15. In this situation The interval over which the product $x(t-\tau)h(\tau)$ is nonzero is the interval $t-1 < \tau < 2$. Therefore, the output is given by

$$y(t) = \int_{t-1}^{2} d\tau = 2 - (t-1) = 3 - t$$

Notice that as t increases in this interval the range of nonzero overlap decreases and hence the value of $y(t)$ decreases as well.

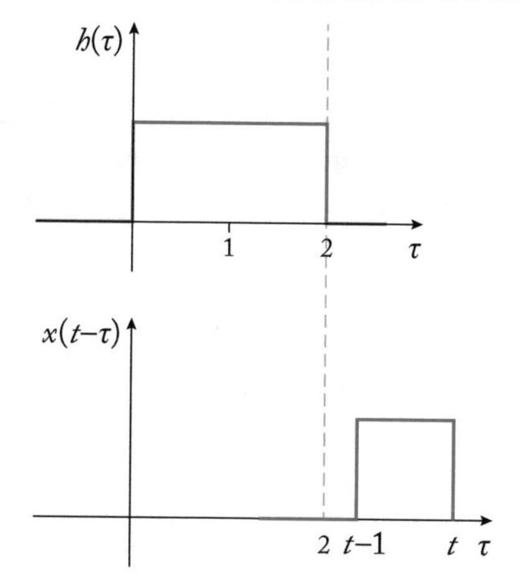

Figure 9.16: The impulse response $h(\tau)$ and the input $x(t-\tau)$ for $t > 3$.

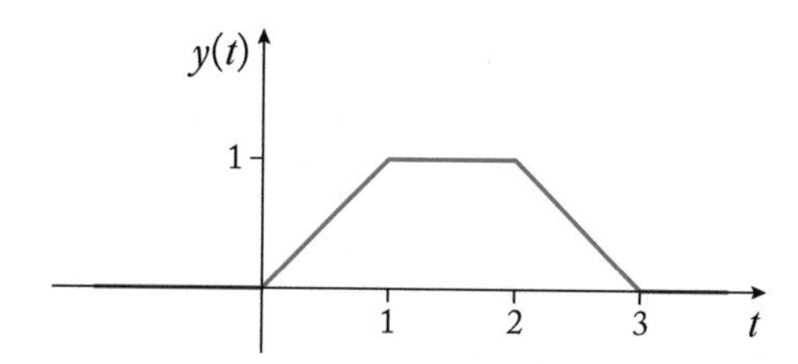

Figure 9.17: The result of the convolution of two pulses of unequal width.

And finally we increase $t-1$, the trailing edge of $x(t-\tau)$ to be greater than 2. This situation is shown in Figure 9.16. Again there is no overlap between the regions in which $h(\tau)$ is nonzero and the region in which $x(t-\tau)$ is nonzero. Therefore, the product $x(t-\tau)h(\tau)$ is zero for all values of τ and the integral, and hence $y(t)$, is zero.

We can combine all of these results together to get

$$y(t) = \begin{cases} 0 & t < 0 \\ t & 0 < t < 1 \\ 1 & 1 < t < 2 \\ 3-t & 2 < t < 3 \\ 0 & t > 3 \end{cases}$$

The output is plotted in Figure 9.17.

9.2 SUMMARY

In this module we worked through several examples of continuous time convolution. As in the previous module I strongly recommend that once you have understood the examples and their solutions you close the book and see if you can solve the examples without looking at the solutions.

9.3 EXERCISES

(Answers on the following page)

1. Given the impulse response $h(t)$ and the input $x(t)$, in each case find the output $y(t)$.
   (a) $x(t) = u(t-2)$, $h(t) = e^{-\alpha t}u(t)$ where $u(t)$ is the unit step function
   (b) $x(t) = u(t)$, $h(t) = e^{-\alpha(t-2)}u(t-2)$ where $u(t)$ is the unit step function
   (c) $x(t) = u(t) - u(t-1)$, $h(t) = u(t-1) - u(t-2)$ where $u(t)$ is the unit step function
   (d) $h(t) = u(t) - u(t-2)$, $x(t) = u(t-1) - u(t-2)$, where $u(t)$ is the unit step function
   (e) $h(t) = e^{-2t}u(t)$, $x(t) = u(t-1) - u(t-2)$, where $u(t)$ is the unit step function
   (f) $h(t) = e^{-2t}u(t)$ and $x(t) = e^{-t}u(t)$, where $u(t)$ is the unit step function
   (g) $h(t) = e^{-2t}u(t)$ and $x(t) = \delta(t-2)$, where $\delta(t)$ is the impulse function
2. The input-output relationship of a linear time-invariant system is given by

$$y(t) = -\frac{1}{\alpha}\int_{-\infty}^{t} x(\tau)d\tau$$

   What is the impulse response of this system? Can you come up with a circuit that would have this impulse response?
3. The input-output relationship of a linear time-invariant system is given by

$$y(t) = \int_{-\infty}^{t} e^{-(t-\tau)}x(\tau-1)d\tau$$

   What is the impulse response of this system?
4. Let

$$\begin{aligned} a(t) &= e^{-2t}u(t) \\ b(t) &= u(t) \\ c(t) &= \delta(t) + \delta(t-1) \end{aligned}$$

   Find $a(t) \circledast b(t) \circledast c(t)$

9.4 ANSWERS

1. (a)

$$y(t) = \frac{1}{\alpha}\left(1 - e^{-\alpha(t-2)}\right) u(t-2)$$

(b)

$$y(t) = \frac{1}{\alpha}\left(1 - e^{-\alpha(t-2)}\right) u(t-2)$$

(c)

$$y(t) = \begin{cases} 0 & t < 1 \\ t-1 & 1 < t < 2 \\ 3-t & 2 < t < 3 \\ 0 & t > 3 \end{cases}$$

(d)

$$y(t) = \begin{cases} 0 & t < 1 \\ t-1 & 1 < t < 2 \\ 1 & 2 < t < 3 \\ 4-t & 3 < t, 4 \\ 0 & t > 4 \end{cases}$$

(e)

$$y(t) = \begin{cases} 0 & t < 1 \\ \frac{1}{2}\left(1 - e^{-2(t-1)}\right) & 1 < t < 2 \\ \frac{1}{2}\left(e^{-2(t-2)} - e^{-2(t-1)}\right) & t > 2 \end{cases}$$

(f)

$$y(t) = \left(e^{-t} - e^{-2t}\right) u(t)$$

(g)

$$y(t) = e^{-2(t-2)} u(t-2)$$

2.

$$h(t) = -\frac{1}{\alpha} u(t)$$

3.

$$h(t) = e^{-(t-1)} u(t-1)$$

4.

$$y(t) = \begin{cases} 0 & t < 0 \\ \frac{1}{2}\left(1 - e^{-2t}\right) & 0 < t < 1 \\ \frac{1}{2}\left(2 - e^{-2} - e^{-2(t-1)}\right) & t > 1 \end{cases}$$

MODULE 10

Fourier Series

We have spent most of our time until now on systems. We have defined their properties and have narrowed our focus to linear time-invariant systems. We have shown that such systems can be completely characterized by their impulse response. We will now pivot to focus mostly on the signals part of the course and develop tools to analyze signals. In order to develop the tools we will use the properties of linear time-invariant systems.

Let's begin with a very simple world in which all signals look something like those shown in Figure 10.1. Suppose I also know that the output of a linear time-invariant system to an input—we will call it a basis function—$b(t)$ shown in Figure 10.2 is $y(t)$.

Comparing $b(t)$ with the first signal $x(t)$ in Figure 10.1 we can see that we can write $x(t)$ in terms of shifted versions of $b(t)$. We can write $x(t)$ as

$$x(t) = b(t) + b(t-2) + b(t-4)$$

Let's define

$$b_k(t) = b(t-k)$$

Then we can write $x(t)$ as

$$x(t) = 1 \cdot b_0(t) + 0 \cdot b_1(t) + 1 \cdot b_2(t) + 0 \cdot b_3(t) + 1 \cdot b_4(t)$$

or in a more succinct form as

$$x(t) = \sum_{k=0}^{4} a_k b_k(t)$$

where

$$a_k = \begin{cases} 1 & k = 0, 2, 4 \\ 0 & k = 1, 3 \end{cases}$$

We can use the same approach to express $v(t)$ in terms of shifted and scaled versions of $b(t)$. Looking at the function we can see that we can write $v(t)$ as

$$v(t) = 0.5b(t+1) - b(t) - 0.5b(t-1) + 0 \cdot b(t-2) + 0.5b(t-3) + 1.5b(t-4)$$

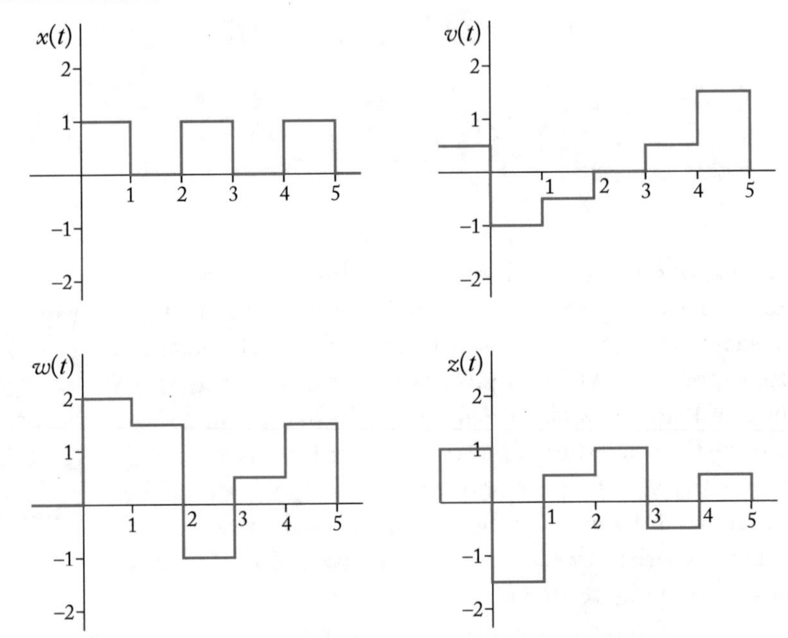

Figure 10.1: Signals in a simple world.

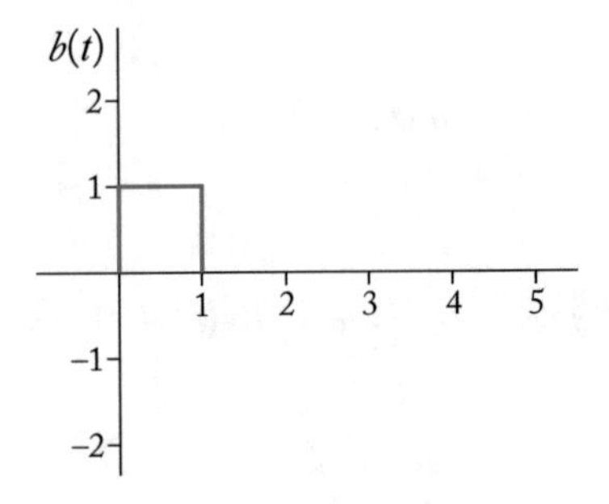

Figure 10.2: The basis function $b(t)$.

or

$$v(t) = \sum_{k=-1}^{4} a_k b_k(t)$$

where

$$a_k = \begin{cases} 0.5 & k = -1 \\ -1 & k = 0 \\ -0.5 & k = 1 \\ 0 & k = 2 \\ 0.5 & k = 3 \\ 1.5 & k = 4 \end{cases}$$

What use is this rewriting of $x(t)$ and $v(t)$ in terms of shifted and scaled versions of $b(t)$? Recall that we said that for a particular linear time invariant system the output of the system for an input of $b(t)$ is $y(t)$. Also recall that a defining property of linear time invariant systems is that if we know the output for specific inputs we also know the outputs for all linear combinations of time-shifted versions of those inputs. As the system we are dealing with is a linear time-invariant system knowing the output for $b(t)$ means we also know the output of the system if the input is $b(t-k)$. The output of the system for this input (because of time-invariance) will be $y(t-k)$. And if the input is $a_k b(t-k)$, or $a_k b_k(t)$, the output is $a_k y(t-k)$. Finally, if the input is $\sum a_k b_k(t)$. The output will be $\sum a_k y(t-k)$. Knowing the output of a linear time invariant system for the input $b(t)$ we can write down the output of the system to any signal from our simple world. The only thing we need to do is to find the values of a_k.

We have been able to find the values of a_k by observation. However, in general, we don't have to rely on our visual acuity to find the values of a_k. We can simply multiply the signal with $b_k(t)$ and integrate. Let's explore this with $w(t)$. Multiplying $w(t)$ with $b_0(t)$ results in a signal which is equal to 2 in the interval [0,1] and 0 outside. Integrating this function we get a value of 2 for a_0. Similarly multiplying $w(t)$ with $b_1(t)$ results in a function which has a value of 1.5 for t between 1 and 2, and 0 otherwise. Integrating this gives us a value of 1.5 for a_1. This is shown in Figure 10.3

We can write this more generally as

$$a_k = \int_{-\infty}^{\infty} w(t) b_k(t) dt$$

We have used limits of $-\infty$ to ∞ here but we could have used any limits that included $b_k(t)$.

If we know the output of a linear time-invariant system for $b(t)$, writing signals in terms of the set $\{b_k(t)\}$ allows us to specify the output of this system for reasonably complex inputs, albeit from our simple world, as long as we know the output when the input is the very simple function $b(t)$. The set $\{b_k(t)\}$ is called a basis set for the signals of our simple world. It would be nice if we could find a basis set for signals which are not limited to our simple world. As it happens there are many such basis sets. However, one set of signals is particularly well suited for linear time-invariant systems. This set is the infinite set consisting of the basis vectors $\{e^{j\omega t}\}$. Why infinite? Well we are talking about a much much more complicated world than our simple

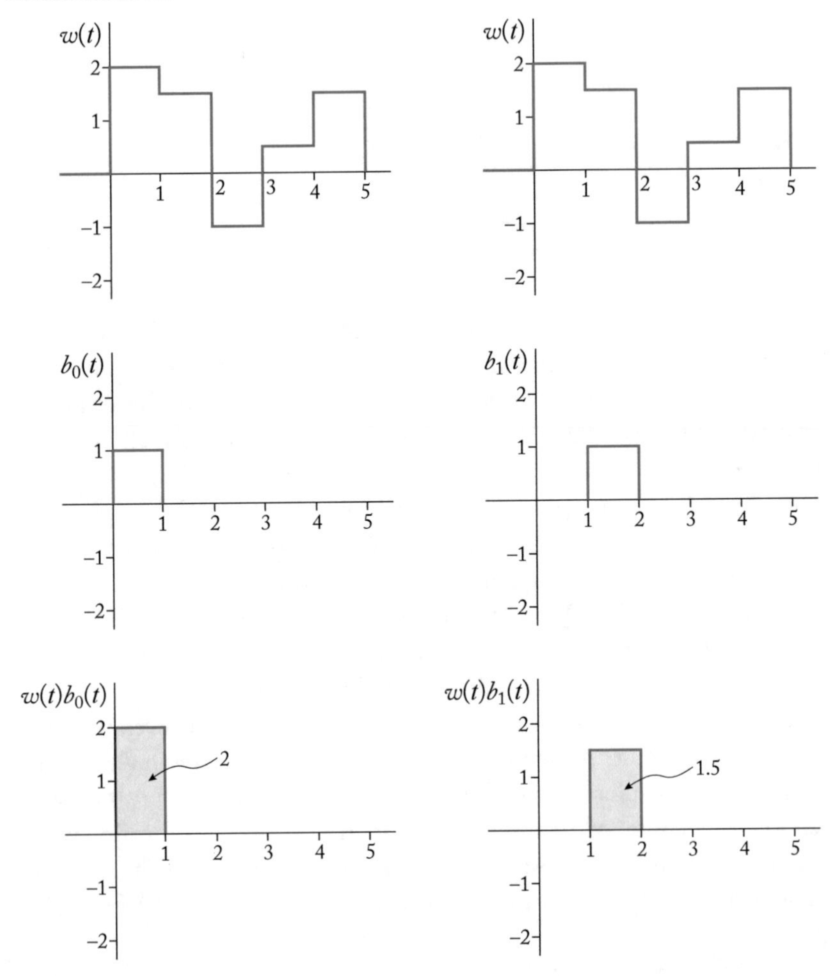

Figure 10.3: Expansion of $w(t)$ in terms of $\{b_k(t)\}$.

world. So what is the very nice property that differentiates this set from other sets? Consider a linear time invariant system with impulse response $h(t)$. We know that the output $y(t)$ is given by the convolution integral

$$y(t) = \int_{-\infty}^{\infty} x(t-\tau)h(\tau)d\tau$$

If the input $x(t)$ is given by

$$x(t) = e^{j\omega t}$$

the output of the system is given by

$$y(t) = \int_{-\infty}^{\infty} e^{j\omega(t-\tau)} h(\tau) d\tau = e^{j\omega t} \int_{-\infty}^{\infty} h(\tau) e^{-j\omega\tau} d\tau$$

Let's define

$$H(\omega) = \int_{-\infty}^{\infty} h(\tau) e^{-j\omega\tau} d\tau$$

then regardless of the value of ω the response of the linear time invariant system to an input of the form $e^{j\omega t}$ will be given by

$$y(t) = e^{j\omega t} H(\omega)$$

But can we write our input in terms of $e^{j\omega t}$? *Si si puede*. We will do this in two steps. First we will restrict the set of signals we are interested in to periodic signals. Once we are comfortable with representing periodic signals in terms of our basis set we will extend our approach to (almost) all signals.

Lets suppose the period of the periodic signal is denoted by T and the fundamental frequency is given by $f_o = 1/T$ in Hz, or $\omega_o = 2\pi f_o = 2\pi/T$ in radians. Our claim for this periodic signal is that it can be represented by linear combinations of $\{e^{jn\omega_o t}\}$. In other words we can write the periodic signal $x(t)$ as

$$x(t) = \sum_{k=-\infty}^{\infty} a_k e^{jk\omega_o t}$$

If we could do this then based on our earlier development we can write $y(t)$ as

$$y(t) = \sum_{k=-\infty}^{\infty} a_k e^{jk\omega_o t} H(k\omega_o)$$

No convolution needed! All we need to do is find $H(\omega)$ once and then whenever we need to find the output for a particular periodic signal all we need to do is find the coefficients $\{a_k\}$. But how are we to do that?

To answer this let's look at the integral over one period of the product of $e^{jk\omega_o t}$ and $e^{-jm\omega_o t}$.

$$\int_T e^{jk\omega_o t} e^{-jm\omega_o t} dt$$

We will look at two cases; when $k \neq m$ and when $k = m$. For convenience we will integrate from 0 to T. First the case where $k \neq m$.

$$
\begin{aligned}
\int_0^T e^{jk\omega_o t} e^{-jm\omega_o t} dt &= \int_0^T e^{j(k-m)\omega_o t} dt \\
&= \frac{1}{(k-m)j\omega_o} e^{j(k-m)\omega_o t} |_0^T \\
&= \frac{1}{(k-m)j\omega_o} \left(e^{j(k-m)\omega_o T} - e^0 \right) \\
&= \frac{1}{(k-m)j\omega_o} \left(e^{j(k-m)\frac{2\pi}{T}T} - 1 \right) \\
&= \frac{1}{(k-m)j\omega_o} \left(e^{j(k-m)2\pi} - 1 \right) \\
&= \frac{1}{(k-m)j\omega_o} (\cos((k-m)2\pi) + j\sin((k-m)2\pi) - 1)
\end{aligned}
$$

where we have used Euler's formula to evaluate $e^{j(k-m)2\pi}$. The cosine of any multiple of 2π is always 1 and the sine of any multiple of 2π is always 0 so this becomes

$$
\frac{1}{(k-m)j\omega_o} (1 + j0 - 1) = 0
$$

Putting all of this together, for $k \neq m$

$$
\int_T e^{jk\omega_o t} e^{-jm\omega_o t} dt = 0
$$

What happens when $k = m$? If $k = m$ then $k - m = 0$ and

$$
\begin{aligned}
\int_0^T e^{jk\omega_o t} e^{-jm\omega_o t} dt &= \int_0^T e^{j(k-m)\omega_o t} dt \\
&= \int_0^T e^0 dt \\
&= \int_0^T dt \\
&= T
\end{aligned}
$$

Combining these we have

$$
\int_T e^{jk\omega_o t} e^{-jm\omega_o t} dt = \begin{cases} 0 & k \neq m \\ T & k = m \end{cases}
$$

If we divide this integral by T we get

$$\frac{1}{T}\int_T e^{jk\omega_o t}e^{-jm\omega_o t}dt = \begin{cases} 0 & k \neq m \\ 1 & k = m \end{cases}$$

Let's use this relationship to find the coefficients $\{a_k\}$ in the expression

$$x(t) = \sum_{k=-\infty}^{\infty} a_k e^{jk\omega_o t}$$

To find a_m we will multiply both sides of this equation by $\frac{1}{T}e^{-jm\omega_o t}$ and integrate over a period.

$$\frac{1}{T}\int_T x(t)e^{-jm\omega_o t}dt = \frac{1}{T}\int_T \sum_{k=-\infty}^{\infty} a_k e^{jk\omega_o t}e^{-jm\omega_o t}dt$$

Interchanging the order of summation and integration (with genuflection to Fubini)

$$\frac{1}{T}\int_T x(t)e^{-jm\omega_o t}dt = \sum_{k=-\infty}^{\infty}\frac{1}{T}\int_T e^{jk\omega_o t}e^{-jm\omega_o t}dt$$

But the integral on the right is zero for all the terms of the summation in which $k \neq m$ and is 1 when $k = m$, so the sum reduces to only one term when $k = m$ and that term is 1. So we get

$$\frac{1}{T}\int_T x(t)e^{-jm\omega_o t}dt = a_m$$

An Alternate Representation (*Optional Reading*)

Think about a point in two or three dimensional space. We can represent this vector as a line from the origin to the point with an arrowhead at the point or we can represent the point with its cartesian coordinates. But what exactly are these coordinates? Let's take a simple example of the point (3, 4) in two dimensional space. We can represent this point as we did with the cartesian coordinates (3, 4) or we can represent it as a vector with magnitude of 5 and an angle of 53.13^o or we can represent the vector as

$$\mathbf{v} = 3u_x + 4u_y$$

or we can represent the vector as

$$\mathbf{v} = \begin{bmatrix} 3 \\ 4 \end{bmatrix}.$$

Consider the representation

$$\mathbf{v} = 3u_x + 4u_y$$

What we are saying here is that the vector representing the point is a linear combination of the *basis vectors* denoted by u_x and u_y. The coordinates are simply the weights of the individual basis vectors. Conceptually this is similar to what we were doing before with the two basic signals—the rectangle and the triangle—back when we were defining a linear time-invariant system. Practically though there are several differences. The basis set $\{u_x, u_y\}$ allow us to represent any point in two dimensional space. We can go in the other direction with the same ease; it is easy to obtain the weights on the individual basis vectors by simply taking the dot product of the vector with the basis vector of interest. So,

$$\mathbf{v} \cdot u_x = (3u_x + 4u_y) \cdot u_x = 3u_x \cdot u_x + 4u_y \cdot u_x = 3 \times 1 + 4 \times 0 = 3$$

and

$$\mathbf{v} \cdot u_y = (3u_x + 4u_y) \cdot u_y = 3u_x \cdot u_y + 4u_y \cdot u_y = 3 \times 0 + 4 \times 1 = 4$$

What makes life easy for us in this case are two facts

1. The basis vectors are orthogonal to each other and have unit magnitude.
2. We have something called the *dot product* or the *inner product* that allows us to both define what orthogonal means and then gives us the machinery to decompose a vector in two dimensional space into components in terms of the basis vectors.

Our task is now clear: we need to find a definition of an inner product which will work for the signals in which we are interested. We need to find a basis set which is orthogonal in terms of this inner product, in which each element has unit magnitude, and the combinations of which will give us all of the signals in which we are interested. Let's deal with the last issue first—what are all the signals of interest to us? Practically speaking these would be all signals with a finite amount of energy. For an aperiodic signal $x(t)$ this means that

$$\int_{-\infty}^{\infty} |x(t)|^2 < \infty$$

For periodic signals we will limit the integral to be over one period. Are there basis vectors that can be used to represent all signals with finite energy? Turns out there are quite a few, and we get to choose which set we want. We will select a set which has the additional very nice property that given any member of this basis set and the impulse response of a linear time invariant system we can easily figure out the output.

As we have discussed previously this set is the infinite set consisting of the basis vectors $\{e^{j\omega t}\}$. How can we use these ideas of inner product and basis sets to write our input in terms of $e^{j\omega t}$? To see this let us first restrict the set of signals to periodic signals. Lets suppose the period of the periodic signal is denoted by T and the fundamental frequency is given by $f_o = 1/T$ in Hz, or $\omega_o = 2\pi f_o = 2\pi/T$ in radians. Our claim for this periodic signal is that it can be represented by linear combinations of $\{e^{jn\omega_o t}\}$. In other words we can write the periodic signal $x(t)$ as

$$x(t) = \sum_{n=-\infty}^{\infty} a_n e^{jn\omega_o t}$$

If we could do this then based on our earlier development we can write $y(t)$ as

$$y(t) = \sum_{n=-\infty}^{\infty} a_n e^{jn\omega_o t} H(n\omega_o)$$

To find the values of a_n let's refer back to our simple example of a vector in two dimensional space. We found the weights of the basis vector by taking the inner product of the vector with the basis functions. We know what the basis functions are in this case—$e^{jn\omega_o t}$. All we need is the inner product. For continuous functions $f(t)$ and $g(t)$ we can define an inner product as

$$\langle f(t), g(t) \rangle = \alpha \int f(t) g^*(t) dt$$

where $g^*(t)$ is the complex conjugate of $g(t)$ and α is a fudge factor we will use to make sure that our basis vectors have a magnitude of one. For our specific case we will pick α to be equal to $1/T$ and we will pick the limits of integration to cover one period. That means we can integrate from 0 to T or from $-T/2$ to $T/2$ or whatever makes the integral more tractable. So our inner product is

$$\langle f(t), g(t) \rangle = \frac{1}{T} \int_0^T f(t) g^*(t) dt$$

Let's use this definition of the inner product to see if the set $\{e^{jn\omega_o t}\}$ is orthonormal. First, lets see if this set is orthogonal. To do this we pick two basis functions $e^{jn\omega_o t}$ and $e^{jm\omega_o t}$ where $m \neq n$ and examine their inner product.

$$\langle e^{jn\omega_o t}, e^{jm\omega_o t} \rangle = \frac{1}{T} \int_0^T e^{jn\omega_o t} e^{-jm\omega_o t} dt$$

which we have already shown to be zero if $n \neq m$. This means that the elements of the set $\{e^{jn\omega_o t}\}$ are orthogonal. To check orthonormality we need to see if the magnitude of these elements is one. We can do that by taking the inner product of $e^{jn\omega_o t}$ with itself

$$\langle e^{jn\omega_o t}, e^{jn\omega_o t} \rangle = \frac{1}{T} \int_0^T e^{jn\omega_o t} e^{-jn\omega_o t} dt$$

which we have already shown to be 1. So the set $\{e^{jn\omega_o t}\}$ is indeed orthonormal. Now if we want to find the components a_n all we need to do is to take the inner product of the signal $x(t)$ with the unit vector in the nth direction—$e^{jn\omega_o t}$.

$$\begin{aligned}
\langle x(t), e^{jn\omega_o t}\rangle &= \left\langle \sum_{k=-\infty}^{\infty} a_k e^{jn\omega_o t}, e^{jn\omega_o t} \right\rangle \\
&= \frac{1}{T}\int_0^T \sum_{k=-\infty}^{\infty} a_k e^{jk\omega_o t} e^{-jn\omega_o t}\, dt \\
&= \sum_{k=-\infty}^{\infty} a_k \left[\frac{1}{T}\int_0^T e^{j(k-n)\omega_o t}\, dt\right] \\
&= a_n
\end{aligned}$$

where the last equality is a result of the fact that the quantity in brackets is zero for all values of k except for $k = n$. So,

$$a_n = \langle x(t), e^{jn\omega_o t}\rangle = \frac{1}{T}\int_T x(t)e^{-jn\omega_o t}\, dt$$

Note that we have replaced the limits of the integral from 0 and T to a more general form. This is to indicate that integral limits can be anything as long as they cover one whole period. So instead of taking the integral from 0 to T we could take the integral from $-T/2$ to $T/2$ or from $-T/4$ to $3T/4$ etc.

Does this integral always exist? No, it doesn't but the conditions under which it does not exist are extremely rare from an engineering point of view. The conditions for the existence of the integral are called the Dirichlet conditions for convergence. They are:

1. The function $x(t)$ should be absolutely integrable.

$$\int |x(t)| < \infty$$

2. The function has bounded variation which means that in any finite interval it has only a finite number of zero crossings.

3. It has a finite number of finite discontinuities.

Pretty much every signal we will ever deal with will satisfy these conditions.

We can speak of a_n as the coefficients of the Fourier series expansion or we can view them as the components of the signal at different frequencies. In particular a_0 is the component of the signal at 0 frequency which we refer to as the DC value or the average value of the function. To check that a_0 is indeed the average value of the function we can plug in $n = 0$ in the equation for obtaining the Fourier series coefficients and we get

$$a_0 = \frac{1}{T}\int_T x(t)e^{-j(0)\omega_0 t}dt = \frac{1}{T}\int_T x(t)dt$$

which is indeed the average value of $x(t)$.

10.1 FOURIER SERIES EXPANSION OF A SQUARE WAVE

Let's find the Fourier series coefficients a_n for the square wave shown in Figure 10.4.

$$\begin{aligned}
a_n &= \frac{1}{T}\int_{-T/2}^{T/2} e^{-jn\omega_0 t}dt \\
&= \frac{1}{T}\int_{-T_0}^{T_0} e^{-jn\omega_0 t}dt \\
&= \frac{1}{T}\frac{1}{-jn\omega_0}\left(e^{-jn\omega_0 T_0} - e^{jn\omega_0 T_0}\right) \\
&= \frac{1}{jn\omega_0 T}\left(e^{jn\omega_0 T_0} - e^{-jn\omega_0 T_0}\right)
\end{aligned}$$

substituting

$$\omega_0 = \frac{2\pi}{T}$$

$$\begin{aligned}
a_n &= \frac{T}{jn2\pi T}\left(e^{jn\omega_0 T_0} - e^{-jn\omega_0 T_0}\right) \\
&= \frac{1}{n\pi}\frac{e^{jn\omega_0 T_0} - e^{-jn\omega_0 T_0}}{2j} \\
&= \frac{\sin(n\omega_0 T_0)}{n\pi}
\end{aligned}$$

Let's see what this coefficients look like for different values of the pulse width. First, let's pick T_0 to be equal to $T/4$. Plugging in for ω_o and T_o in the expression for a_n we get

$$a_n = \frac{\sin\left(n\frac{2\pi}{T}\frac{T}{4}\right)}{n\pi} = \frac{\sin(n\pi/2)}{n\pi}$$

Plugging in for $n = 0$ in this expression we get

$$a_0 = \frac{\sin(0)}{0} = \frac{0}{0}$$

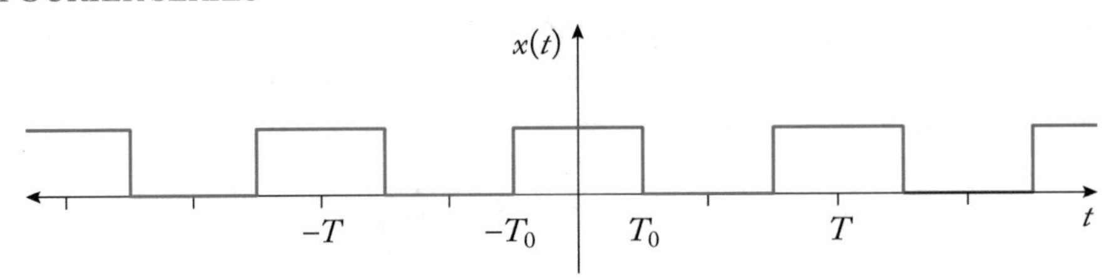

Figure 10.4: A square wave with period T and pulse width $2T_0$.

Because we have a $0/0$ form we can use l'Hopital's rule to evaluate a_0. A quick reminder: the way we use l'Hopital's rule is to take the derivative of the numerator and the denominator with respect to the variable being substituted for and then evaluate the resulting expression. In this case the variable is n so taking the derivative with respect to n and then evaluating the result at $n = 0$ we get

$$a_0 = \frac{\frac{d}{dn}\sin(n\pi/2)}{\frac{d}{dn}n\pi}\Big|_{n=0} = \frac{\pi/2\cos(n\pi/2)}{\pi}\Big|_{n=0} = \frac{\pi/2}{\pi} = \frac{1}{2}$$

a_1 is much simpler to evaluate

$$a_1 = \frac{\sin(n\pi/2)}{n\pi}\Big|_{n=1} = \frac{\sin(\pi/2)}{\pi}\Big|_{n=1} = \frac{1}{\pi}$$

What about a_{-1}?

$$a_{-1} = \frac{\sin(n\pi/2)}{n\pi}\Big|_{n=-1} = \frac{\sin(-\pi/2)}{-\pi}\Big|_{n=1} = \frac{-1}{-\pi} = \frac{1}{\pi}$$

which is the same as a_1. In fact the expression for a_n is even in terms of n

$$a_{-n} = \frac{\sin(-n\pi/2)}{-n\pi} = \frac{-\sin(n\pi/2)}{-n\pi} = \frac{\sin(n\pi/2)}{n\pi} = a_n$$

Continuing with our calculation of the values of a_n

$$\begin{aligned}
a_2 &= \frac{\sin(n\pi/2)}{n\pi}\Big|_{n=2} = \frac{\sin(\pi)}{2\pi} = 0 \\
a_3 &= \frac{\sin(n\pi/2)}{n\pi}\Big|_{n=3} = \frac{\sin(3\pi/2)}{3\pi} = -\frac{1}{3\pi} \\
a_4 &= \frac{\sin(n\pi/2)}{n\pi}\Big|_{n=4} = \frac{\sin(2\pi)}{4\pi} = 0 \\
a_5 &= \frac{\sin(n\pi/2)}{n\pi}\Big|_{n=5} = \frac{5\pi/2}{5\pi} = \frac{1}{5\pi} \\
&\vdots
\end{aligned}$$

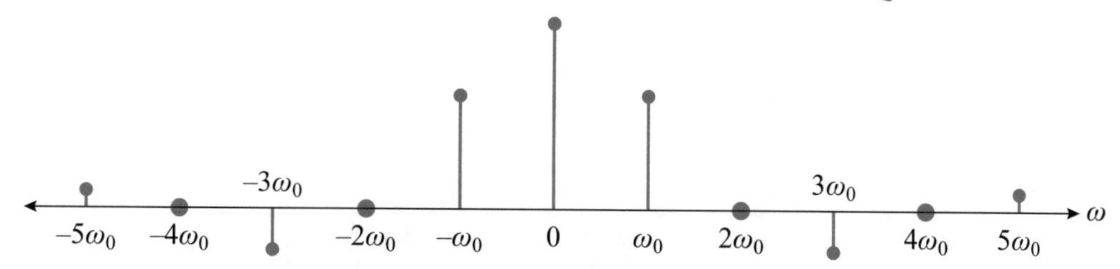

Figure 10.5: Fourier series coefficients for a square wave with period T and pulse width $T/2$.

We don't need to compute the values of a_n for negative values of n because the expression for a_n is even in terms of n. Plotting these coefficients we obtain Figure 10.5

10.2 TIME FREQUENCY DUALITY

Let's continue with our previous example of a square wave but his time let's view it as a train of pulses and let's play a bit with the width of the pulse. In the previous example the width of the pulse was $T/2$. We will make the pulse narrower by setting T_o to $T/8$ so the width of the pulse is $T/4$. Plugging in for ω_o and T_o in the expression for a_n we get

$$a_n = \frac{\sin\left(n\frac{2\pi}{T}\frac{T}{8}\right)}{n\pi} = \frac{\sin(n\pi/4)}{n\pi}$$

Plugging in $n = 0$ will again give us a $0/0$ form so we again use l'Hopital's rule to get

$$a_0 = \frac{\frac{d}{dn}\sin(n\pi/4)}{\frac{d}{dn}n\pi}|_{n=0} = \frac{\pi/4\cos(n\pi/4)}{\pi}|_{n=0} = \frac{\pi/4}{\pi} = \frac{1}{4}$$

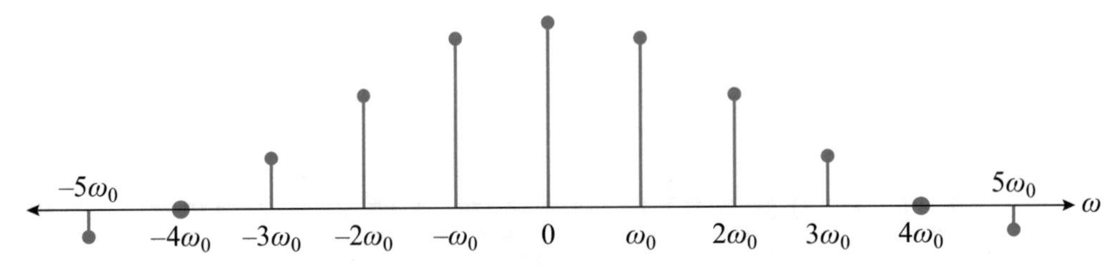

Figure 10.6: Fourier series coefficients for a square wave with period T and pulse width $T/4$.

Computing a_n for other values of n we obtain

$$
\begin{aligned}
a_1 &= \frac{\sin(n\pi/4)}{n\pi}\Big|_{n=1} = \frac{\sin(\pi/4)}{\pi} = \frac{1/\sqrt{2}}{\pi} = \frac{1}{\sqrt{2}\pi} \\
a_2 &= \frac{\sin(n\pi/4)}{n\pi}\Big|_{n=2} = \frac{\sin(\pi/2)}{2\pi} = \frac{1}{2\pi} \\
a_3 &= \frac{\sin(n\pi/4)}{n\pi}\Big|_{n=3} = \frac{\sin(3\pi/4)}{3\pi} = \frac{1/\sqrt{2}}{3\pi} = \frac{1}{3\sqrt{2}\pi} \\
a_4 &= \frac{\sin(n\pi/4)}{n\pi}\Big|_{n=4} = \frac{\sin(\pi)}{4\pi} = 0 \\
a_5 &= \frac{\sin(n\pi/4)}{n\pi}\Big|_{n=5} = \frac{\sin(5\pi/4)}{\pi} = \frac{-1/\sqrt{2}}{5\pi} = -\frac{1}{5\sqrt{2}\pi} \\
&\vdots
\end{aligned}
$$

Plotting these in Figure 10.6 we see that there is a qualitative difference between the spread of the Fourier series coefficients. If we think of the energy in the signal as the sum of the squares of the coefficients, the energy in the coefficients for the narrower pulse is spread much broader than the energy in the coefficients of the broader pulse. A measure of the spread of energy in the frequency components is the *bandwidth* of the signal. Without giving a precise definition of bandwidth we can see that the bandwidth of the narrower pulse train is larger than the bandwidth of the broader pulse train. We will find that this is a basic rule. The narrower a signal is in the time domain the broader will be its spectral footprint corresponding to a broader bandwidth. When we think about digital communication we can see that bandwidth is very closely related to rate. The more bits we need to send the narrower the pulses representing the bits have to be. If we want to send information at 1000 bits per second and we are using one pulse per bit the width of the pulse has to be less than one millisecond. If we want to up the rate to a 1 megabits per second the width of the pulse has to be less than .001 milliseconds. The higher the rate, the narrower the pulse and hence the more bandwidth we need.

The narrower a signal is in the time domain the wider will be its frequency profile and vice versa.

For our next example let's try a really simple periodic function

$$x(t) = \cos(\omega_o t)$$

For this particular example we don't actually need to evaluate an integral. That is because we already know a way to write the cosine function in terms of complex exponentials

$$\cos(\omega_o t) = \frac{1}{2}e^{j\omega_o t} + \frac{1}{2}e^{-j\omega_o t}$$

So, the Fourier series for $x(t) = \cos(\omega_o t)$ has only two nonzero coefficients, a_1 and a_{-1} with

$$a_1 = \frac{1}{2} = a_{-1}$$

Let's look at a the Fourier series expansion, or the frequency representation of a couple more signals. Let's begin with

$$x(t) = \sin(\omega_o t)$$

We can find the expansion for this in the same way we did for $\cos(\omega_o t)$ by using Euler's Excellent Formula.

$$x(t) = \sin(\omega_o t) = \frac{e^{j\omega_o t} - e^{-j\omega_o t}}{2j} = -\frac{1}{2j}e^{-j\omega_o t} + \frac{1}{2j}e^{j\omega_o t}$$

Comparing this with the standard Fourier series representation

$$x(t) = \sum_{n=-\infty}^{\infty} a_n e^{jn\omega_o t}$$

we can see that the Euler's formula does indeed give us the signal as a weighted sum of $\{e^{jn\omega_o t}\}$ where

$$a_n = \begin{cases} -\frac{1}{2j} & n = -1 \\ \frac{1}{2j} & n = 1 \\ 0 & \text{otherwise} \end{cases}$$

Let's get a bit more adventurous and find the frequency representation, or the Fourier series representation of the sawtooth wave shown in Figure 10.7.

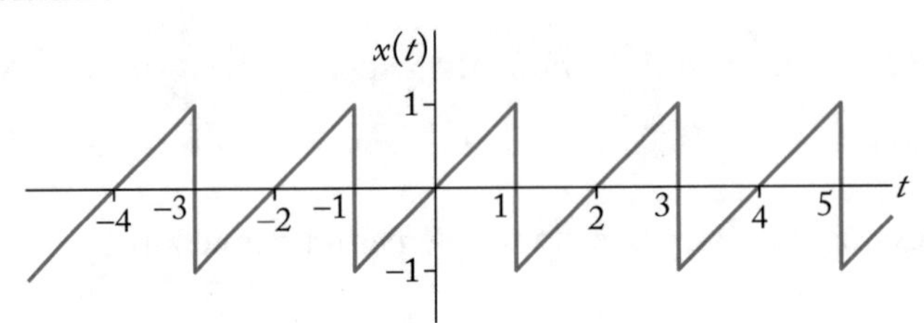

Figure 10.7: A sawtooth wave.

Clearly this is a periodic wave. The period is 2, and therefore, $\omega_o = \pi$. The Fourier series representation is given by

$$x(t) = \sum_{n=-\infty}^{\infty} a_n e^{jn\pi t}$$

with the coefficients a_n given by

$$\begin{aligned}
a_n &= \frac{1}{2}\int_{-1}^{1} te^{-jn\pi t}dt \\
&= \frac{1}{2}\left(\frac{t}{-jn\pi} - \frac{1}{(jn\pi)^2}\right)e^{-jn\pi t}|_{-1}^{1} \\
&= \frac{1}{2}\left[\left(\frac{1}{-jn\pi} + \frac{1}{(n\pi)^2}\right)e^{-jn\pi} - \left(\frac{-1}{-jn\pi} + \frac{1}{(n\pi)^2}\right)e^{jn\pi}\right] \\
&= \frac{1}{2}\left[\frac{-1}{jn\pi}\left(e^{-jn\pi} + e^{jn\pi}\right) + \frac{1}{(n\pi)^2}\left(e^{-jn\pi} - e^{jn\pi}\right)\right] \\
&= -\frac{1}{jn\pi}\left(\frac{e^{jn\pi} + e^{-jn\pi}}{2}\right) - \frac{1}{(n\pi)^2}\left(\frac{e^{jn\pi} - e^{-jn\pi}}{2}\right)
\end{aligned}$$

The first term in parentheses is $\cos(n\pi)$ while the second term in parentheses is $j\ \sin(n\pi)$.

$$\cos(n\pi) = \begin{cases} -1 & n \text{ odd} \\ 1 & n \text{ even} \end{cases} = (-1)^n$$

and $\sin(n\pi)$ is zero for all n. Therefore

$$a_n = -\frac{1}{jn\pi}(-1)^n = \frac{j}{n\pi}(-1)^n$$

10.3 SUMMARY

In this module we learned how to represent a periodic function in terms of complex exponentials as

$$x(t) = \sum_{n=-\infty}^{\infty} a_n e^{jn\omega_o t}$$

where the coefficients a_n are given by

$$a_n = \frac{1}{T} \int_T x(t) e^{-jn\omega_o t} dt$$

We computed the Fourier series for a couple of square waves and a sawtooth wave and looked at the effect of narrowing the pulse width of the square wave on the Fourier series coefficients.

10.4 EXERCISES

(Answers on the following page)

1. Given the signal

$$x(t) = 2\cos\left(\frac{\pi}{4}t\right) + 3\cos\left(\frac{3\pi}{4}t\right)$$

determine the fundamental frequency ω_o and the Fourier series coefficients $\{a_k\}$ such that

$$x(t) = \sum_{k=-\infty}^{\infty} a_k e^{jk\omega_o t}$$

2. One period of a periodic signal $x(t)$ with period $T = 2$ is given by

$$x(t) = \begin{cases} 1 & 0 < t < 1 \\ 0 & 1 < t < 2 \end{cases}$$

Find the Fourier series coefficients $\{a_k\}$ such that

$$x(t) = \sum_{k=-\infty}^{\infty} a_k e^{jk\pi t}$$

Can you relate this to what we did in class?

3. One period of a periodic signal $x(t)$ with period $T = 2$ is given by

$$x(t) = t, \quad -1 < t < 1$$

Find the Fourier series coefficients $\{a_k\}$ such that

$$x(t) = \sum_{k=-\infty}^{\infty} a_k e^{jk\pi t}$$

4. One period of a periodic signal $x(t)$ with period $T = 2$ is given by

$$x(t) = \begin{cases} t+1 & -1 < t < 0 \\ 1-t & 0 < t < 1 \end{cases}$$

Find the Fourier series coefficients $\{a_k\}$ such that

$$x(t) = \sum_{k=-\infty}^{\infty} a_k e^{jk\pi t}$$

10.5 ANSWERS

1.

$$\omega_o = \frac{\pi}{4}$$

$$a_1 = a_{-1} = -1; \quad a_3 = a_{-3} = \frac{3}{2}$$

2.

$$a_n = e^{-jn\frac{\pi}{2}} \frac{\sin\left(\frac{n\pi}{2}\right)}{n\pi}$$

This waveform is a shifted version of the square wave in the example with $T_0 = 1/2$ and the shift $t_0 = 1/2$.

3.

$$a_n = \frac{j}{n\pi}(-1)^n$$

4.

$$a_n = \frac{1}{(n\pi)^2}(1 - \cos(n\pi))$$

MODULE 11

Fourier Series – Properties and Interpretation

We know how to compute the Fourier series. While this is important the Fourier series is useful only if we understand how this representation of the Fourier series relates to the time function from which we calculated it. And what it can tell us about the function that we could perhaps have not seen so easily before. Let's start with the fact that the Fourier series coefficients are a unique representation of the periodic time function. Given a set of Fourier series coefficients and the fundamental frequency there is one and only one periodic function that we can recover from them. And, vice versa. We could write this as

$$x(t) \Leftrightarrow \{a_n\}, \omega_o$$

Another way of thinking about this is that the time function $x(t)$ and the Fourier coefficients $\{a_n\}$ are two ways of looking at the same thing.

One of the obvious things it implies is that the transformation from one representation to the other is linear; the Fourier coefficients of the sum of two signals are the sum of the Fourier coefficients of individual signals. In other words, if we have two signals $x(t)$ and $y(t)$ which are both periodic with the same period T and have Fourier series coefficients $\{a_n\}$ and $\{b_n\}$ then the Fourier series coefficients of any linear combination of $x(t)$ and $y(t)$ will be a linear combination of $\{a_n\}$ and $\{b_n\}$. Suppose we construct a signal

$$z(t) = \alpha x(t) + \beta y(t)$$

The Fourier series coefficients $\{c_n\}$ corresponding to $z(t)$ are given by

$$c_n = \alpha a_n + \beta b_n$$

This is a very nice property for obvious reasons. There are many times when this property will make life easy for us and let us work with linear combinations of signals without requiring us to compute complicated integrals.

We know how to compute the Fourier series coefficients. Remember that the Fourier series coefficients make up an alternate representation of the time function. Therefore, the properties of the Fourier series are connected to the properties of the time function. In this module we look at some of those connections. We have computed the Fourier series coefficients for different square waves and a sawtooth wave. You might have noticed that while all of these functions are real valued, the Fourier series coefficients for the square waves were real while the Fourier series coefficients for the sawtooth waveforms were purely imaginary. So what distinguishes the first two waveforms from the last one? You might have noticed that the two square waves we used were even functions of time while the sawtooth was an odd functions of time. Let's connect the property of evenness and oddness of a time function to the values of the coefficients.

11.1 EVEN AND ODD FUNCTIONS AND THEIR COEFFICIENTS

The coefficients $\{a_n\}$ are even functions of n when $x(t)$ is an even function of t. To show this let's begin with the general expression for computing the Fourier series coefficients and then use the fact that $x(t) = x(-t)$ to get the following:

$$\begin{aligned}
a_n &= \frac{1}{T}\int_T x(t)e^{-jn\omega_0 t}\,dt \\
&= \frac{1}{T}\int_T x(-t)e^{-jn\omega_0 t}\,dt \\
&= \frac{1}{T}\int_T x(t)e^{jn\omega_0 t}\,dt \\
&= \frac{1}{T}\int_T x(t)e^{-j(-n)\omega_0 t}\,dt \\
&= a_{-n}
\end{aligned}$$

(In the third equation above we have used a variable substitution).

The Fourier series coefficients are even functions of n when the periodic function $x(t)$ is an even function of t.

What if $x(t)$ was an odd function? Then following the same logic and using the fact that for $x(t)$ odd, $x(t) = -x(-t)$ we get

$$\begin{aligned} a_n &= \frac{1}{T}\int_T x(t)e^{-jn\omega_0 t}\,dt \\ &= \frac{1}{T}\int_T -x(-t)e^{-jn\omega_0 t}\,dt \\ &= -\frac{1}{T}\int_T x(t)e^{jn\omega_0 t}\,dt \\ &= \frac{1}{T}\int_T x(t)e^{-j(-n)\omega_0 t}\,dt \\ &= -a_{-n} \end{aligned}$$

The Fourier series coefficients are odd functions of n when the periodic function $x(t)$ is an odd function of t.

We can look at this in a slightly different way using Euler's formula.

$$\begin{aligned} a_n &= \frac{1}{T}\int_{-T/2}^{T/2} x(t)e^{-jn\omega_0 t}\,dt \\ &= \frac{1}{T}\int_{-T/2}^{T/2} x(t)\left(\cos(n\omega_0 t) - j\,\sin(n\omega_0 t)\right)dt \\ &= \frac{1}{T}\int_{-T/2}^{T/2} x(t)\cos(n\omega_0 t)dt - j\frac{1}{T}\int_{-T/2}^{T/2} x(t)\sin(n\omega_0 t)dt \end{aligned}$$

If $x(t)$ is even then the integrand in the first integral is also even as $\cos(n\omega_0 t)$ is even and even $\times$ even $=$ even. The integrand in the second integral is odd as $\sin(n\omega_0 t)$ is odd and even $\times$ odd $=$ odd. We are integrating over a symmetric interval and the integral over a symmetric interval of an odd function is identically zero, therefore, if $x(t)$ is even the Fourier series coefficients are given by

$$a_n = \frac{1}{T}\int_{-T/2}^{T/2} x(t)\cos(n\omega_0 t)dt$$

We can see from this that a_n is an even function of n as

$$a_{-n} = \frac{1}{T}\int_{-T/2}^{T/2} x(t)\cos(-n\omega_0 t)dt = \frac{1}{T}\int_{-T/2}^{T/2} x(t)\cos(n\omega_0 t)dt = a_n$$

We can also see something else from this equation. If $x(t)$ besides being even is also real, then a_n will be real.

The Fourier series coefficients are real even functions of n when the periodic function $x(t)$ is a real even function of t.

What if $x(t)$ is odd? In this case the first integrand will be odd and the second integrand will be even so we will have

$$a_n = -j\frac{1}{T}\int_{-T/2}^{T/2} x(t)\sin(n\omega_0 t)dt$$

If $x(t)$ is real besides being odd then a_n will be purely imaginary.

The Fourier series coefficients are purely imaginary, odd functions of n when the periodic function $x(t)$ is a real and odd function of t.

So what happens if $x(t)$ is neither even nor odd? We explore that next in the context of time shifts.

11.2 TIME SHIFTS ARE PHASE SHIFTS

Let's begin exploring that by rewriting the Fourier series of the even square wave.

$$x(t) = \sum_{n=-\infty}^{\infty} a_n e^{jn\omega_o t} = \sum_{n=-\infty}^{-1} a_n e^{jn\omega_o t} + a_0 + \sum_{n=1}^{\infty} a_n e^{jn\omega_o t}$$

If we replace n by $-n$ in the first summation we get

$$\sum_{n=-\infty}^{-1} a_n e^{jn\omega_o t} = \sum_{-n=-\infty}^{-1} a_{-n} n e^{-jn\omega_o t} = \sum_{n=\infty}^{1} a_{-n} e^{-jn\omega_o t} = \sum_{n=1}^{\infty} a_{-n} e^{-jn\omega_o t}$$

So now we have

$$x(t) = \sum_{n=1}^{\infty} a_{-n} e^{-jn\omega_o t} + a_0 + \sum_{n=1}^{\infty} a_n e^{jn\omega_o t}$$

The two summations have the same limits so we can put them together

$$x(t) = a_0 + \sum_{n=1}^{\infty} \left[a_{-n} e^{-jn\omega_o t} + a_n e^{jn\omega_o t}\right]$$

Recall that the coefficients we obtained were even functions of n, in other words $a_n = a_{-n}$. Therefore, we can replace a_{-n} with a_n in the summation

$$x(t) = a_0 + \sum_{n=1}^{\infty} \left[a_n e^{-jn\omega_o t} + a_n e^{jn\omega_o t}\right]$$

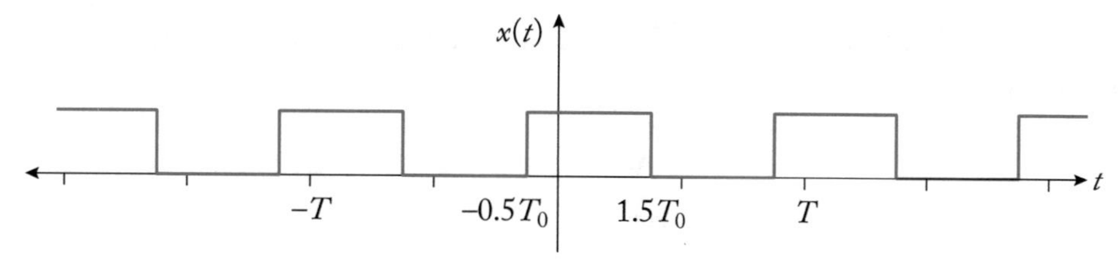

Figure 11.1: The square wave with period T and pulse width $2T_0$ shifted by $T_0/2$.

or

$$x(t) = a_0 + \sum_{n=1}^{\infty} a_n \left[e^{-jn\omega_o t} + e^{jn\omega_o t}\right]$$

but the quantity in brackets is just $2\cos(n\omega_o t)$. Therefore,

$$x(t) = a_0 + \sum_{n=1}^{\infty} 2a_n \cos(n\omega_o t)$$

The square wave in our example is a sum of sinusoids oscillating at frequencies which are integer multiples of the fundamental frequency of the square wave.

What happens if we take the square wave and shift it in time by $T_o/2$ so that it is neither even nor odd as shown in Figure 11.1. Clearly the coefficient a_0 will not change as the area under the curve over one period remains the same. Let's take a look at what happens to the other coefficients.

$$\begin{aligned} a_n &= \int_{-T/2}^{T/2} x(t)e^{-jn\omega_o t}\,dt \\ &= \int_{-T/2}^{-T_0/2} x(t)e^{-jn\omega_o t}\,dt + \int_{-T_0/2}^{3T_0/2} x(t)e^{-jn\omega_o t}\,dt + \int_{3T_0/2}^{T/2} x(t)e^{-jn\omega_o t}\,dt \end{aligned}$$

In the interval corresponding to the first and third integrals $x(t)$ is zero therefore, the integral is zero. So we are left with

$$\begin{aligned} a_n &= \int_{-T_0/2}^{3T_0/2} x(t)e^{-jn\omega_o t}\,dt \\ &= -\frac{1}{jn\omega_o T}\left[e^{-jn\omega_o \frac{3T_0}{2}} - e^{jn\omega_o \frac{T_0}{2}}\right] \\ &= -\frac{1}{jn2\pi}\left[e^{-jn\omega_o \frac{3T_0}{2}} - e^{jn\omega_o \frac{T_0}{2}}\right] \end{aligned}$$

where we have replaced ω_o with $2\pi/T$ in the last equation. We can now take $e^{-jn\omega_o T_0/2}$ in common to get

$$
\begin{aligned}
a_n &= -\frac{e^{-jn\omega_o T_0/2}}{jn2\pi}\left[e^{-jn\omega_o T_0} - e^{jn\omega_o T_0}\right] \\
&= \frac{e^{-jn\omega_o T_0/2}}{n\pi}\left[\frac{e^{jn\omega_o T_0} - e^{-jn\omega_o T_0}}{2j}\right] \\
&= e^{-jn\omega_o T_0/2}\frac{\sin(n\omega_o T_0)}{n\pi}
\end{aligned}
$$

This looks like the expression we had for the even square wave with an additional complex exponential factor. This complex exponential prevents the coefficients from being even functions of n. So how does this effect our sinusoidal representation? We could replace this expression for a_n in our Fourier series expansion, break up the summation into three intervals—from negative infinity to -1, 0, and from 1 to infinity as we did previously and see the effect. It is a lot simpler to just take a look at the effect for one value of n and a numerical value for T_0. As before let's set $T_0 = T/4$. Then

$$
a_n = e^{-jn\omega_o T/8}\frac{\sin(n\pi/2)}{n\pi}
$$

Let's plug this in to the Fourier series expression

$$
\begin{aligned}
x(t) &= \sum_{n=-\infty}^{-1} e^{-jn\omega_o T/8}\frac{\sin(n\pi/2)}{n\pi}e^{jn\omega_o t} + a_0 + \sum_{n=1}^{\infty} e^{-jn\omega_o T/8}\frac{\sin(n\pi/2)}{n\pi}e^{jn\omega_o t} \\
&= a_0 + \sum_{n=1}^{\infty} e^{jn\omega_o T/8}\frac{\sin(-n\pi/2)}{-n\pi}e^{-jn\omega_o t} + \sum_{n=1}^{\infty} e^{-jn\omega_o T/8}\frac{\sin(n\pi/2)}{n\pi}e^{jn\omega_o t} \\
&= a_0 + \sum_{n=1}^{\infty}\frac{\sin(n\pi/2)}{n\pi}\left[e^{jn\omega_o T/8}e^{-jn\omega_o t} + e^{-jn\omega_o T/8}e^{jn\omega_o t}\right] \\
&= a_0 + \sum_{n=1}^{\infty}\frac{\sin(n\pi/2)}{n\pi}\left[e^{-jn\omega_o(t-T/8)} + e^{jn\omega_o(t-T/8)}\right] \\
&= a_0 + \sum_{n=1}^{\infty}\frac{\sin(n\pi/2)}{n\pi}2\cos(n\omega_o(t-T/8))
\end{aligned}
$$

We again have a sum of sinusoids. The only difference between the sinusoids here and the sinusoids we got before for the even square wave is a phase shift.

A delay in a periodic function results in a phase shift in the Fourier representation of that function.

We will see this effect over and over again in our various representations—a time shift in the time domain representation will translate to a phase shift in the frequency domain representation.

Before we leave this let's take a look at how the coefficients change for a general periodic function $x(t)$ if we shift it in time. Let's begin with the coefficients for $x(t)$

$$a_n = \frac{1}{T}\int_T x(t)e^{-jn\omega_o t}\,dt$$

Now let's look at what happens to the coefficients when we shift the periodic function to get $x(t - t_0)$.

$$b_n = \frac{1}{T}\int_T x(t - t_0)e^{-jn\omega_o t}\,dt$$

Let $\tau = t - t_0$. Then, $t = \tau + t_0$ and

$$\begin{aligned}
b_n &= \frac{1}{T}\int_T x(\tau)e^{-jn\omega_o(\tau+t_0)}\,d\tau \\
&= \frac{1}{T}\int_T x(\tau)e^{-jn\omega_o\tau}e^{-jn\omega_o t_0}\,d\tau \\
&= e^{-jn\omega_o t_0}\,\frac{1}{T}\int_T x(\tau)e^{-jn\omega_o\tau}\,d\tau \\
&= e^{-jn\omega_o t_0}a_n
\end{aligned} \tag{11.1}$$

So a time shift of t_0 results in the Fourier series coefficients being multiplied by $e^{-jn\omega_o t_0}$. For specific values of t_0 this quantity can be real or purely imaginary. For example if $t_0 = T/2$. Then

$$e^{-jn\omega_o t_0} = e^{-jn\frac{2\pi}{T}\frac{T}{2}} = e^{-jn\pi} = (-1)^n$$

11.3 THE FOURIER COEFFICIENTS – MEANING AND EXTRACTION

Let's for a moment consider the physical meaning of the Fourier coefficients. The sinusoidal representation is useful to understand the physical meaning so let's begin with that. The general sinusoidal form is

$$x(t) = a_0 + \sum_{n=1}^{\infty} \alpha_n \cos(n\omega_o(t - t_0))$$

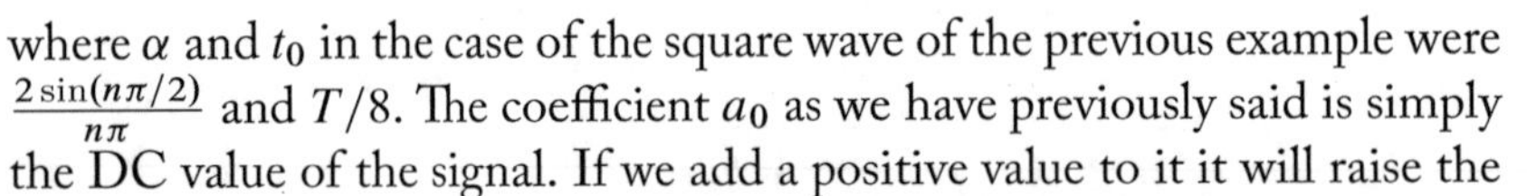

where α and t_0 in the case of the square wave of the previous example were $\frac{2\sin(n\pi/2)}{n\pi}$ and $T/8$. The coefficient a_0 as we have previously said is simply the DC value of the signal. If we add a positive value to it it will raise the signal level and if we subtract from it the signal level will be lowered; both without effecting the shape of the signal. The coefficients α_n—very closely related to a_n—are the multipliers of sinusoids of frequency $n\omega_o$. These then are the frequency components of a signal. A signal which has a lot of low frequencies—think of a jackhammer—will have larger magnitudes of a_n for smaller values of n. A squeal like you would get from a worn out brake pad will have larger values for higher values of n and thus the higher frequencies.

If we wanted to isolate these components we could do that by using linear time invariant systems called filters. A filter that lets through low frequency components while blocking high frequency components is called a *low pass filter*. A filter that blocks low frequency components and lets through high frequency components is called a *high pass filter* while a filter that blocks both high and low frequency components letting through only a band of frequency components is called a *bandpass filter*.

How these filters work can be easily understood by revisiting our reason for using $e^{j\omega t}$ as the basis signal for Fourier analysis. Recall that if the input to a linear time invariant system with impulse response $h(t)$ is $e^{j\omega_o t}$ the output is given by

$$y(t) = e^{j\omega_o t} H(\omega_o)$$

where

$$H(\omega_o) = \int_{-\infty}^{\infty} h(t) e^{-j\omega_o t}\, dt$$

The function $H(\omega)$ is called the *transfer function* of the linear time invariant system. Because this is a linear time invariant system we can easily find the output of any linear combination of complex exponentials. So if the input to this system is

$$x(t) = \alpha_0 e^{j\omega_0 t} + \alpha_1 e^{j\omega_1 t}$$

The output of this system will be

$$y(t) = \alpha_0 e^{j\omega_0 t} H(\omega_0) + \alpha_1 e^{j\omega_1 t} H(\omega_1)$$

Generalizing this if the input is

$$x(t) = \sum_{n=-\infty}^{\infty} a_n e^{jn\omega_o t}$$

then the output is

$$y(t) = \sum_{n=-\infty}^{\infty} a_n e^{jn\omega_o t} H(n\omega_o)$$

Suppose we wanted to extract the DC and the first frequency component from a periodic signal. We could do this by using a filter with transfer function

$$H(\omega) = \begin{cases} 1 & |\omega| < 1.5\omega_o \\ 0 & \text{otherwise} \end{cases}$$

(The 1.5 in this equation is arbitrary—we could have picked any number greater than one but less than two). A filter with such a transfer function is physically impossible to build for reasons we shall see later on (though we can try to approximate it) but it is useful for us right now so we will ignore its physical impossibility.

If we apply this filter to the input signal we will zero out all components other than for $n = -1, 0, 1$. So the output of the filter will be

$$y(t) = a_{-1}e^{-j\omega_o t} + a_0 + a_1 e^{j\omega_o t}$$

If we combine the first and third term we will get a raised cosine

$$y(t) = a_0 + \alpha \cos(\omega_o(t - t_0))$$

where α and t_0 can be obtained from a_1 and a_{-1}. For the even square wave with a period of 0.1 seconds and $T_0 = .0125$ this is plotted in Figure 11.2. If your eyes are good you can extract the period and frequency of this component from the graph in Figure 11.2.

If we wanted instead to extract the second and third components we could use a filter with transfer function

$$H(\omega) = \begin{cases} 1 & 1.5\omega_o < |\omega| < 3.5\omega_o \\ 0 & \text{otherwise} \end{cases}$$

The output of this filter would be the sum of two cosines

$$y(t) = \alpha_1 \cos(2\omega_o(t - t_0)) + \alpha_2 \cos(3\omega_o(t - t_0))$$

For the even square wave with a period of 0.1 seconds and $T_0 = .0125$ this is plotted in Figure 11.3. You can see that this is not a pure sinusoid but a sum of two sinusoids.

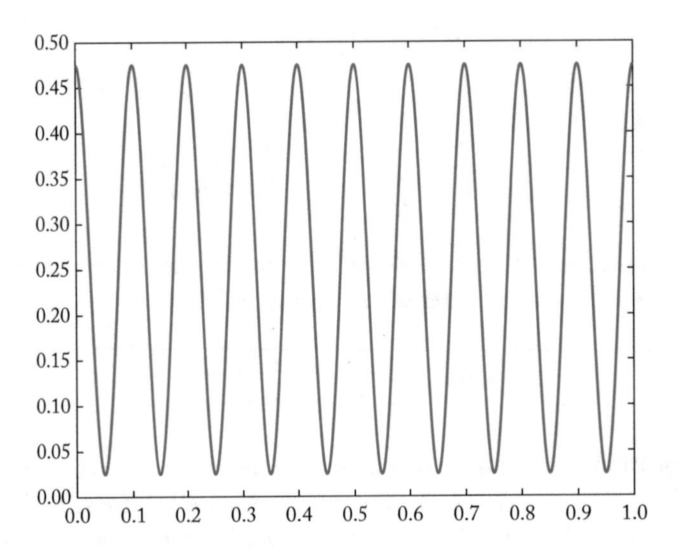

Figure 11.2: Output of the lowpass filter which extracts the DC and first component of the square wave signal.

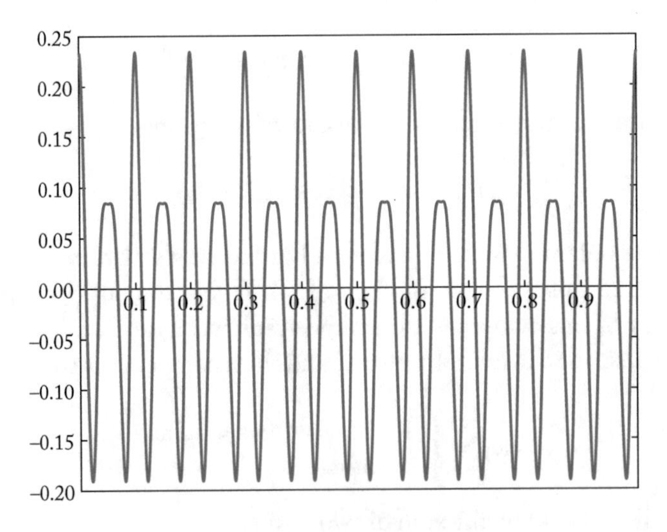

Figure 11.3: Output of the lowpass filter which extracts the second and third frequency component of the square wave signal.

11.4 THE ENERGY IN A SIGNAL DOES NOT CHANGE BASED ON ITS REPRESENTATION

Parseval's Identity The last property we will look at is a kind of conservation of energy property. Both the time description of the signal $x(t)$ and the frequency description obtained using the Fourier series coefficients are the description of the same physical signal. One of the properties of this signal is the amount of average power in the signal. Because both descriptions are descriptions of the same signal it makes sense that the signal power computed in either the time or frequency domain should be identical. The average power for a periodic signal in the time domain is computed as

$$P = \frac{1}{T}\int_T |x(t)|^2 dt$$

where

$$|x(t)|^2 = x(t)x^*(t)$$

In the frequency domain the power is the sum of the magnitude squares of the frequency coefficients

$$P = \sum_{n=-\infty}^{\infty} |a_n|^2$$

To show that these two are the same we begin with our usual expression for the Fourier series expansion.

$$x(t) = \sum_{n=-\infty}^{\infty} a_n e^{jn\omega_o t}$$

Therefore,

$$x^*(t) = \left[\sum_{n=-\infty}^{\infty} a_n e^{jn\omega_o t}\right]^* = \sum_{n=-\infty}^{\infty} a_n^* e^{-jn\omega_o t}$$

Substituting for $x^*(t)$ in the integral for the energy we get

$$\begin{aligned}
\frac{1}{T}\int_T |x(t)|^2 dt &= \frac{1}{T}\int_T x(t)x^*(t)dt \\
&= \frac{1}{T}\int_T x(t)\sum_{n=-\infty}^{\infty} a_n^* e^{-jn\omega_o t} dt \\
&= \sum_{n=-\infty}^{\infty} a_n^* \frac{1}{T}\int_T x(t)e^{-jn\omega_o t} dt \\
&= \sum_{n=-\infty}^{\infty} a_n^* a_n \\
&= \sum_{n=-\infty}^{\infty} |a_n|^2
\end{aligned}$$

11.5 SUMMARY

In this module we examined several useful properties of the Fourier series coefficients.

1. The Fourier series coefficients are even functions of n when the periodic function is an even function of t.
2. The Fourier series coefficients are odd functions of n when the periodic function is an odd function of t
3. The Fourier series coefficients are real and even functions of n when the periodic function is a real and even function of t.
4. The Fourier series coefficients are purely imaginary and odd functions of n when the periodic function is a real and odd function of t.
5. A delay in the periodic function corresponds to a phase shift in the Fourier series coefficients.
6. Because both the time function and the Fourier series represent the same physical waveform if we calculate physical quantities such as power the result will be the same regardless of the representation we use.
7. Because the Fourier series coefficients each correspond to a specific frequency we can see how we can design filters to block, attenuate, or enhance particular frequencies.

11.6 EXERCISES

(Answers on the following page)

1. Find the Fourier series expansion of

$$x(t) = |\cos(2\pi t)|$$

2. You are given a periodic function $x(t)$ with Fourier series

$$x(t) = \sum_{n=-\infty}^{\infty} a_n e^{jn\pi t}$$

   (a) If $y(t) = x(t-1)$ find the coefficients $\{b_n\}$ for the Fourier series expansion of $y(t)$ in terms of $\{a_n\}$.

   (b) If $y(t) = x(t-1/2)$ find the coefficients $\{b_n\}$ for the Fourier series expansion of $y(t)$ in terms of $\{a_n\}$.

3. For each of the following determine if $x(t)$ is real, if $x(t)$ is even, if $x(t)$ is odd.

$$(a) \qquad x(t) = \sum_{n=0}^{\infty} \left(\frac{1}{2}\right)^n e^{jn\pi t}$$

$$(b) \qquad x(t) = \sum_{n=-\infty}^{\infty} \left(\frac{1}{2}\right)^{|n|} e^{jn\frac{\pi}{2}t}$$

$$(c) \qquad x(t) = \sum_{n=-10}^{10} \left(\frac{-jn}{2}\right) e^{jnt}$$

4. If

$$\begin{aligned} x(t) &= j\sin(\omega_o t) \\ y(t) &= j\sin(2\omega_o t) \end{aligned}$$

   and

$$z(t) = x(t)y(t)$$

   (a) Find the Fourier series coefficients for $z(t)$.

   (b) Express $z(t)$ as a sum of cosines.

5. Given the square wave $x(t)$ shown in Figure 11.4 which can be represented by the Fourier series coefficients $\{a_n\}$.

   Sketch $y(t)$ if $\{b_n\}$ the Fourier series coefficients for $y(t)$ are given by

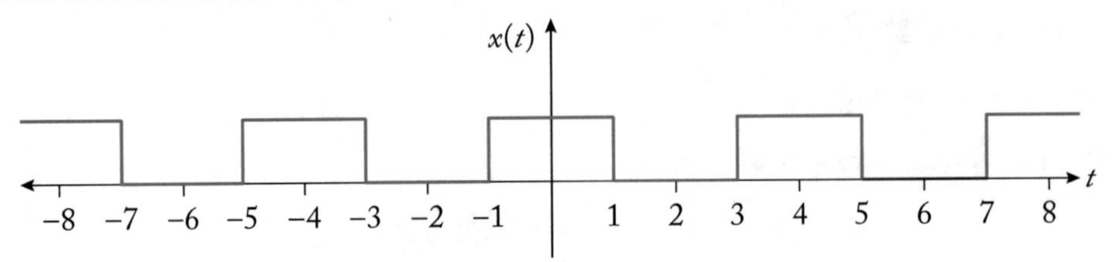

Figure 11.4: Square wave.

(a)

$$b_n = \begin{cases} a_n & n = -1, 0, 1 \\ 0 & \text{otherwise} \end{cases}$$

(b)

$$b_n = \begin{cases} a_n & n = -1, 1 \\ 0 & \text{otherwise} \end{cases}$$

11.7 ANSWERS

1.

$$a_n = \frac{2\sin\left(\frac{2-4n}{4}\right)\pi}{2-4n} + \frac{2\sin\left(\frac{2+4n}{4}\right)\pi}{2+4n}$$

2. $\omega_o = \pi$

   (a)

$$b_n = e^{-jn\pi} a_n$$

   (b)

$$b_n = e^{-jn\frac{\pi}{2}} a_n$$

3. (a) $x(t)$ is real, but it is neither even nor odd

   (b) $x(t)$ is real and even

   (c) $x(t)$ is real and odd

4. (a)

$$a_1 = a_{-1} = -\frac{1}{4}; \quad a_3 = a_{-3} = \frac{1}{4}$$

   (b)

$$z(t) = \frac{1}{2}\cos(3\omega_o t) - \frac{1}{2}\cos(\omega_o t)$$

5. See Figure 11.5.

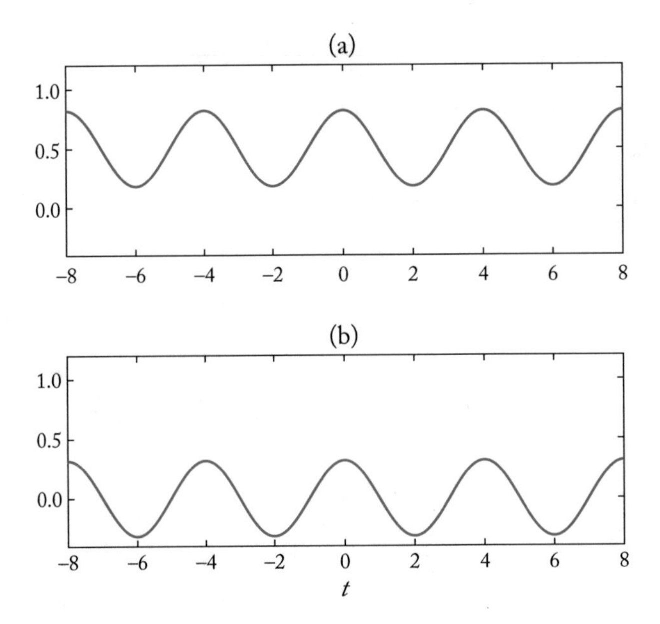

Figure 11.5: Sketch.

MODULE 12

The Fourier Transform

We have been looking at the frequency representation of periodic signals. While periodic signals are important for life, the universe, and everything we are also interested in the many aperiodic signals that we encounter daily. For this we need to extend our analysis from periodic to aperiodic signals.

12.1 EXTENDING THE FREQUENCY VIEW TO APERIODIC FUNCTIONS

So what happens if $x(t)$ is aperiodic? Let's assume for the moment that $x(t)$ is non-zero over a finite interval $[0, T_o]$. Let's define a periodic extension

$$x_p(t) = \sum_{k=-\infty}^{\infty} x(t - kT)$$

where $T > T_o$. An example for this function using only three terms is shown in Figure 12.1.

As we increase the number of terms we can see that the final result $x_p(t)$ will be a periodic function. Because $x_p(t)$ is a periodic function with period T we can write a Fourier series expansion for it with coefficients

$$a_k = \frac{1}{T} \int_{-T/2}^{T/2} x_p(t) e^{-jk\omega_0 t}\, dt$$

Let's make a few modification to this equation. Let's move the scale factor to the left and lets replace ω_0 with $\Delta\omega$, so

$$a_k T = \int_{-T/2}^{T/2} x_p(t) e^{-jk\Delta\omega t}\, dt$$

When we integrate the right hand side we are left with a function of $k\Delta\omega$. Let's make this explicit by rewriting the left hand side as $X(k\Delta\omega)$. Now let's take the limit as $T \to \infty$. When $T \to \infty$, $\Delta\omega$ will go to zero, and $k\Delta\omega$ will go to ω.

$$X(\omega) = \int_{-\infty}^{\infty} x(t) e^{-j\omega t}\, dt$$

where

$$\lim_{T \to \infty} x_p(t) = x(t)$$

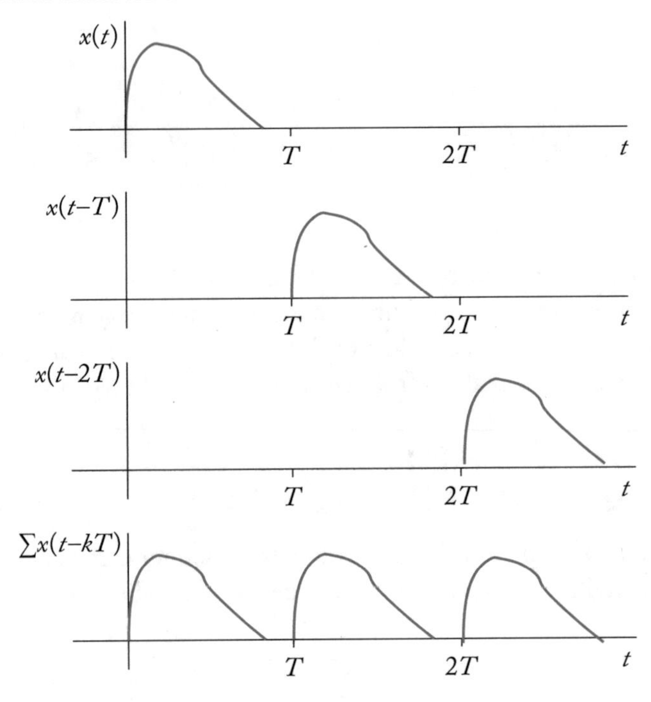

Figure 12.1: A sum of three shifted versions of a time-limited function.

Now

$$x_p(t) = \sum_{k=-\infty}^{\infty} \frac{X(k\Delta\omega)}{T} e^{jk\Delta\omega t}$$

but,

$$T = \frac{2\pi}{\Delta\omega}$$

so

$$x_p(t) = \frac{1}{2\pi} \sum_{k=-\infty}^{\infty} X(k\Delta\omega) e^{jk\Delta\omega t} \Delta\omega$$

Taking the limit as $T \to \infty$ we get

$$x(t) = \frac{1}{2\pi} \int_{-\infty}^{\infty} X(\omega) e^{j\omega t} d\omega$$

These two equations are called the Fourier transform. The equation

$$X(\omega) = \int_{-\infty}^{\infty} x(t) e^{-j\omega t} dt$$

is sometimes called the forward Fourier transform. A shorthand notation for the forward Fourier transform is

$$\mathcal{F}[x(t)] = X(\omega) \text{ or } X(j\omega)$$

I prefer using $X(\omega)$ so that is what we will use in these modules, but you may find others using $X(j\omega)$ instead. The equation which can be used to "recover" $x(t)$ from $X(\omega)$

$$x(t) = \frac{1}{2\pi}\int_{-\infty}^{\infty} X(\omega)e^{j\omega t}\,d\omega$$

is often called the inverse Fourier transform. Using the same operator notation as the forward Fourier transform the inverse Fourier transform is often shown as

$$\mathcal{F}^{-1}[X(\omega)] = x(t)$$

Does the Fourier integral always exist? Without proof we will assert that the Fourier transform of a function $x(t)$ exists if $x(t)$ satisfies the Dirichlet conditions—the same conditions we needed for the Fourier series to exist. This makes sense as the way we obtained the Fourier transform was by making the aperiodic function $x(t)$ one period of a periodic function $x_p(t)$. Remember that these conditions were:

1. The function $x(t)$ should be absolutely integrable.

$$\int |x(t)| < \infty$$

2. The function has bounded variation which means that in any finite interval it has only a finite number of zero crossings.

3. It has a finite number of finite discontinuities.

Most signals that we will ever deal with will always satisfy these conditions.

There are several properties of the Fourier transform that we can immediately see from the definition.

The Fourier transform is Linear The Fourier transform being linear means that if the Fourier transform of $x(t)$ is $X(\omega)$ and the Fourier transform of $y(t)$ is $Y(\omega)$ then the Fourier transform of $\alpha x(t) + \beta y(t)$ is $\alpha X(\omega) + \beta Y(\omega)$. The property is easy to prove as the Fourier transform is an integral transform.

$$\begin{aligned}
\mathcal{F}[\alpha x(t) + \beta y(t)] &= \int_{-\infty}^{\infty} (\alpha x(t) + \beta y(t))\, e^{-j\omega t} dt \\
&= \alpha \int_{-\infty}^{\infty} x(t)e^{-j\omega t} dt + \beta \int_{-\infty}^{\infty} y(t)e^{-j\omega t} dt \\
&= \alpha\mathcal{F}[x(t)] + \beta\mathcal{F}[y(t)] \\
&= \alpha X(\omega) + \beta Y(\omega)
\end{aligned}$$

We can show the linearity of the inverse Fourier transform in exactly the same way.

The Fourier Transform is Unique The Fourier transform $X(\omega)$ of the signal $x(t)$ and the signal $x(t)$ form a unique pair. That is if we know $x(t)$ we exactly know $X(\omega)$ and if we know $X(\omega)$ we exactly know $x(t)$. If we come across a spectral profile $Y(\omega)$ which is the same as $X(\omega)$ then we know that $y(t)$, the inverse transform of $Y(\omega)$ is equal to $x(t)$. We can show this using the linearity property of the Fourier transform. Given two signals $x(t)$ and $y(t)$ with Fourier transforms $X(\omega)$ and $Y(\omega)$, the Fourier transform of the difference is given by

$$\mathcal{F}[x(t) - y(t)] = X(\omega) - Y(\omega)$$

If $x(t)$ equals $y(t)$ then $x(t) - y(t)$ is equal to zero. As the Fourier transform of zero is zero, the left-hand side of this equation is zero, which then implies that

$$X(\omega) - Y(\omega) = 0$$

or $X(\omega)$ equals $Y(\omega)$.

A real signal has a Fourier transform with even magnitude and odd phase To see this we begin with a general—not necessarily real—function $x(t)$ and examine the Fourier transform of $x^*(t)$, the complex conjugate of $x(t)$. The signal $x(t)$ can be written as the inverse Fourier transform of $X(\omega)$

$$x(t) = \frac{1}{2\pi}\int_{-\infty}^{\infty} X(\omega)e^{j\omega t}\,d\omega$$

Taking the complex conjugate of both sides

$$\begin{aligned}
x^*(t) &= \left[\frac{1}{2\pi}\int_{-\infty}^{\infty} X(\omega)e^{j\omega t}\,d\omega\right]^* \\
&= \frac{1}{2\pi}\int_{-\infty}^{\infty} X^*(\omega)e^{-j\omega t}\,d\omega \\
&= \frac{1}{2\pi}\int_{-\infty}^{\infty} X^*(\omega)e^{j(-\omega)t}\,d\omega \\
&= \mathcal{F}^{-1}X^*(-\omega)
\end{aligned}$$

or

$$\mathcal{F}\left[x^*(t)\right] = X^*(-\omega)$$

If $x(t)$ is real

$$x^*(t) = x(t)$$

Therefore,

$$X^*(-\omega) = X(\omega)$$

Writing $X(\omega)$ in polar form

$$X(\omega) = |X(\omega)|\, e^{j\theta(\omega)}$$

Therefore,

$$X^*(\omega) = |X(\omega)|\, e^{-j\theta(\omega)}$$

and

$$X^*(-\omega) = |X(-\omega)|\, e^{-j\theta(-\omega)}$$

For a real $x(t)$, $X^*(-\omega) = X(\omega)$ or

$$|X(-\omega)|\, e^{-j\theta(\omega)} = |X(\omega)|\, e^{j\theta(\omega)}$$

which means that

$$|X(\omega)| = |X(-\omega)|$$

and

$$\theta(\omega) = -\theta(-\omega)$$

That is, if $x(t)$ is real the magnitude of the transform $X(\omega)$ is real and the phase of $X(\omega)$ is odd.

What if $x(t)$ is real and even. If $x(t)$ is real

$$X(-\omega) = X^*(\omega)$$

and if $x(t)$ is even

$$X(\omega) = X(-\omega)$$

Therefore, if $x(t)$ is real and even

$$X(\omega) = X^*(\omega)$$

and $X(\omega)$ is real.

If $x(t)$ is real but not even we can write it as

$$x(t) = \mathcal{E}v(x(t)) + \mathcal{O}d(x(t))$$

Also

$$X(\omega) = \mathcal{R}e(X(\omega)) + j\mathcal{I}m(X(\omega))$$

Therefore,

$$\mathcal{F}[\mathcal{E}v(x(t))] = \mathcal{R}e(X(\omega))$$

and

$$\mathcal{F}[\mathcal{O}d(x(t))] = j\mathcal{I}m(X(\omega))$$

Parseval's relation $x(t)$ and $X(\omega)$ are two representations of the same signal. Therefore, the physical properties of the signal, such as the energy of the signal is the same in both domains.

$$\begin{aligned}
\int_{-\infty}^{\infty} |x(t)|^2 dt &= \int_{-\infty}^{\infty} x(t)x^*(t)dt \\
&= \int_{-\infty}^{\infty} x(t)\left[\frac{1}{2\pi}\int_{-\infty}^{\infty} X(\omega)e^{j\omega t}d\omega\right]^* dt \\
&= \frac{1}{2\pi}\int_{-\infty}^{\infty} x(t)\int_{-\infty}^{\infty} X^*(\omega)e^{-j\omega t}d\omega dt \\
&= \frac{1}{2\pi}\int_{-\infty}^{\infty} X^*(\omega)\int_{-\infty}^{\infty} x(t)e^{-j\omega t}dt d\omega \\
&= \frac{1}{2\pi}\int_{-\infty}^{\infty} X^*(\omega)X(\omega)d\omega \\
&= \frac{1}{2\pi}\int_{-\infty}^{\infty} |X(\omega)|^2 d\omega
\end{aligned}$$

Example 12.1 Let's find the Fourier transform of one of our favorite signals the decaying exponential.

$$x(t) = e^{-2t}u(t)$$

$$\begin{aligned}
X(\omega) &= \int_{-\infty}^{\infty} e^{-2t}u(t)e^{-j\omega t}dt \\
&= \int_{0}^{\infty} e^{-(2+j\omega)t}dt \\
&= -\frac{1}{2+j\omega}e^{-(2+j\omega)t}|_0^\infty
\end{aligned}$$

We know what happens when we evaluate $e^{-(2+j\omega)t}$ at the lower limit of $t = 0$—we get e^0 which is 1. Let's take a closer look at what happens at the upper limit. In order to evaluate $e^{-(2+j\omega)t}$ at the upper limit of infinty let's break $e^{-(2+j\omega)t}$ into its factors e^{-2t} and $e^{-j\omega t}$. When we evaluate e^{-2t} at infinity we get zero because this is number smaller than one raised to a power of infinity. What about $e^{-j\omega t}$? The quantity $e^{-j\omega}$ is a complex number. What does it mean to raise a complex number to the power of infinity? Thanks to Euler we know that

$$e^{-j\omega t} = \cos(\omega t) - j\sin(\omega t)$$

If we valuate the right hand side at infinity we will get $\cos(\infty) - j\sin(\infty)$. We don't know the value of that expression, however, we do know whatever it is, its finite as both cosine and sine are bounded by one. So when we evaluate $e^{-(2+j\omega)t}$ as $t \to \infty$ we get a product of two terms—zero from e^{-2t} and an unknown but finite value from $e^{-j\omega t}$. As the product of zero with any finite value is zero the result is zero. So

$$\begin{aligned} X(\omega) &= -\frac{1}{2+j\omega} e^{-(2+j\omega)t}\Big|_0^\infty \\ &= -\frac{1}{2+j\omega}[0-1] \\ &= \frac{1}{2+j\omega} \end{aligned} \tag{12.1}$$

The function $x(t)$ was a real function so let's see if the magnitude and phase of $X(\omega)$ follow the rule we described above. The magnitude of $X(\omega)$ is given by

$$\begin{aligned} |X(\omega)| &= \sqrt{X(\omega)X^*(\omega)} \\ &= \sqrt{\frac{1}{2+j\omega}\cdot\frac{1}{2-j\omega}} \\ &= \sqrt{\frac{1}{4+\omega^2}} \\ &= \frac{1}{\sqrt{4+\omega^2}} \end{aligned}$$

The phase of $X(\omega)$ is given by

$$X(\omega) = \tan^{-1}\frac{Im[X(\omega]}{Re[X(\omega)]}$$

where $Im[X(\omega]$ is the imaginary part of $X(\omega)$ and $Re[X(\omega)]$ is the real part of $X(\omega)$. We can write the real and imaginary parts of $X(\omega)$ as

$$X(\omega) = \frac{2}{4+\omega^2} - j\frac{\omega}{4+\omega^2}$$

so the phase of $X(\omega)$ is

$$\angle X(\omega) = \tan^{-1}\left(\frac{-\omega}{2}\right)$$

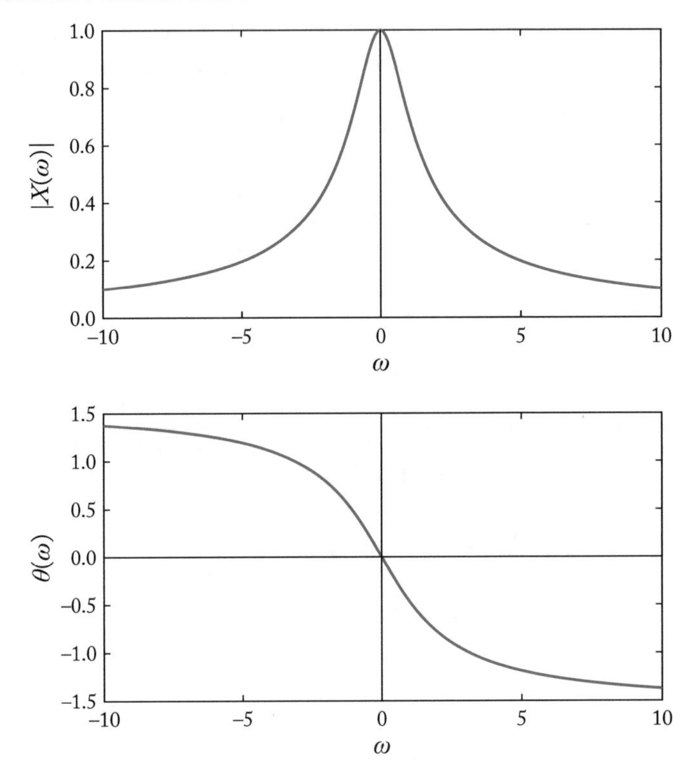

Figure 12.2: The magnitude and phase of the Fourier transform of $e^{-2t}u(t)$.

The magnitude and phase are plotted in Figure 12.2.
Notice that the magnitude is even and the phase is odd.

12.2 SUMMARY

In this module we extended the idea behind the Fourier series to obtain a frequency domain representation of aperiodic functions. We showed that the Fourier transform of a real function has a magnitude which is even and a phase which is odd. We also showed that if the time function is real and even the corresponding Fourier transform is real. As with the Fourier series coefficients the time function and the Fourier transform represent the same physical signal. Therefore, if we compute a physical property such as power from either representation we will get the same result.

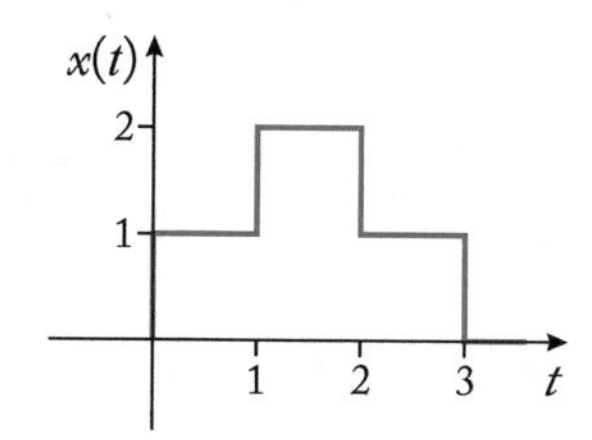

Figure 12.3

12.3 EXERCISES

(Answers on the following page)

1. Given

$$x(t) = \begin{cases} 1 & 0 \leq t < 1 \\ 2 & 1 \leq t < 2 \\ 1 & 2 \leq t < 3 \\ 0 & \text{otherwise} \end{cases}$$

   shown in Figure 12.3. Without using integration:

   (a) Find $X(0)$

   (b) Find $\int_{-\infty}^{\infty} X(\omega) d\omega$

2. Find the Fourier transform of

   (a)

$$x(t) = e^{-4t} u(t)$$

   (b)

$$x(t) = e^{-4|t|}$$

   (c)

$$x(t) = \begin{cases} 1 & |t| < 1 \\ 0 & \text{otherwise} \end{cases}$$

   (d)

$$x(t) = \delta(t-3)$$

   (e)

$$x(t) = \sum_{n=-\infty}^{\infty} \delta(t - nT)$$

3. Find the inverse Fourier transform of

(a)

$$X(\omega) = \frac{1}{1 + j\omega}$$

(b)

$$X(\omega) = \delta(\omega - \omega_o) + \delta(\omega + \omega_o)$$

(c)

$$X(\omega) = \sum_{n=-\infty}^{\infty} a_n \delta(\omega - n\omega_o)$$

12.4 ANSWERS

1. (a) $X(0) = 4$

   (b) $\int_{-\infty}^{\infty} X(\omega)d\omega = 2\pi$

2. (a)

$$X(\omega) = \frac{1}{4 + j\omega}$$

   (b)

$$X(\omega) = \frac{8}{16 + \omega^2}$$

   (c)

$$X(\omega) = 2\frac{\sin(\omega)}{\omega}$$

   (d)

$$X(\omega) = e^{-j3\omega}$$

   (e)

$$X(\omega) = \sum_{n=-\infty}^{\infty} e^{-jn\omega T}$$

3. (a)

$$x(t) = e^{-t}u(t)$$

   (b)

$$\frac{1}{\pi}\cos(\omega_o t)$$

   (c)

$$x(t) = \frac{1}{2\pi}\sum_{n=-\infty}^{\infty} a_n e^{jn\omega_o t}$$

MODULE 13

Properties of the Fourier Transform

Let's acquaint ourselves with this transform and its properties. We will alternate between properties and examples so we can get a physical feel for this transform. In this module we begin with the property that in some sense was our motivation for the development of the Fourier transform and the one property of the Fourier transform we will use over and over again—to the point we almost forget where it came from—is the convolution property. We will then introduce other properties as we need them.

13.1 CONVOLUTION PROPERTY

The convolution property should maybe more properly be named the convolution avoidance property as it is the property of the Fourier transform that allows us to avoid convolutions when we are relating the input and output of a linear time invariant system.

For a linear time invariant system with impulse response $h(t)$ and input $x(t)$, the output is given by

$$y(t) = \int_{-\infty}^{\infty} x(\tau)h(t-\tau)d\tau$$

Taking the Fourier transform of the output $y(t)$ we get

$$\begin{aligned} Y(\omega) &= \int_{-\infty}^{\infty} y(t)e^{-j\omega t}dt \\ &= \int_{-\infty}^{\infty}\int_{-\infty}^{\infty} x(\tau)h(t-\tau)d\tau e^{-j\omega t}d\tau dt \\ &= \int_{-\infty}^{\infty} x(\tau)\left[\int_{-\infty}^{\infty} h(t-\tau)d\tau e^{-j\omega t}dt\right]d\tau \end{aligned}$$

set $u = t - \tau$ then $t = u + \tau$ and

$$Y(\omega) = \int_{-\infty}^{\infty} x(\tau)\left[\int_{-\infty}^{\infty} h(u)e^{-j\omega(u+\tau)}du\right]d\tau$$

or

$$Y(\omega) = \int_{-\infty}^{\infty} x(\tau)e^{-j\omega\tau}\left[\int_{-\infty}^{\infty} h(u)e^{-j\omega u}du\right]d\tau$$

or

$$Y(\omega) = X(\omega)H(\omega)$$

This means convolution in the time domain becomes multiplication in the frequency domain. This is a very useful property as it gives us a much simpler view of what is going on between the input and output of a linear time invariant system. If we know the input and have a desired output, this also allows us to design the linear time invariant system. As, given $X(\omega)$ and $Y(\omega)$,

$$H(\omega) = \frac{Y(\omega)}{X(\omega)}$$

Filters The most common linear time invariant system for which we use this relationship is a filter. A filter is a device or process which allows us to selectively remove or enhance certain frequency components. In the very first module we used a filter to remove unwanted interference. Often it is not possible to entirely remove unwanted interference in which case we can use the filter to attenuate the unwanted parts of the input.

We have been using brick wall filters with transfer function of the form

$$H(\omega) = \begin{cases} 1 & \omega < W \\ 0 & \text{otherwise} \end{cases}$$

in some of our examples with the caveat that these filters are not particularly realistic. Let's take a look at the corresponding impulse response to see why this might be the case.

$$\begin{aligned} h(t) &= \frac{1}{2\pi}\int_{-\infty}^{\infty} H(\omega)e^{j\omega t}\,d\omega \\ &= \frac{1}{2\pi}\int_{-W}^{W} e^{j\omega t}\,d\omega \\ &= \frac{1}{2\pi}\frac{1}{jt}e^{j\omega t}\Big|_{-W}^{W} \\ &= \frac{1}{2\pi jt}\left(e^{jWt} - e^{-jWt}\right) \\ &= \frac{\sin(Wt)}{\pi t} \end{aligned}$$

We have plotted this impulse response in Figure 13.1.

Notice a couple of things. First note the duality at work here. The Fourier transform of a pulse in the time domain was a $\sin(x)/x$ function in the frequency domain. Here we have essentially a pulse in the frequency domain and it's inverse Fourier transform is a $\sin(x)/x$ function. Looking at the impulse response as the response to an impulse notice that it is nonzero for $t < 0$. As we know this means that the filter is noncausal. Furthermore, as the impulse response

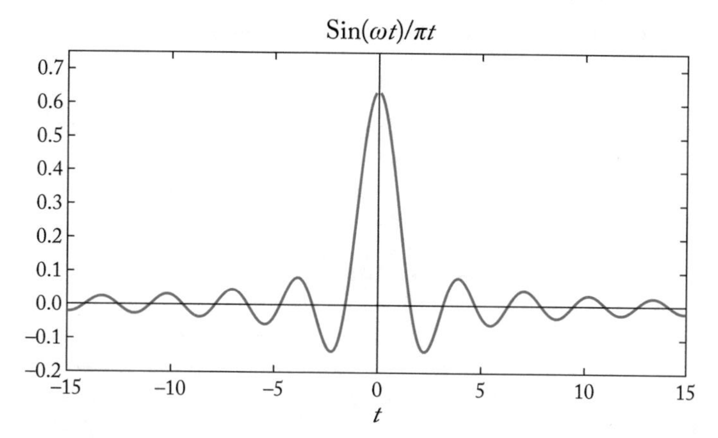

Figure 13.1: The impulse response of a brick wall filter.

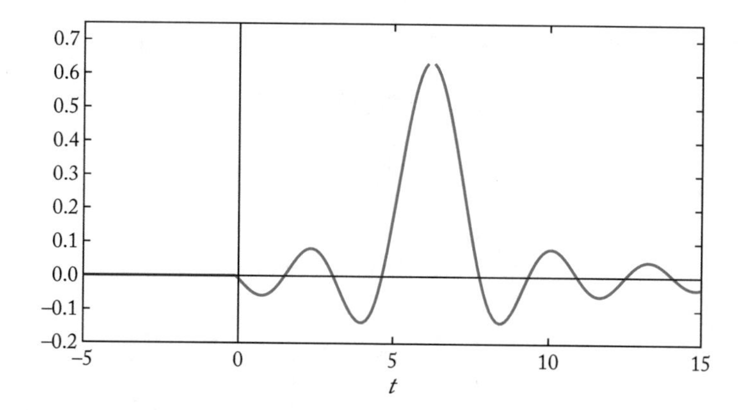

Figure 13.2: A causal impulse response.

is nonzero for $t \rightarrow -\infty$, we cannot implement it by simply delaying the response by a little bit. The best that we could do is to try and approximate this response by delaying it a bit and then setting the response to zero for $t < 0$—something like the response shown in Figure 13.2. If we take the Fourier transform of this we will of course not get our ideal brick wall filter characteristic back.

So what does the transfer function of a realistic filter look like. If you recall we had found the impulse response of a simple RC filter to be

$$h(t) = \frac{1}{RC} e^{-t/RC} u(t)$$

Let's find the transfer function for this filter. For convenience we replace $1/RC$ with α.

$$\begin{aligned} H(\omega) &= \int_{-\infty}^{\infty} h(t)e^{-j\omega t}\,dt \\ &= \int_{-\infty}^{\infty} \alpha e^{-\alpha t}u(t)e^{-j\omega t}\,dt \\ &= \int_{0}^{\infty} \alpha e^{-(\alpha+j\omega)t}\,dt \\ &= -\frac{\alpha}{\alpha+j\omega}e^{-(\alpha+j\omega)t}\,|_0^{\infty} \\ &= -\frac{\alpha}{\alpha+j\omega}[0-1] \\ &= \frac{\alpha}{\alpha+j\omega} \end{aligned}$$

We can write the transfer function in terms of its magnitude and phase

$$H(\omega) = |H(\omega)|\,e^{j\theta(\omega)}$$

where the magnitude is given by

$$|H(\omega)| = \frac{\alpha}{\sqrt{\alpha^2+\omega^2}}$$

and the phase is given by

$$\theta(\omega) = -\tan^{-1}\left(\frac{\omega}{\alpha}\right)$$

The magnitude and the phase are plotted in Figure 12.2 for $\alpha = 2$. Clearly the magnitude is nothing like what we plotted for the ideal filter. However, remember that this is a very simple filter.

If we are willing to accept a little more complexity we can come closer to the ideal characteristics—without ever really achieving it. Some of the known classes of filters include the Butterworth, Chebyshev, and Elliptic filters. They are outside of the scope of this class but they are really simple to understand and implement. The magnitude of the transfer function for a Butterworth filter with the same shape as the filters we have been looking at is given by

$$|H(\omega)| = \frac{1}{\sqrt{1+\omega^{2n}}}$$

Where n denotes the order of the filter. We have plotted the magnitude of the transfer function for a third and fifth order Butterworth filter in Figure 13.3.

An implementation of a third order Butterworth filter is shown in Figure 13.4.

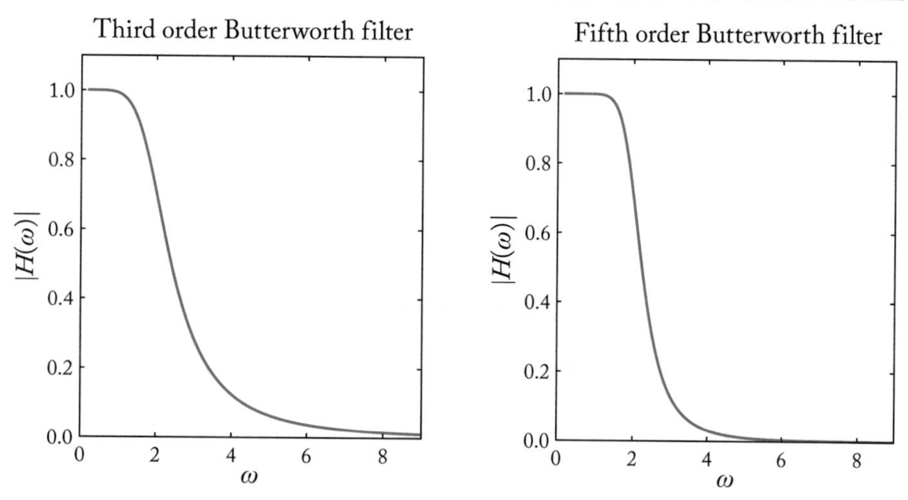

Figure 13.3: Magnitude of the transfer function of a third order Butterworth filter and a Fifth order Butterworth filter.

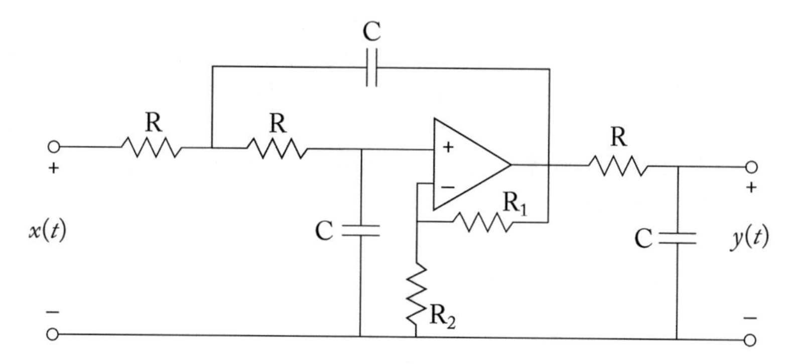

Figure 13.4: An implementation of a third order Butterworth filter.

Example 13.1 Let's take a look at our RC circuit again and examine the response of the circuit to different inputs. Let's suppose the impulse response of the circuit is

$$h(t) = e^{-t}u(t)$$

and the input is another decaying exponential

$$x(t) = e^{-2t}u(t)$$

The transforms $H(\omega)$ and $X(\omega)$, based on our earlier calculations, are given by

$$\begin{aligned} H(\omega) &= \frac{1}{1+j\omega} \\ X(\omega) &= \frac{1}{2+j\omega} \end{aligned}$$

We know that

$$y(t) = x(t) \circledast h(t)$$

Therefore, by the convolution property

$$\begin{aligned} Y(\omega) &= H(\omega)X(\omega) \\ &= \left(\frac{1}{1+j\omega}\right)\left(\frac{1}{2+j\omega}\right) \\ &= \frac{1}{(1+j\omega)(2+j\omega)} \end{aligned}$$

We have $Y(\omega)$. How do we find $y(t)$? There is a reason we have left the denominator in its factored form. We know that the inverse Fourier transform of expressions of the form $1/(a+j\omega)$ is $e^{-at}u(t)$. By linearity this means that the inverse Fourier transform of $\alpha/(a+j\omega)$ is $\alpha e^{-at}u(t)$. If we can expand $Y(\omega)$ using partial fraction expansion into terms of the form $\alpha/(a+j\omega)$ we can use the linearity of the Fourier transform to easily find $y(t)$. Using partial fraction expansion we can write $Y(\omega)$ as

$$Y(\omega) = \frac{1}{(1+j\omega)(2+j\omega)} = \frac{1}{1+j\omega} - \frac{1}{2+j\omega}$$

and, therefore,

$$y(t) = e^{-t}u(t) - e^{-2t}u(t)$$

How does this compare with the result using convolution. The plot of $x(\tau)$ and $h(t-\tau)$or $t<0$ and $t>0$ is shown in Figure 13.5.

We can see from the figure that for $t<0$ $y(t)$ is equal to zero. For $t>0$

$$\begin{aligned} y(t) &= \int_0^t e^{-(t-\tau)}e^{-2\tau}d\tau \\ &= e^{-t}\int_0^t e^{-\tau}d\tau \\ &= e^{-t}\left[1-e^{-t}\right] \\ &= e^{-t}-e^{-2t} \end{aligned}$$

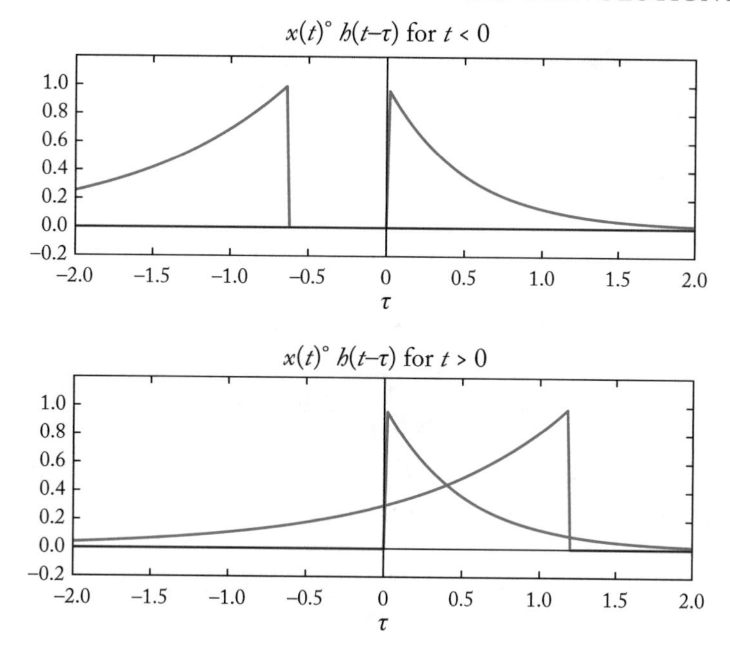

Figure 13.5: A plot of $x(\tau)$ and $h(t-\tau)$or $t < 0$ and $t > 0$.

or

$$y(t) = \left(e^{-t} - e^{-2t}\right) u(t)$$

which is the same result as that obtained from using the Fourier transform.

Example 13.2 In our previous module we looked at how the RC circuit responds to a unit step. Let's try to solve the same problem using Fourier transforms. The transfer function is still the same

$$H(\omega) = \frac{1}{1 + j\omega}$$

However, when we try to compute $X(\omega)$ we run into a problem. The input $x(t)$ does not satisfy the Dirichlet condition

$$\int_{-\infty}^{\infty} u(t)dt \to \infty$$

This means we cannot use the Fourier integral to obtain the Fourier transform of the unit step function. The unit step is a rather important function. The usefulness of the Fourier transform would be diminished if we cannot find its transform. We need to reach into our bag of tricks and pull out something that will allow us to find the Fourier transform of such problematic functions. The trick we will use is the differentiation property of the Fourier transform.

13.2 THE DIFFERENTIATION PROPERTY AND THE FOURIER TRANSFORM OF THE UNIT STEP FUNCTION

If the Fourier transform of $x(t)$ is $X(\omega)$ the derivative property states that

$$\mathcal{F}\left[\frac{d}{dt}x(t)\right] = j\omega X(\omega)$$

Before we go about deriving this let's emphasize something. If you see

$$Bill = \frac{1}{2\pi}\int_{-\infty}^{\infty} Bob\ e^{j\omega t}\,dt$$

then regardless of what *Bob* or *Bill* may say *Bob* is the Fourier transform of *Bill*. Keeping this in mind the derivative property is easy to derive

$$x(t) = \frac{1}{2\pi}\int_{-\infty}^{\infty} X(\omega)e^{j\omega t}\,d\omega$$

Taking the derivative of both sides and changing order of integration and differentiation on the right we get

$$\frac{d}{dt}x(t) = \int_{-\infty}^{\infty} j\omega X(\omega)e^{j\omega t}\,d\omega$$

Now, remembering *Bill* and *Bob*

$$\mathcal{F}\left[\frac{d}{dt}x(t)\right] = j\omega X(\omega)$$

or

$$\mathcal{F}\left[\frac{d}{dt}x(t)\right] = j\omega\mathcal{F}\left[x(t)\right]$$

or

$$\mathcal{F}\left[x(t)\right] = \frac{1}{j\omega}\mathcal{F}\left[\frac{d}{dt}x(t)\right] \tag{13.1}$$

So we can find the Fourier transform of a $x(t)$ by first finding the Fourier transform of the derivative of $x(t)$ and dividing that by $j\omega$. We have to be a bit careful though. Using the derivative property to find a Fourier transform can be touchy unless you have a signal whose dc value is 0 because the derivative will kill the dc value. The unit step definitely has a dc value (of 0.5). However, the signum function $sgn(t)$, which is -1 for $t < 0$ and 1 for $t > 0$ does not. We can write the unit step in terms of the signum function as

$$u(t) = \frac{1}{2}sgn(t) + \frac{1}{2}$$

One more thing before we proceed. The Fourier transform of a constant (forward or inverse) is a delta function. We can see that easily by using the sifting property of the delta function. So

$$\mathcal{F}[\delta(t)] = \int_{-\infty}^{\infty} \delta(t)e^{-j\omega t}dt = e^0 = 1$$

and

$$\mathcal{F}^{-1}[\delta(\omega)] = \frac{1}{2\pi}\int_{-\infty}^{\infty} \delta(\omega)e^{j\omega t}d\omega = \frac{1}{2\pi}e^0 = \frac{1}{2\pi}$$

Given that we can write the unit step as a sum of $\frac{1}{2}sgn(t)$ and $\frac{1}{2}$ we can use the fact that the Fourier transform is linear to write

$$\mathcal{F}[u(t)] = \mathcal{F}\left[\frac{1}{2}sgn(t)\right] + \mathcal{F}\left[\frac{1}{2}\right]$$

The derivative of the signum function is zero everywhere except at zero where it is $2\delta(t)$ So we can find $\mathcal{F}\left[\frac{1}{2}sgn(t)\right]$ using Equation (13.1) as

$$\begin{aligned}
\mathcal{F}\left[\frac{1}{2}sgn(t)\right] &= \frac{1}{j\omega}\mathcal{F}\left[\frac{1}{2}\frac{d}{dt}sgn(t)\right] && (13.2)\\
&= \frac{1}{j\omega}\mathcal{F}[\delta(t)] && (13.3)\\
&= \frac{1}{j\omega} && (13.4)
\end{aligned}$$

and the Fourier transform of $\frac{1}{2}$ is

$$\mathcal{F}\left[\frac{1}{2}\right] = 2\pi \cdot \frac{1}{2} \cdot \delta(\omega) = \pi\delta(\omega)$$

which gives us the Fourier transform of $u(t)$.

$$\mathcal{F}[u(t)] = \frac{1}{j\omega} + \pi\delta(\omega)$$

Example 13.3 Let's return now to our example where we were investigating the output of the RC filter when the input is a step function. Using the convolution property of the Fourier transform we multiply the Fourier transform of the unit step with the transfer function of the filter to obtain

$$\begin{aligned}
Y(\omega) &= \left[\frac{1}{j\omega} + \pi\delta(\omega)\right] \cdot \frac{1}{1+j\omega}\\
&= \frac{1}{j\omega(1+j\omega)} + \pi\delta(\omega)\frac{1}{1+j\omega}\\
&= \frac{1}{j\omega(1+j\omega)} + \pi\delta(\omega)
\end{aligned}$$

where we have used the fact that in the second term $\delta(\omega)$ is nonzero only when $\omega = 0$. Expanding the first term using partial fraction expansion we get

$$Y(\omega) = \frac{1}{j\omega} - \frac{1}{1 + j\omega} + \pi\delta(\omega)$$

Using the linearity and uniqueness of the Fourier transform we get

$$y(t) = \frac{1}{2}sgn(t) - e^{-t}u(t) + \frac{1}{2} = u(t) - e^{-t}u(t)$$

Or, to put it in a more familiar form

$$y(t) = \left[1 - e^{-t}\right]u(t)$$

One other thing to note is that the differentiation property also clearly shows something we already know—namely the derivate of a function is proportional to the rate of change of the function. Let's take a look at the Fourier transform of the derivative of a function in terms of its physical meaning.

$$\mathcal{F}\left[\frac{dx(t)}{dt}\right] = j\omega X(\omega)$$

The DC or constant value of the signal is $X(0)$ and as we can see for $\omega = 0$ we get $j0X(0) = 0$. In other words the derivative of the constant term in $x(t)$ is zero. For ω small or low frequencies, or the slow variations in the signal $j\omega X(\omega)$ will be small, and for ω large the high frequency components of $X(\omega)$ will be multiplied with this larger value of ω.

13.3 SUMMARY

In this module we learned

1. Convolution in the time domain gets transformed to multiplication in the Fourier domain

$$\mathcal{F}\left[x(t) \circledast h(t)\right] = X(\omega)H(\omega)$$

2. There is a reason we cannot have an ideal brick wall filter.

3. The Fourier transform of the derivative of a function is given by

$$\mathcal{F}\left[\frac{d}{dt}x(t)\right] = j\omega X(\omega)$$

The derivative operation weights low frequency components of $x(t)$ less and the high frequency components of $x(t)$ more.

4. The Fourier transform of the unit step is given by

$$\mathcal{F}[u(t)] = \frac{1}{j\omega} + \pi\delta(\omega)$$

We will continue in the next module with properties of the Fourier transform and connect them to various applications.

13.4 EXERCISES

(Answers on the following page)

1. The signal $y(t)$ is the output from a linear time invariant system with impulse response $h(t)$ when the input is $x(t)$. Given that $X(\omega)$ the Fourier transform of the input and $y(t)$ are given by

$$\begin{aligned} X(\omega) &= 1 \\ y(t) &= e^{-2t}u(t) \end{aligned}$$

   find $h(t)$.

2. Given the input $x(t)$

$$x(t) = e^{-3t}u(t)$$

   to a linear time invariant system with transfer function

$$H(\omega) = \frac{1}{4 + j\omega}$$

   Find the output $y(t)$.

3. Find the Fourier transform of

   (a) $x(t) = \delta(t)$
   (b) $x(t) = \delta(t-2)$
   (c) $x(t) = \delta(t-2) + \delta(t+2)$

4. Find the inverse Fourier transform of

   (a) $X(\omega) = \delta(\omega)$
   (b) $X(\omega) = \delta(\omega - 2)$
   (c) $X(\omega) = \delta(\omega - 2) + \delta(\omega + 2)$

5. The input output relationship of a linear time invariant system is given by

$$y(t) = \frac{dx(t)}{dt}$$

   What is the transfer function for this system?

6. A linear time invariant system is described by the differential equation

$$\frac{d^2y(t)}{dt^2} + 3\frac{dy(t)}{dt} + 2y(t) = x(t)$$

   (a) Use the differentiation property of the Fourier transform to find the transfer function of this system.
   (b) Find the impulse response of this system.

13.5 ANSWERS

1.
$$h(t) = e^{-2t}u(t)$$

2.
$$y(t) = e^{-3t}u(t) - e^{-4t}u(t)$$

3. (a) $X(\omega) = 1$

   (b) $X(\omega) = e^{-j\omega}$

   (c) $X(\omega) = 2\cos(2\omega)$

4. (a)
$$x(t) = \frac{1}{2\pi}$$

   (b)
$$x(t) = \frac{1}{2\pi}e^{j2t}$$

   (c)
$$x(t) = \frac{1}{\pi}\cos(2t)$$

5.
$$H(\omega) = j\omega$$

6. (a)
$$H(\omega) = \frac{1}{2 + 3j\omega - \omega^2}$$

   (b)
$$h(t) = \left[e^{-t} - e^{-2t}\right]u(t)$$

MODULE 14

Some More Useful Properties of the Fourier Transform

Let's look at a few more properties of the Fourier transform which will come in handy when dealing with applications we are interested in. We examined the differentiation property in the last module so let's begin with the integration property.

14.1 INTEGRATION PROPERTY

If we know that the Fourier transform of $x(t)$ is $X(\omega)$ then the integration property says that

$$\mathcal{F}\left[\int_{-\infty}^{t} x(\tau)d\tau\right] = \frac{1}{j\omega}X(\omega) + \pi X(0)\delta(\omega)$$

There are a number of different ways of showing the integration property. We will use what we learned in the previous module. Notice that the limits of integration are $-\infty$ to t instead of from $-\infty$ to ∞. We can get this form if we convolve $x(t)$ with $u(t)$

$$x(t) \circledast u(t) = \int_{-\infty}^{\infty} x(\tau)u(t-\tau)d\tau = \int_{-\infty}^{t} x(\tau)d\tau$$

By the convolution property we know that

$$\mathcal{F}[x(t) \circledast u(t)] = X(\omega)\mathcal{F}[u(t)]$$

And in the previous module we found that

$$\mathcal{F}[u(t)] = \frac{1}{j\omega} + \pi\delta(\omega)$$

Therefore,

$$\mathcal{F}[x(t) \circledast u(t)] = \mathcal{F}\left[\int_{-\infty}^{t} x(\tau)d\tau\right] = X(\omega)\cdot\left[\frac{1}{j\omega} + \pi\delta(\omega)\right]$$

or

$$\mathcal{F}\left[\int_{-\infty}^{t} x(\tau)d\tau\right] = \frac{1}{j\omega}X(\omega) + \pi\delta(\omega)X(\omega)$$

$\delta(\omega)$ is nonzero only for $\omega = 0$. Therefore,

$$\mathcal{F}\left[\int_{-\infty}^{t} x(\tau)d\tau\right] = \frac{1}{j\omega}X(\omega) + \pi\delta(\omega)X(0)$$

14.2 TIME AND FREQUENCY SCALING

By time scaling we mean the broadening or narrowing of the function by multiplying the argument with a scale factor. We can see how this works by considering a pulse in time

$$x(t) = \begin{cases} 1 & -1 \leq t \leq 1 \\ 0 & \text{otherwise} \end{cases}$$

Let's scale the argument by a scale factor α to generate $x(\alpha t)$. Examining $x(\alpha t)$,

$$x(\alpha t) = \begin{cases} 1 & -1 \leq \alpha t \leq 1 \\ 0 & \text{otherwise} \end{cases}$$

For the moment let's assume that α is positive. Then we get

$$x(\alpha t) = \begin{cases} 1 & -\frac{1}{\alpha} \leq t \leq \frac{1}{\alpha} \\ 0 & \text{otherwise} \end{cases}$$

The pulse width went from 2 to $2/\alpha$. If α is greater than 1 this will mean a narrowing of the pulse. If α is less than 1 this will mean a broadening of the pulse. Now let's see how scaling in the time domain effects the spectral profile of the signal. Taking the Fourier transform

$$\mathcal{F}[x(\alpha t)] = \int_{-\infty}^{\infty} x(\alpha t)e^{-j\omega t}dt$$

substitute $\tau = \alpha t \Rightarrow t = \tau/\alpha$. If α is positive the limits stay the same. However if α is negative the limits flip—when t goes to infinity αt goes to negative infinity and when t goes to negative infinity αt goes to positive infinity. Let's work through both cases. For $\alpha > 0$ we get

$$\begin{aligned}
\mathcal{F}[x(\alpha t)] &= \int_{-\infty}^{\infty} x(\tau)e^{-j\omega\tau/\alpha}\frac{1}{\alpha}d\tau \\
&= \frac{1}{\alpha}\int_{-\infty}^{\infty} x(\tau)e^{-j(\frac{\omega}{\alpha})\tau}d\tau \\
&= \frac{1}{\alpha}X\left(\frac{\omega}{\alpha}\right)
\end{aligned}$$

For $\alpha < 0$ we get

$$\begin{aligned} \mathcal{F}[x(\alpha t)] &= \int_{\infty}^{-\infty} x(\tau)e^{-j\omega\tau/\alpha}\frac{1}{\alpha}d\tau \\ &= -\frac{1}{\alpha}\int_{-\infty}^{\infty} x(\tau)e^{-j(\frac{\omega}{\alpha})\tau}d\tau \\ &= -\frac{1}{\alpha}X\left(\frac{\omega}{\alpha}\right) \end{aligned}$$

Combining both we get

$$\mathcal{F}[x(\alpha t)] = \frac{1}{|\alpha|}X\left(\frac{\omega}{\alpha}\right)$$

where we have used the fact that

$$|x| = \begin{cases} x & \text{for } x > 0 \\ -x & \text{for } x < 0 \end{cases}$$

Notice now that instead of getting αt as the argument we have ω/α. This means the effect of the magnitude of α is reversed. When α is greater than one $|X(\omega)|$ will be broader and when α is less than 1 it will be narrower. This agrees with our earlier discussions. Narrowing of a signal in the time domain results in a broadening of its spectral profile and vice versa.

When $x(t)$ is even $X(\omega)$ is even Let's take a look at the special case where $\alpha = -1$. Substituting this value for α we get

$$\mathcal{F}[x(-t)] = \frac{1}{|-1|}X\left(\frac{\omega}{-1}\right) = X(-\omega)$$

If $x(t)$ is even, $x(t) = x(-t)$ and given the fact that the Fourier transform of $x(-t)$ is $X(-\omega)$ and the uniqueness of the Fourier transform, this means that $X(\omega) = X(-\omega)$.

Example 14.1 For our first example let's pick a square pulse—what we would get if we started with the square wave and let the period go to infinity.

$$x(t) = \begin{cases} 1 & |t| < T_o \\ 0 & \text{otherwise} \end{cases}$$

The Fourier transform is given by

$$\begin{aligned} X(\omega) &= \int_{-\infty}^{\infty} x(t)e^{-j\omega t}\,dt \\ &= \int_{-T_o}^{T_o} e^{-j\omega t}\,dt \\ &= -\frac{1}{j\omega}e^{-j\omega t}\Big|_{-T_o}^{T_o} \\ &= -\frac{1}{j\omega}\left(e^{-j\omega T_o} - e^{-j\omega(-T_o)}\right) \\ &= \frac{1}{j\omega}\left(e^{j\omega T_o} - e^{-j\omega T_o}\right) \\ &= \frac{\sin(\omega T_o)}{\omega/2} \end{aligned} \tag{14.1}$$

This form is very similar to the form of the Fourier series coefficients of the square wave. Recall that for for a square wave the Fourier series coefficients were given by

$$a_n = \frac{\sin(n\omega_o T_o)}{n\omega_o T/2}$$

$X(\omega)$ is plotted in Figure 14.1 for $T_o = 1$.

What happens if we pick T_o to be equal to a half? Making T_o smaller narrows our pulse. In the frequency domain, as seen in Figure 14.2 the effect is to broaden the frequency profile. In practice this duality has a major impact. If we are using pulses to transmit binary data—positive pulse for a one and negative pulse for a zero, or a pulse for a one and no pulse for a zero, the width of the pulse depends on the rate at which we want to send the bits. If we want to send 1000 bits per second the width of the pulse has to be less than a millisecond. If we want to send 10,000 bits per second the width of the pulse has to be less than 0.1 milliseconds. The higher the rate, the narrower the pulse, and, therefore, based on what we see here the broader will be the frequency profile. A measure of the width of the frequency profile is referred to as the bandwidth of the signal. Thus, a higher rate will require a higher bandwidth.

14.3 FOURIER TRANSFORM OF PERIODIC SIGNALS

We know that we can find the frequency components of a periodic signal using the Fourier series representation. We motivated the Fourier transform by our need to find the frequency representation of aperiodic signals. It would be nice if we could also include periodic signals within the Fourier transform framework. That is what we are going to do in this section.

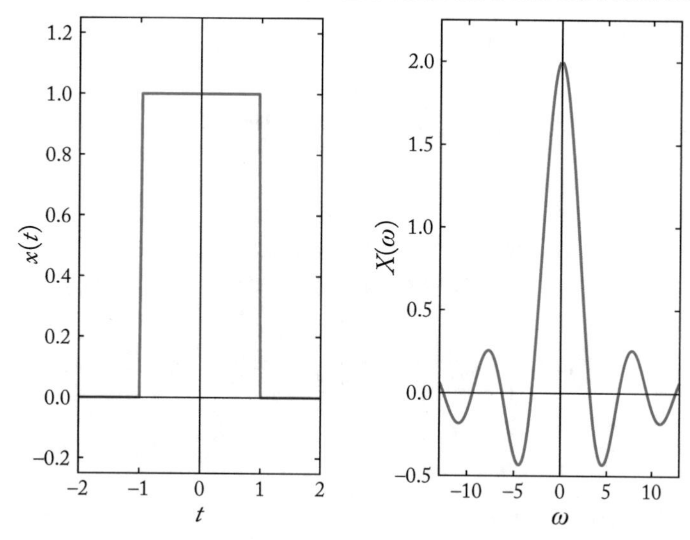

Figure 14.1: The Fourier transform of a rectangular pulse with pulse width 2.

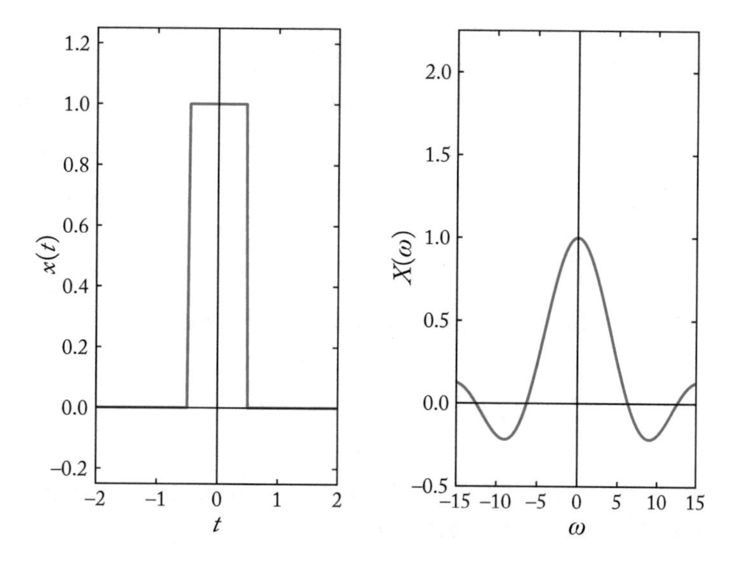

Figure 14.2: The Fourier transform of a rectangular pulse with pulse width 1.

We begin by using the linearity of the Fourier transform. Suppose $x(t)$ is a periodic signal. Then we can write it as

$$x(t) = \sum_{n=-\infty}^{\infty} a_n e^{jn\omega_o t}$$

Using the Linearity of the Fourier transform

$$\mathcal{F}[x(t)] = \sum_{n=-\infty}^{\infty} a_n \mathcal{F}\left[e^{jn\omega_o t}\right]$$

In order to evaluate this transform we need to find the Fourier transform of $e^{jn\omega_o t}$. We will make use of the uniqueness of the Fourier transform to find this.

We begin by evaluating the inverse Fourier transform of $\delta(\omega - \omega_o)$.

$$\begin{aligned}\mathcal{F}^{-1}[\delta(\omega - \omega_0)] &= \frac{1}{2\pi}\int_{-\infty}^{\infty} \delta(\omega - \omega_0) e^{j\omega t} d\omega \\ &= \frac{1}{2\pi} e^{j\omega_o t}\end{aligned}$$

where we have used the sifting property of the delta function. Multiplying both sides by $1/2\pi$ we get

$$\mathcal{F}^{-1}[2\pi\delta(\omega - \omega_0)] = e^{j\omega_o t}$$

Because the Fourier transform is unique, this means that if the inverse transform of $2\pi\delta(\omega - \omega_0)$ is $e^{j\omega_o t}$, then the Fourier transform of $e^{j\omega_o t}$ is $2\pi\delta(\omega - \omega_0)$.

Returning to our original quest for the Fourier transform of a periodic function $x(t)$ we get

$$\mathcal{F}[x(t)] = \sum_{n=-\infty}^{\infty} a_n \mathcal{F}\left[e^{jn\omega_o t}\right] = 2\pi \sum_{n=-\infty}^{\infty} a_n \delta(\omega - n\omega_0)$$

Example 14.2 Let's apply this to a particular periodic function—the impulse train.

$$x(t) = \sum_{k=-\infty}^{\infty} \delta(t - kT)$$

The impulse train is clearly periodic and can therefore be written as a Fourier series. To find the series coefficient take the integral over one period.

$$\begin{aligned}a_n &= \frac{1}{T}\int_{-\frac{T}{2}}^{\frac{T}{2}} \sum_{n=-\infty}^{\infty} \delta(t - kT) e^{-jk\omega_0 t} dt \\ &= \frac{1}{T}\int_{-\frac{T}{2}}^{\frac{T}{2}} \delta(t) e^{-jk\omega_0 t} dt \\ &= \frac{1}{T}\end{aligned}$$

Thus,

$$x(t) = \sum_{n=-\infty}^{\infty} \frac{1}{T} e^{jkn\omega_0 t}$$

Therefore,

$$X(\omega) = \frac{2\pi}{T} \sum_{n=-\infty}^{\infty} \delta(\omega - k\omega_0)$$

Therefore, Fourier transform of an impulse train is an impulse train. Notice that the separation between impulses in the time and frequency domains are inversely related. In the time domain they are separated by T while in the frequency domain they are separated by $\omega_0 = 2\pi/T$. Therefore, if the impulses in the time domain are brought closer the impulses in the frequency domain will move apart.

14.4 A SHIFT IN TIME IS A PHASE CHANGE IN FREQUENCY

As in the case of the Fourier series a time shift in the time domain results in a phase shift in the frequency domain. This is easy to show. Let $x(t)$ and $X(\omega)$ be a Fourier transform pair

$$\mathcal{F}[x(t)] = X(\omega) = \int_{-\infty}^{\infty} x(t)e^{-j\omega t}dt$$

The Fourier transform of $x(t - t_o)$ is

$$\mathcal{F}[x(t - t_o)] = \int_{-\infty}^{\infty} x(t - t_o)e^{-j\omega t}dt$$

Substituting $\tau = t - t_o$

$$\begin{aligned} \mathcal{F}[x(t - t_o)] &= \int_{-\infty}^{\infty} x(\tau)e^{-j\omega(\tau + t_o)}d\tau \\ &= e^{-j\omega t_o} \int_{-\infty}^{\infty} x(\tau)e^{-j\omega\tau}d\tau \\ &= e^{-j\omega t_o} X(\omega) \end{aligned}$$

If we write $X(\omega)$ in polar form

$$X(\omega) = |X(\omega)|\, e^{j\theta(\omega)}$$

we can see that

$$\mathcal{F}[x(t - t_o)] = |X(\omega)|\, e^{j\theta(\omega)}e^{-j\omega t_o} = |X(\omega)|\, e^{j(\theta(\omega) - \omega t_o)}$$

Example 14.3 Let's find the Fourier transform of

$$x(t) = \cos(2t - 2)$$

First, notice that we can write $x(t)$ as $\cos(2(t-1))$ so we can find the Fourier transform of $\cos(2t)$ and then apply the shifting property to find the Fourier transform of $\cos(2t-2)$. To find the Fourier transform of $\cos(2t)$ let's write the expression using Euler's formula

$$\cos(2t) = \frac{1}{2}e^{j2t} + \frac{1}{2}e^{-j2t}$$

Using the linearity of the Fourier transform we can take the Fourier transform of each individual term and then combine them together to get the Fourier transform of $\cos(2t)$.

$$\begin{aligned} \mathcal{F}\left[\frac{1}{2}e^{j2t}\right] &= \frac{1}{2}\mathcal{F}\left[e^{j2t}\right] \\ &= \frac{1}{2}2\pi\delta(\omega-2) \\ &= \pi\delta(\omega-2) \\ \mathcal{F}\left[\frac{1}{2}e^{-j2t}\right] &= \frac{1}{2}\mathcal{F}\left[e^{-j2t}\right] \\ &= \frac{1}{2}2\pi\delta(\omega+2) \\ &= \pi\delta(\omega+2) \end{aligned}$$

Therefore,

$$\mathcal{F}[\cos(2t)] = \pi\delta(\omega-2) + \pi\delta(\omega+2)$$

Now using the shifting property

$$\mathcal{F}[\cos(2(t-1))] = \pi\delta(\omega-2)e^{-j\omega} + \pi\delta(\omega+2)e^{-j\omega}$$

We could go one step further and noting that $\delta(\omega-2)$ is nonzero only at $\omega = 2$ and $\delta(\omega+2)$ is nonzero only at $\omega = -2$,

$$\mathcal{F}[\cos(2(t-1))] = \pi\delta(\omega-2)e^{-2j} + \pi\delta(\omega+2)e^{2j}$$

14.5 SUMMARY

In this module we looked at the following properties of the Fourier transform:

1. **Integration property:** If we integrate a time domain function we tend to smooth out the high frequency variations. We can see that in the first term below

$$\mathcal{F}\left[\int_{-\infty}^{t} x(\tau)d\tau\right] = \frac{1}{j\omega}X(\omega) + \pi X(0)\delta(\omega)$$

   The second term reflects the situation where the time signal has a non-zero average value or DC bias.

2. **Time and Frequency scaling:** Narrowing a function in time broadens its frequency profile while broadening a function in time narrows its frequency profile.

3. **Time shift:** If we introduce a time shift in a signal this appears as a phase shift in the frequency domain.

We also extended the Fourier transform to periodic signals.

14.6 EXERCISES

(Answers on the following page)

1. Find the Fourier transform of

   (a) $x(t) = \cos(3t)$.

   (b) $x(t) = \sin(3t)$

   (c) $x(t) = \cos(3t + 1)$

   (d) $x(t) = \cos(2t + 1)$

2. The Fourier transform of

$$a(t) = u\left(t + \frac{1}{2}\right) - u\left(t - \frac{1}{2}\right)$$

   has been shown in the writeup to be

$$A(\omega) = 2\sin(\omega/2)/\omega$$

   Without using the Fourier integral find the Fourier transform of the following

   (a)

$$x(t) = u(t) - u(t - 1)$$

   (b)

$$y(t) = u(t + 1) - u(t - 1)$$

   (c)

$$z(t) = u(t) - u(t - 2)$$

3. Find and sketch the Fourier transform of

$$x(t) = \sum_{k=-\infty}^{\infty} p(t - kT)$$

   where

   (a)

$$p(t) = \begin{cases} 1 & |t| < T/4 \\ 0 & \text{otherwise} \end{cases}$$

   (b)

$$p(t) = \begin{cases} 1 & |t| < T/8 \\ 0 & \text{otherwise} \end{cases}$$

4. Find the Fourier transform of

   (a)

$$x(t) = e^{-2|t|}$$

   (b)

$$x(t) = e^{-2|t|} \cos(5t)$$

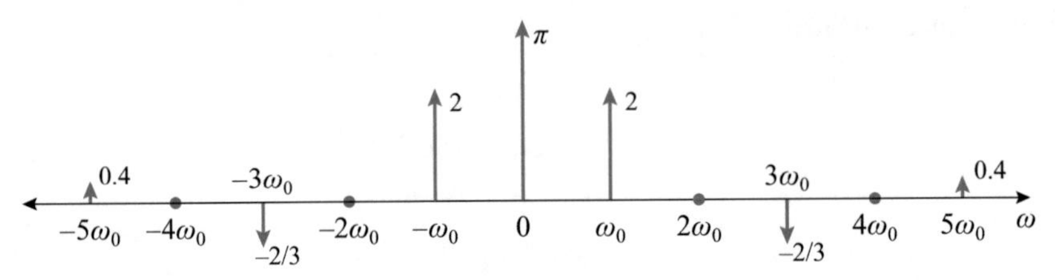

Figure 14.3: The Fourier transform of a train of rectangular pulses with pulse width $T/4$.

14.7 ANSWERS

1. (a)

$$\mathcal{F}\left[\cos(3t)\right] = \pi\delta(\omega - 3) + \pi\delta(\omega + 3)$$

(b)

$$\mathcal{F}\left[\sin(3t)\right] = -j\pi\delta(\omega - 3) + j\pi\delta(\omega + 3)$$

(c)

$$X(\omega) = e^{j}\pi\delta(\omega - 3) + e^{-j}\pi\delta(\omega + 3)$$

(d)

$$X(\omega) = e^{j}\pi\delta(\omega - 2) + e^{-j}\pi\delta(\omega + 2)$$

2. (a)

$$X(\omega) = 2e^{-j\omega/2}\frac{\sin(\omega/2)}{\omega}$$

(b)

$$Y(\omega) = 2\frac{\sin(\omega)}{\omega}$$

(c)

$$Z(\omega) = 2e^{-j\omega}\frac{\sin(\omega)}{\omega}$$

3. (a) (Figure 14.3)

$$X(\omega) = \sum_{n=-\infty}^{\infty} 2\,\frac{\sin(n\pi/2)}{n}\delta(\omega - n\omega_o)$$

(b) (Figure 14.4)

$$X(\omega) = \sum_{n=-\infty}^{\infty} 2\,\frac{\sin(n\pi/4)}{n}\delta(\omega - n\omega_o)$$

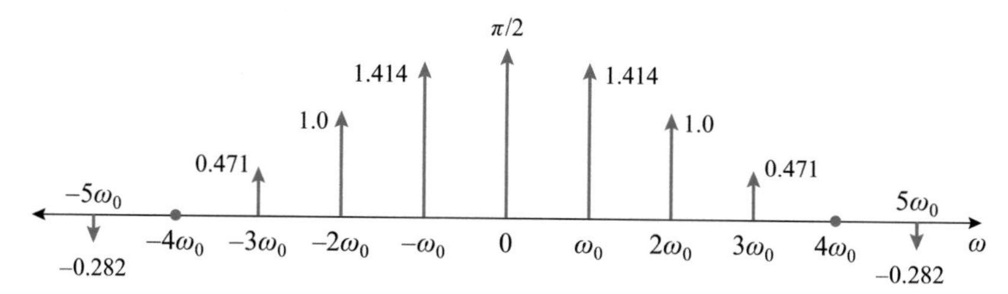

Figure 14.4: The Fourier transform of a train of rectangular pulses with pulse width $T/8$.

4. (a)

$$X(\omega) = \frac{4}{4 + \omega^2}$$

(b)

$$X(\omega) = \frac{4\pi}{4 + (\omega - 5)^2} + \frac{4\pi}{4 + (\omega + 5)^2}$$

MODULE 15

Sampling

Much of the processing nowadays is conducted in the digital domain. However, the signals being processed, voice, music, video, etc., are all analog. To go from analog signals to the digital signals we first have to move from the continuous time domain to the discrete time domain. The operation which allows us to do this is sampling. Processing the signal in the digital domain is not the end goal of many applications especially if the processed signal is for human consumption. We often need to convert the discrete time signal back into a continuous time signal which can be perceived by humans. In this module we introduce the process of sampling and the reconstruction of signals from the sampled values. We have almost all the tools we need to understand this process. The only tool we are missing is the multiplication property of the Fourier transform which is what we look at next.

15.1 THE MULTIPLICATION PROPERTY OF THE FOURIER TRANSFORM

Consider for a moment the forward and inverse Fourier transforms

$$\begin{aligned} X(\omega) &= \int_{-\infty}^{\infty} x(t)e^{-j\omega t}\,dt \\ x(t) &= \frac{1}{2\pi}\int_{-\infty}^{\infty} X(\omega)e^{j\omega t}\,d\omega \end{aligned}$$

Notice the similarity of the forward and inverse Fourier transforms. As one would expect this similarity is reflected in the properties of the Fourier transform. In particular the convolution property of the Fourier transform is that the Fourier transform of the convolution of signals in the time domain is the product of the Fourier transform of the signals. You would then expect that the Fourier transform of the product of signals in the time domain would be the convolution of the Fourier transform of the signals. You would be right in this expectation. Let's validate the expectation Suppose

$$g(t) = a(t)b(t)$$

Then

$$G(\omega) = \int_{-\infty}^{\infty} g(t)e^{j\omega t}\,dt = \int_{-\infty}^{\infty} a(t)b(t)e^{j\omega t}\,dt$$

Writing $b(t)$ as the inverse Fourier transform of $B(\omega)$

$$\begin{aligned} G(\omega) &= \int_{-\infty}^{\infty} a(t)\left[\frac{1}{2\pi}\int_{-\infty}^{\infty} B(\sigma)e^{j\sigma t}d\sigma\right]e^{-j\omega t}dt \\ &= \frac{1}{2\pi}\int_{-\infty}^{\infty} B(\sigma)\left[\int_{-\infty}^{\infty} a(t)e^{-j\omega t}e^{j\sigma t}d\omega\right]d\sigma \\ &= \frac{1}{2\pi}\int_{-\infty}^{\infty} B(\sigma)\left[\int_{-\infty}^{\infty} a(t)e^{-j(\omega-\sigma)t}d\omega\right]d\sigma \\ &= \frac{1}{2\pi}\int_{-\infty}^{\infty} B(\sigma)A(\omega-\sigma)d\sigma \end{aligned}$$

or

$$G(\omega) = \frac{1}{2\pi}A(\omega) \circledast B(\omega)$$

Let's now use this property to explore sampling.

15.2 IDEAL SAMPLING

When we sample a signal in time can we recover the original signal exactly from the sampled signal? At first sight the answers seems to be a clear no. By the process of sampling we have thrown away all the information about the signal between the samples. However, surprisingly, the answer to the question is a qualified yes. Even more surprisingly the answer is still yes when all our intuition yells no. Consider the portion of the sinusoidal signal shown in top panel of Figure 15.1. In the bottom panel of the figure we show just the samples themselves. It is intuitively obvious that given the samples we cannot accurately recover the original signal. After all there are an infinite number of signals that could have resulted in those samples. Intuition in this case is completely and utterly wrong.

We owe this humbling of our intuitive self ostensibly to Harry Nyquist and Claude Shannon. Nyquist published a version of the result we will now look at in the *Bell Systems Technical Journal* in 1928 where it languished until Claude Shannon working in the same lab resurrected and expanded the result into (almost) the current form in 1949. (Ostensibly because as with almost all ideas attributed to particular people this idea was also discovered by others—in this case going back to at least 1897.)

Suppose we have a signal $x(t)$ with a Fourier transform which is zero for ω greater than some W. In other words $x(t)$ is a bandlimited signal. Just as an example we could visualize $X(\omega)$ as shown in Figure 15.2. We can model a sampled signal $x_S(t)$ as a product of the original signal $x(t)$ with an impulse train.

$$x_S(t) = x(t) \cdot \sum_{k=-\infty}^{\infty} \delta(t - kT_S)$$

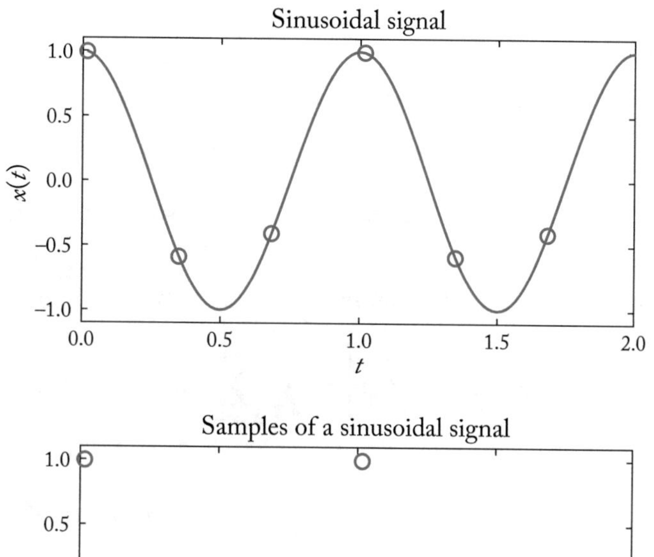

Figure 15.1: Samples of the sinusoidal signal.

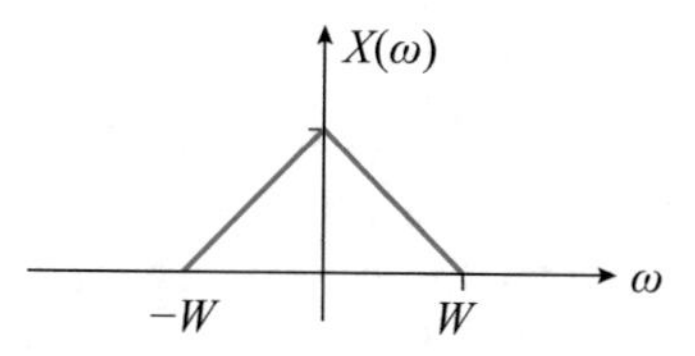

Figure 15.2: An example signal.

where T_S is the sampling interval. Using the multiplication property we can write the Fourier transform of the sampled signal as the convolution of the original signal and the Fourier transform of the impulse train.

$$X_S(\omega) = \frac{1}{2\pi} X(\omega) \circledast \mathcal{F}\left[\sum_{k=-\infty}^{\infty} \delta(t - kT_S)\right]$$

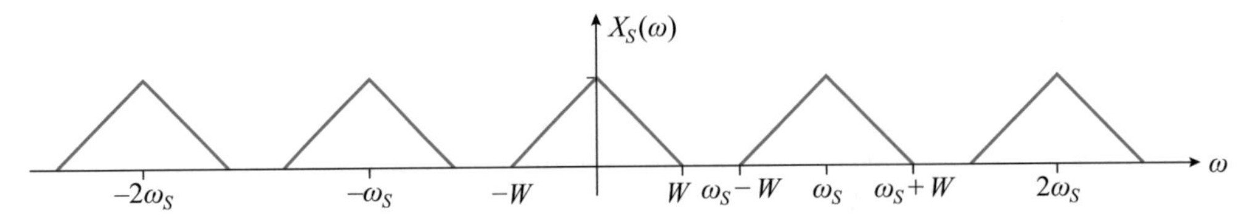

Figure 15.3: Fourier transform of the sampled version of the signal shown in Figure 15.2.

But as we have previously noted the Fourier transform of an impulse train is an impulse train.

$$\mathcal{F}\left[\sum_{k=-\infty}^{\infty} \delta(t - kT_S)\right] = \frac{2\pi}{T_S} \sum_{n=-\infty}^{\infty} \delta(\omega - k\omega_S)$$

so

$$X_S(\omega) = \frac{1}{2\pi} X(\omega) \circledast \frac{2\pi}{T_S} \sum_{n=-\infty}^{\infty} \delta(\omega - k\omega_S)$$

We can switch the order of convolution and multiplication to obtain

$$\begin{aligned} X_S(\omega) &= \frac{1}{T_S} \sum_{n=-\infty}^{\infty} X(\omega) \circledast \delta(\omega - k\omega_S) \\ &= \frac{1}{T_S} \sum_{n=-\infty}^{\infty} X(\omega - k\omega_S) \end{aligned}$$

The Fourier transform of the sampled signal is shown in Figure 15.3. It is transform of the original signal replicated at intervals of length ω_S.

Clearly we can recover the original signal perfectly from the sampled signal. All we need to do is use a low pass filter with a bandwidth of W and we have our original signal back. There is however one little detail we glossed over. Take another look at Figure 15.3. The only reason we can recover our original signal is that the various replicas (known as images) are distinctly separate from each other. If the sampling frequency ω_S had been lower the various replicas would add to each other and it would be impossible to recover the original signal. The picture in Figure 15.3 actually tells us exactly what the lower limit for the sampling frequency has to be in order to be able to recover the original signal. Because the original signal is bandlimited to W radians/sec, the lower limit of the first replica is at $\omega_S - W$. In order for the two signals not to overlap we need

$$\omega_S - W > W \Rightarrow \omega_S > 2W$$

W is the highest frequency in the original signal. So, the requirement for perfect reconstruction from the samples, at least theoretically, is to sample at a rate which is greater than twice the highest frequency. This frequency is called the *Nyquist frequency* after its discoverer Harry

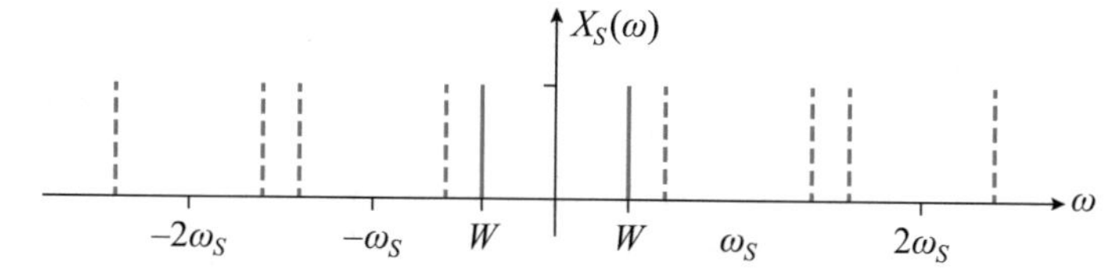

Figure 15.4: A 440 Hz signal sampled at 1100 samples per second.

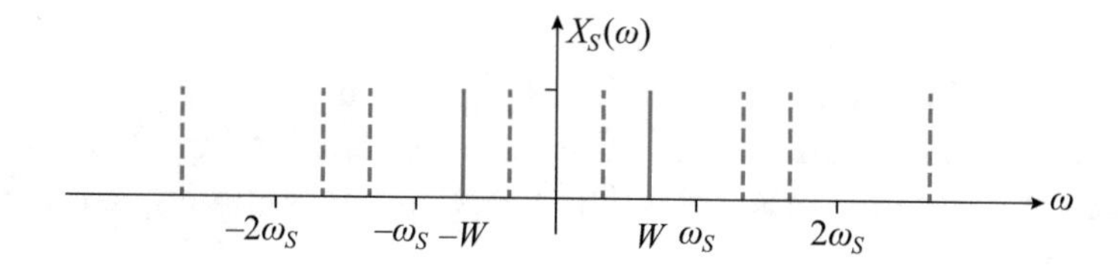

Figure 15.5: A 440 Hz signal sampled at 660 samples per second.

Nyquist. In practice we would want to sample at a higher rate as the larger the value of ω_S, the further apart the replicas will be. The further apart the replicas are the more relaxed the requirements on the reconstruction filter.

We can see what happens when the sampling frequency is less than the Nyquist frequency if we sample a pure tone. Let's sample a sinusoid at a frequency of 440 Hz. If we sample it at 1100 samples per second we will get a sampled signal with a frequency profile which looks something like that shown in Figure 15.4. In the figure the replicas are shown with dotted lines. We can use a low pass filter with a 450 Hz cutoff and recover the original tone.

Now let's take the same signal and sample it at 660 samples per second. Now the spectral profile of the sampled signal looks like that shown in Figure 15.5. If we tried to use the same low pass filter to recover the original signal we would instead get two tones back—one at 440 Hz and one at 220 Hz.

This distortion due to undersampling is called *aliasing* and in most real world situations getting rid of it is impossible. There are times when we are stuck with a particular sampling frequency which is less than twice the highest frequency of the signal being sampled because of processor or other constraints. This is the situation when sampling voice signals for transmission over telephones. Because of various historical reasons the voice signal transmitted over cellphones is usually sampled at 8000 samples per second. This would be fine if all frequency components in the voice signal were at frequencies below 4 kHz. However, the voice contains components at least up to 8 kHz. Sampling this signal at 8000 samples per second would result in a huge amount of aliasing. Fortunately for us most of the information in the voice signal is contained in frequencies below 4 kHz. Therefore, in order to avoid the aliasing distortion we first filter the voice signal using an *antialiasing filter* with a cutoff at 3.6 kHz. Note that this still introduces distortion—the components above 4 kHz have much to do with the audio qual-

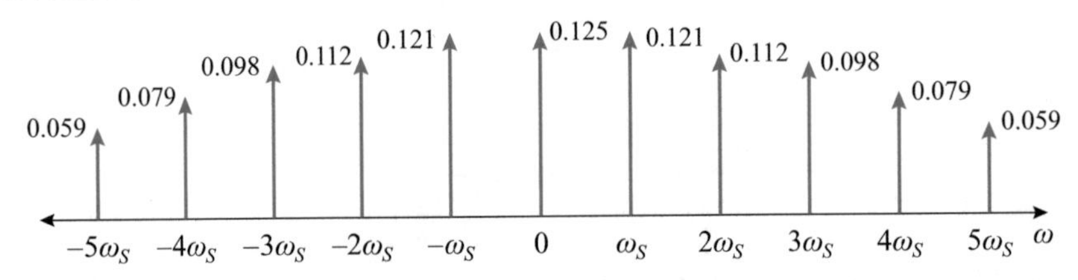

Figure 15.6: The Fourier transform of a pulse train with $T_o = T/16$.

ity of the voice signal. However, this distortion is significantly less than the aliasing noise. In addition it is controlled distortion. Basically the use of the antialiasing filter is there to cut our losses. Some voice-over-internet protocols (VoIP) uses a higher sampling rate and can therefore provide a higher quality voice.

Where quality is important—for example for music signals—we use a much higher sampling rate—usually 44,100 samples per second.

Why not just use a higher sampling rate all the time? First, increased sampling rate requires increased processor power which may not necessarily be available or convenient. Second, each sample has to be represented by number of bits—usually eight bits per sample for voice and sixteen bits for other audio signals. Increasing the sampling rate results in increasing the bit rate which in turn means a higher bandwidth for transmission. In the end it is a matter of tradeoffs.

15.3 NONIDEAL SAMPLING

In the discussion above we modeled the sampling process using an impulse train. An impulse train is a mathematical abstraction and not particularly realistic. A more realistic model would use a pulse train in place of the impulse train. Does doing that mess up our development? Not really. Suppose we used a version of the square wave we found the Fourier transform for in the previous module. Let's make the pulse a bit narrower and pick $T_o = T/16$. This would give us values for a_n as

$$a_n = \frac{\sin(n\pi/8)}{n\pi}$$

The Fourier transform of this square wave is shown in Figure 15.6. Convolving this with our original signal we again get a set of replicas of the original signal. The only difference is that the replicas each have a different gain. As we are only interested in the replica at the origin this is not really much of an issue. In fact we could use another shape for the pulse and we would get pretty much the same result.

Finally, let's go back to our first example. Why is it that against all intuition we can get our original signal back despite the fact that there are an infinite number of possible signals that would generate the samples shown in Figure 15.1. Let's take a look at the Fourier transform of the sampled signal. The signal itself is a 1 Hz signal being sampled three times each cycle—or

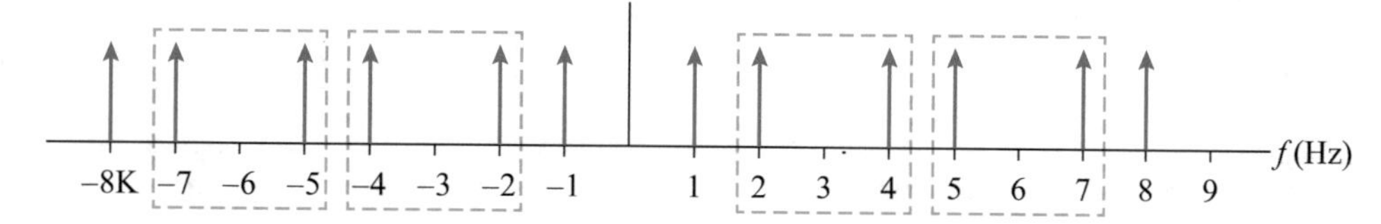

Figure 15.7: A 1 Hz signal sampled at 3 samples per second.

three samples per second. The Fourier transform of the sampled signal is shown in Figure 15.7 where we have put dashed boxes around the replicas. There are infinitely many sinusoids here and by picking and choosing we could generate an infinite number of different signals. However, once we apply a low pass filter to extract only the replica around the origin we get the original signal back.

15.4 SUMMARY

In this module we looked at what happens to the frequency domain representation of a signal when we sample it in the time domain as an application of the multiplication property of Fourier transforms. We introduced the counterintuitive result that as long as we sample a signal at a rate greater than twice the highest frequency in the signal we can recover the original signal without distortion. This result forms the basis of much of our digital world today.

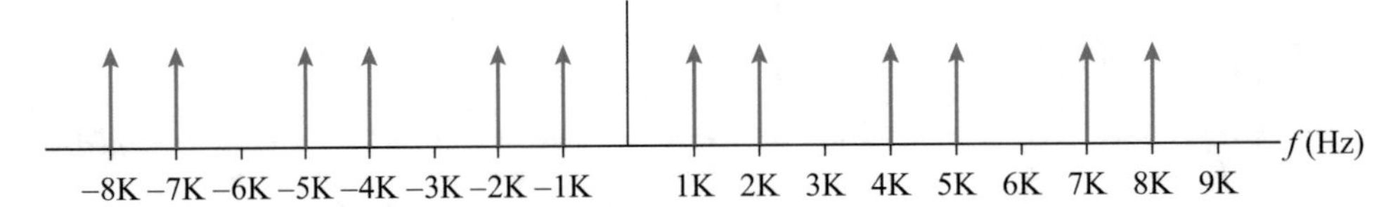

Figure 15.8: Fourier transform.

15.5 EXERCISES

(Answers on the following page)

1. The signal $x(t)$ is a 2 KHz tone

$$x(t) = \cos(4000\pi t)$$

   (a) In theory what is the minimum number of samples per seconds (the Nyquist sampling rate) we need in order to be able to reconstruct the original signal from its samples.

   (b) This 2 KHz signal is undersampled with the resulting Fourier transform shown in Figure 15.8.
   What was the sampling rate used in this case?

2. Suppose we have a signal $x(t)$ which can be exactly recovered from its samples. If the sampling rate is 10,000 samples/second (or $\omega_S = 2\pi \times 10{,}000$ radians/sec). What is the highest nonzero frequency component of this signal?

3. Suppose we generate a new signal

$$y(t) = \frac{dx(t)}{dt}$$

   where $x(t)$ is the signal described in the previous question. What is the highest nonzero frequency component of this signal?

4. Suppose we have a signal $x(t)$ with a Nyquist sampling rate of 10,000 samples/second (or $\omega_S = 2\pi \times 10{,}000$ radians/sec). Suppose we generate a new signal $z(t)$

$$z(t) = x(t - 1)$$

   What is the Nyquist sampling rate for $z(t)$?

5. Suppose we have a signal $x(t)$ with a Nyquist sampling rate of 10,000 samples/second (or $\omega_S = 2\pi \times 10{,}000$ radians/sec). We generate a new signal $w(t)$

$$w(t) = x(2t)$$

   What is the Nyquist sampling rate for $w(t)$?

6. A signal $x(t)$ has a Fourier transform which is nonzero for $|\omega| < 2\pi 5000$. We sample this signal at a rate of 15,000 samples per second. The original signal is recovered from the sampled signal using an ideal low pass filter with cutoff frequency $\omega_c = 2\pi f_c$. What is the acceptable range of values for f_c.

15.6 ANSWERS

1. (a) 4 kHz
   (b) 3 kHz
2. 5000 Hz (or $2\pi \times 5000$ radians/sec)
3. 5000 Hz (or $2\pi \times 5000$ radians/sec)
4. 10,000 Hz (or $2\pi \times 10{,}000$ radians/sec)
5. 20,000 Hz (or $2\pi \times 20{,}000$ radians/sec)
6. f_c can range from 5–10 kHz

MODULE 16

Amplitude Modulation

Let's look at one more application of the multiplication property. This has to do with broadcasting signals over a shared communication channel. This shared channel can be the atmosphere as is the case with broadcast radio and television signals, or it can be a a coaxial or optical cable such as that used to distribute television programs. Basically, any communication medium that is being used by a number of users at the same time. If the signals generated by the users occupy the same frequency band sending them all over the same channel at the same time will cause them all to be distorted. We have to separate them somehow. We can do this by separating them in time, or we could separate them in frequency, or as the Austrian-American actress Hedy Lamar demonstrated you could do both (remember her the next time you use Bluetooth). In this module we take a look at one way of moving the signals into separate slots in the frequency domain. At first sight separating similar signals in frequency does not seem like a viable option because generally the signals we send over a common broadcast medium are similar signals and therefore have similar spectral profiles. However, remember from our discussion of sampling that we can move a signal around in the frequency domain by convolving it with a delta function.

$$X(\omega) \circledast \delta(\omega - \omega_o) = X(\omega - \omega_o)$$

If $x(t)$ is real it has an even magnitude and odd phase. To preserve this symmetry we need to shift the signal to both $\pm\omega_o$. So we convolve it with two delta functions

$$X(\omega) \circledast (\delta(\omega - \omega_o) + \delta(\omega + \omega_o)) = X(\omega - \omega_o) + X(\omega + \omega_o)$$

For different signals we can pick different values of ω_o and as long as these are far enough apart we can accommodate a host of signals over a common channel. But how do we go about convolving $X(\omega)$ with a delta function? To see this let's take the inverse Fourier transform of the shifted signal. In the frequency domain we convolved $X(\omega)$ with the pair of delta function. In the time domain the convolution becomes a multiplication where we use the multiplication property.

$$\mathcal{F}[a(t)b(t)] = \frac{1}{2\pi} A(\omega) \circledast B(\omega)$$

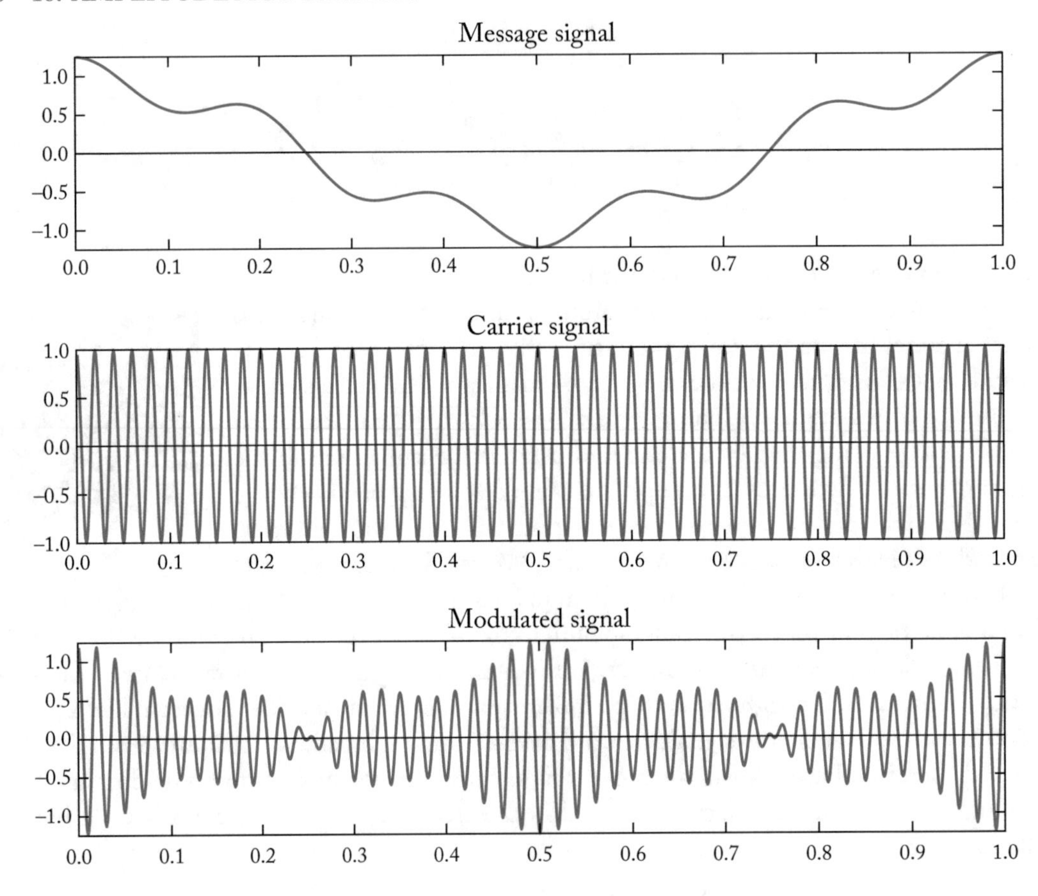

Figure 16.1: An example of a message, carrier, and modulated signal.

We can show that the inverse Fourier transform of $(\delta(\omega - \omega_o) + \delta(\omega + \omega_o))$ is $(1/2\pi)[e^{j\omega_o t} + e^{-j\omega_o t}]$ which is simply $(1/\pi)\cos(\omega_o t)$. Putting all of this together we get

$$\mathcal{F}^{-1}\left[X(\omega - \omega_o) + X(\omega + \omega_o)\right] = 2x(t)\cos(\omega_o t)$$

This process of multiplying a message bearing signal with a sinusoid to move it around in frequency is called *amplitude modulation* (or more technically Double Sideband—Suppressed Carrier Modulation). The sinusoid is essentially carrying the message signal, therefore, it is called the *carrier signal*. The reason for the names becomes clearer if we look at an example. In Figure 16.1 we show a (sinusoidal) message signal, a carrier signal and the modulated signal. You can see that the *amplitude* of the carrier signal is being *modulated* by the message signal, hence, amplitude modulation.

So, now that we have moved the signal to a higher frequency how do we recover it at the receiver? One way is to do exactly what we did to move it up—multiply it again by the carrier signal $\cos(\omega_o t)$. We can see what happens in the time domain using a trigonometric identity.

$$\begin{aligned} x(t)\cos(\omega_o t)\cdot\cos(\omega_o t) &= x(t)\cos^2(\omega_o t) \\ &= x(t)\left[\frac{1}{2}+\frac{1}{2}\cos(2\omega_o t)\right] \\ &= \frac{1}{2}x(t)+\frac{1}{2}x(t)\cos(2\omega_o t) \end{aligned}$$

The second term is the signal moved out to $2\omega_o$ and can be easily removed by a low pass filter. This method of recovery or *demodulation* is called *coherent demodulation*. In order to use this at the receiver we need to generate a signal with the exact same frequency and phase as the carrier signal. What happens if we cannot get the phase right? Suppose at the transmitter we used the carrier signal $\cos(\omega_o t)$, but at the receiver we only had $\cos(\omega_o t+\theta)$ available. If we multiplied the modulated signal with this phase shifted signal we would get

$$\begin{aligned} x(t)\cos(\omega_o t)\cdot\cos(\omega_o t+\theta) &= x(t)\left[\frac{1}{2}\cos(2\omega_o t+\theta)+\frac{1}{2}\cos(\theta)\right] \\ &= \frac{1}{2}x(t)\cos(\theta)+\frac{1}{2}x(t)\cos(2\omega_o t+\theta) \end{aligned}$$

Here we have used the trigonometric identity (which you could prove using Euler's excellent formula)

$$\cos(\alpha)\cos(\beta)=\frac{1}{2}cos(\alpha+\beta)+\frac{1}{2}\cos(\alpha-\beta)$$

Once again we can filter out the signal at $2\omega_o t$ but we are left with not $x(t)$ but $x(t)\cos(\theta)$. If θ is close to $\pi/2$ this would mean that the signal would disappear as $\cos(\pi/2)=0$. In a sense a worse situation is when θ is not fixed but varies continuously. The received signal then would fade in and out—not a very desirable situation.

Generating a signal with the exact same frequency is not necessarily a difficult task, however, the phase requirement can be onerous and expensive.

To overcome this requirement we can do something really simply. In Figure 16.2 we have plotted the modulated signal with the original signal overlayed and the modulated signal with the positive envelope emphasized. The positive envelope looks like the message signal when the message signal is positive and is flipped when the message signal is negative

The positive envelope is something we can get using a simple rectifying filter shown in Figure 16.3. If we could modify our scheme so that the positive envelope of the modulated signal is the message signal we would not need to get the frequency or phase of the carrier. Looking at Figure 16.2 we can see a way to do this. All we need to do is to ensure the original signal is always positive. We can do that with a DC shift.

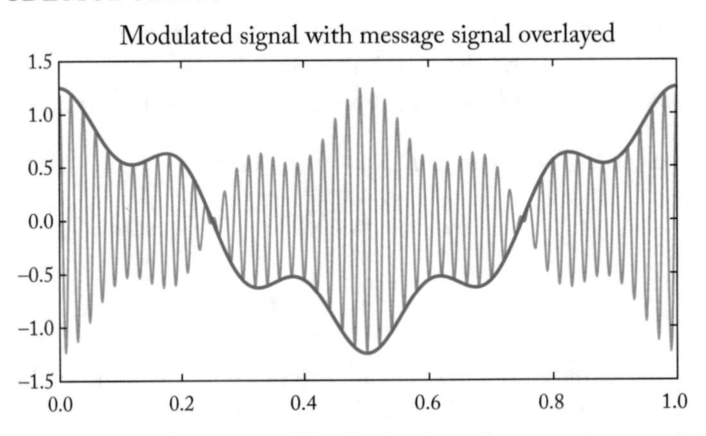

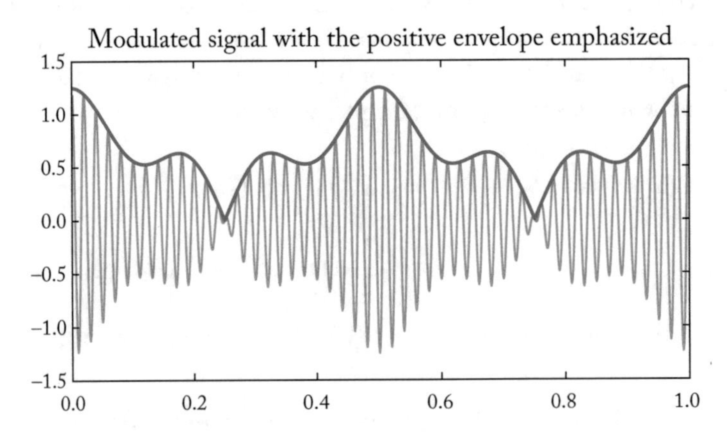

Figure 16.2: The modulated signal and the message signal.

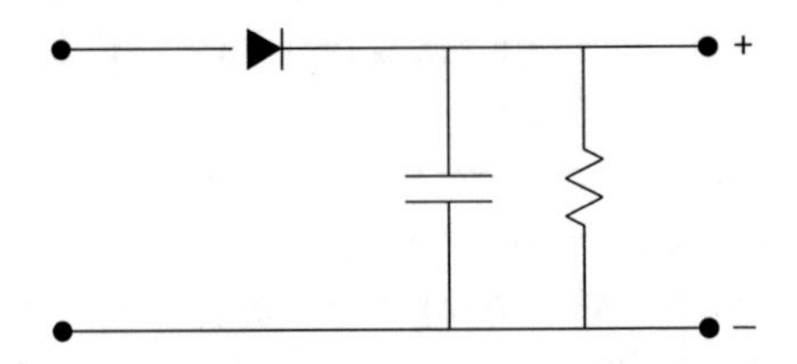

Figure 16.3: A simple envelope detector.

We have plotted this situation in Figure 16.4. You can see that the positive envelope of the modulated signal is the level shifted message signal. To emphasize this we plot the modulated and message signal overlayed in Figure 16.5. After putting the modulated signal through an envelope detector we can remove the level shift by using a blocking capacitor.

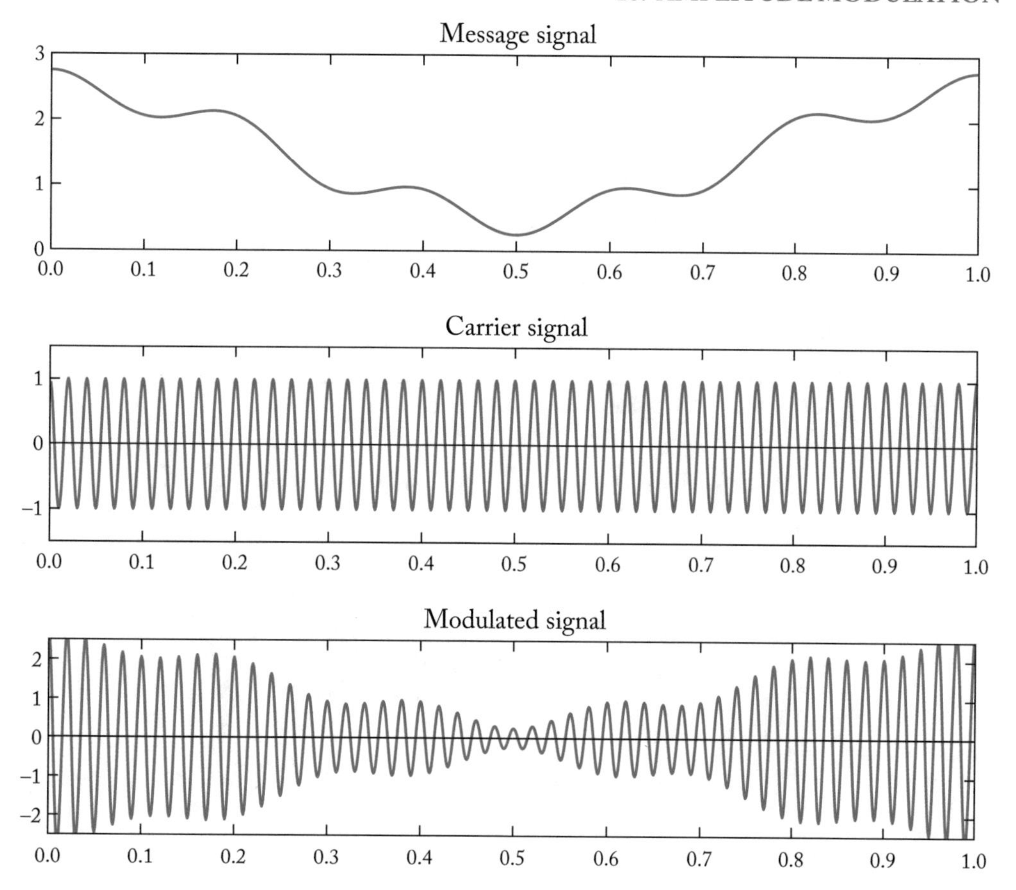

Figure 16.4: An example of a level shifted message, carrier, and modulated signal.

This latter method of level shifting the signal and then multiplying (or mixing) with a sinusoidal carrier is used in commercial AM systems. Here are some acronyms for those interested in such things. The method used by the commercial broadcaster where we add a DC value prior to modulation is called Double Sideband-Large Carrier (DSB-LC). The LC part is because the addition of the DC value prior to modulation can also be seen as the addition of a carrier term after modulation.

$$(x(t) + A)\cos(\omega_c t) = x(t)\cos(\omega_c t) + A\cos(\omega_c t)$$

The amount of additional carrier is measured by a quantity called the *modulation index* μ which is defined as

$$\mu = \frac{x_m}{A}$$

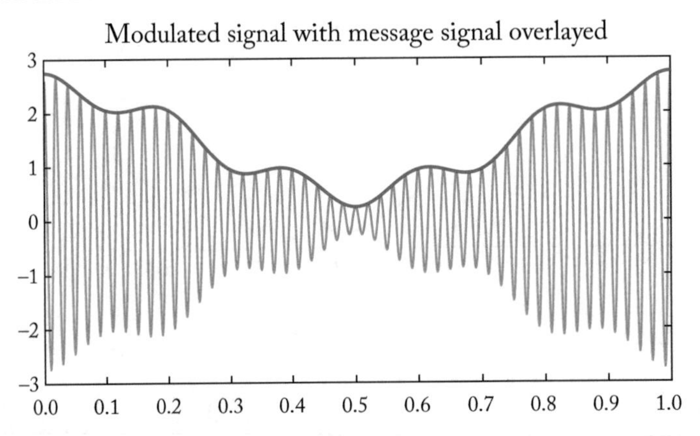

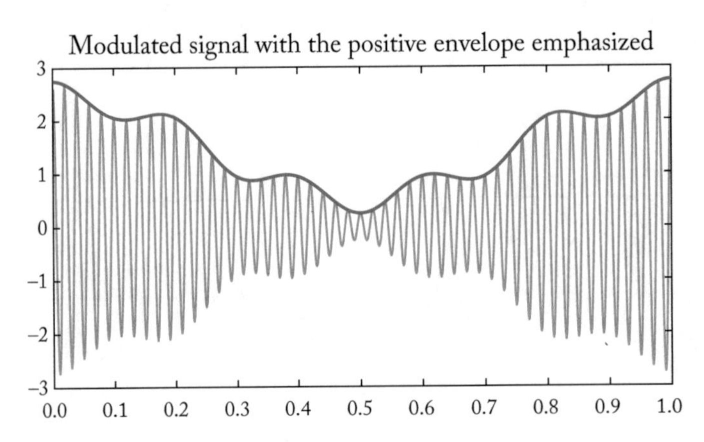

Figure 16.5: The modulated signal and the message signal overlayed.

where x_m is the maximum value of the absolute value of the message signal $|x(t)|$. To understand the Double Sideband part consider the Fourier transform of a real signal. We know that the magnitude of the Fourier transform has even symmetry and the phase has odd symmetry. This symmetry means that we actually only need to know half of the transform. Either the transform for positive ω, or the transform for negative ω. We label the transform for positive ω as the upper sideband and the transform for negative ω as the lower sideband. During the process of modulation both sidebands get shifted to the carrier frequency and transmitted —hence the "double sideband" in the name. For the same reason the amplitude modulation scheme without the level shift is called Double Sideband-Suppressed Carrier (DSB-SC). There are techniques that transmit only one sideband and are known as (you guessed it) Single Sideband (SSB) systems. Finally, there are techniques (mostly used for the transmission of TV signals) in which one sideband plus a part of another sideband is transmitted. This method is called Vesitigial Sideband (VSB) signaling.

16.1 AM RECIEVER

We now know how to demodulate a DSB-LC signal. We put it through an envelope detector, remove the DC bias and Bob's your uncle. But there are multiple AM stations each transmitting using the same method. In the U.S. the AM band ranges from 530–1700 kHz with any particular radio station taking up 10 kHz. How can we be sure we will get the station we want? The answer to that requires a number of steps. Let's briefly describe them.

It seems the easy way out would be to use a tunable bandpass filter with a bandwidth of 10 kHz tuned to the station we want. Unfortunately, sharp tunable bandpass filters are expensive to make. Furthermore the signals we want to receive are at high frequencies and the higher the frequency the more difficult and hence more expensive the amplifier design. Instead the strategy used is to move the signal of interest to a lower *intermediate frequency (IF)*, which in the U.S. is 455 kHz. At this intermediate frequency we filter the signal using a sharp (non-tunable) bandpass filter rejecting all the signals we are not interested in. This filter is referred to as the IF filter. We still have the problem of moving the signal of interest to the intermediate frequency. As we have seen moving signals in frequency can be accomplished by multiplying the signal with a sinusoid. The signal we are moving is a modulated sinusoid so the effect of multiplying this with another sinusoid is going to be two modulated signals—one at the sum of the frequencies of the two sinusoids and one at the difference. We want the locally generated sinusoid to be at a frequency that results in one of these modulated signals to be at the intermediate frequency. The standard is to use the difference (or hetrodyne) signal be at the intermediate frequency. We generate a sinusoid using a local oscillator with a frequency ω_{lo} which is 455 kHz greater than the frequency of the signal of interest. Multiplying the signal of interest with a the output of the local oscillator we get a copy of our signal of interest moved to the intermediate frequency and another copy to a frequency equal to the sum of the carrier and local oscillator frequency.

$$\begin{aligned}(x(t)+A)\cos(\omega_c t)\cos(\omega_{lo}t) =& \frac{1}{2}(x(t)+A)\cos((\omega_c-\omega_{lo})t)\\ &+\frac{1}{2}(x(t)+A)\cos((\omega_c+\omega_{lo})t)\end{aligned}$$

There is one slight problem. If there is a signal at a frequency which is the sum of the local oscillator frequency and 455 kHz, it will also be moved to the intermediate frequency (can you see why). We remove this *image frequency* using a tunable bandpass filter called the RF (for radio frequency) filter. But wasn't this whole rigmarole to avoid using a tunable bandpass filter? No, it was to avoid using a *sharp* tunable bandpass filter. In the case of the RF filter the signal that has to be knocked out is far away in frequency from the signal of interest. So a poor filter will suffice.

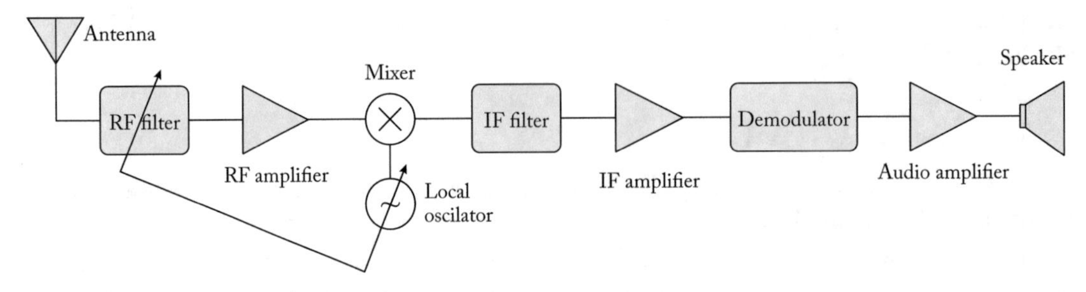

Figure 16.6: A superhetrodyne receiver.

This particular receiver is called a superheterodyne receiver shown in Figure 16.6.

Let's go through the process using an example. Our example station is KFOR which broadcasts from Lincoln, Nebraska at a frequency of 1240 kHz. To bring this signal down to the intermediate frequency of 455 kHz we need the local oscillator signal to be at 1,240 + 455 = 1,695 kHz. This local oscillator frequency will also bring the image frequency of 1695 + 455 = 2,150 kHz down to 455 kHz as the difference between 2,150 kHz and 1,695 kHz is also 455 kHz. It is the job of the RF filter to knock out this image frequency.

16.2 AM STATIONS IN THE U.S.

The AM stations in the U.S. are limited to 530–1700 kHz with a station required to limit its broadcast to a 10 kHz band. This gives us about 118 slots. However there are more than 4,500 stations currently operating in the U.S. The reason we can do this is because radio waves in these frequencies have a limited daytime range so the same frequency can be assigned to multiple stations which are geographically separated. For example, there are close to 150 stations that broadcast at the same frequency as KFOR. During night the amplitude modulated signals can bounce off the ionosphere vastly extending their range. This also increases the possibility of stations interfering with each other. To prevent this from happening the Federal Communication Commission has a number of requirements—one of them being that certain stations either have to shut down transmission at their assigned frequency at night or transmit at a much reduced power level. These stations used to be called daytimers.

The FCC classifies AM channels as Clear Channel, Regional, and Local as well as Class A, B, C, and D. Class A stations are generally clear channel stations and can transmit using 10kW to 50kW transmitters and are allowed to transmit for 24 hours a day. Class B stations which are generally regional stations may transmit from 250 W to 50 KW and can transmit for 24 hours a day. Class C stations are local stations and are limited to power levels of between 250 W to 1 kW and can transmit at all hours. Class D stations have a wide power range—from 250 W to 50 kW but are restricted in terms of time. Many of the class D stations transmit only during the day (the daytimers). If these stations transmit during the night they are limited to 250 W. There are currently twenty AM stations in Nebraska. Of these KFAB transmitting at 1110 kHz is the only Clear Channel station transmitting at 50 KW. KCRO transmits at 660 kHz which is the frequency used by a Clear Channel station in New York City. Therefore, while it transmits at 1 kW during the day it has to reduce it's night time transmission power to 54 Watts. There are three AM stations in Lincoln, KFOR, KLIN, and KLMS. KFOR and KLIN are class C stations and transmit using 1 kW during both day and night. KLMS is the local ESPN station broadcasting at 1480 kHz. It transmits using a 1 kW during the day and 750 W during the night.

16.3 SUMMARY

In this module we looked at amplitude modulation, another practical application of the multiplication property of Fourier transforms. We looked at how it is used in AM radio broadcast and reception. We also looked at some of the features of AM radio broadcast in the US.

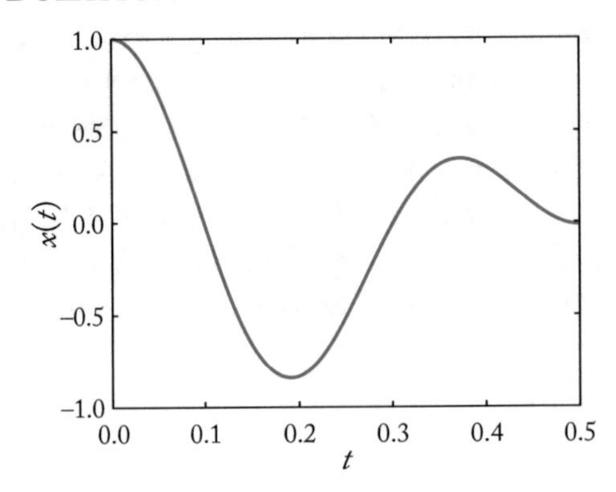

Figure 16.7

Table 16.1: Match the figures with the signal

Signal	Figure
$x(t)\cos(\omega_o t)$	
$x(t) + \cos(\omega_o t)$	
$x(t)\cos^2(\omega_o t)$	
$(1 + x(t))\cos(\omega_o t)$	

16.4 EXERCISES

(Answers on the following page)

1. The signal $x(t)$ shown in Figure 16.7 combined with a sinusoid in a number of ways to generate the signals shown in Figure 16.8.

   Match the figures with the signal (see Table 16.1).

2. If we wanted to receive KLIN on our AM receiver

   (a) What would be the frequency of the local oscillator?

   (b) What is the image frequency that the RF filter has to block out?

3. If we wanted to receive KFAB on our AM receiver

   (a) What would be the frequency of the local oscillator?

   (b) What is the image frequency that the RF filter has to block out?

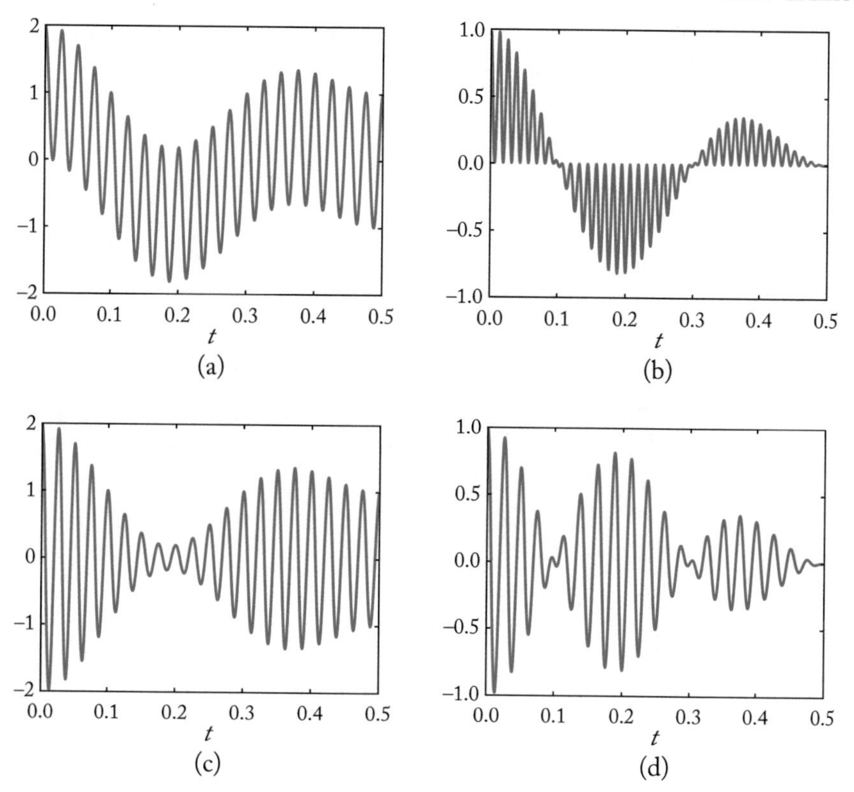

Figure 16.8

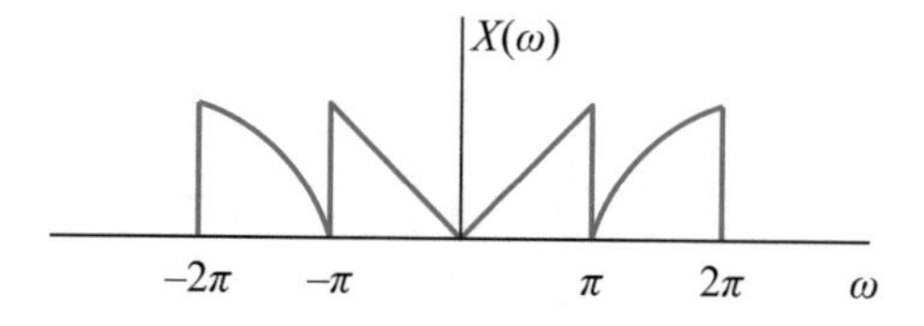

Figure 16.9: Fourier transform of the message signal.

4. Suppose I want to transmit the message signal

$$x(t) = \cos(\omega_o t)$$

using amplitude modulation and a modulation index of 0.5. What is the value of the dc shift A?

5. The Fourier transform of the message signal is shown in Figure 16.9

If we multiply the message signal with a carrier signal at a frequency of 2 Hz sketch the transform of the modulated signal.

Table 16.2: Match the figures with the signal

Signal	Figure
$x(t)\cos(\omega_o t)$	D
$x(t) + \cos(\omega_o t)$	A
$x(t)\cos^2(\omega_o t)$	B
$(1 + x(t))\cos(\omega_o t)$	C

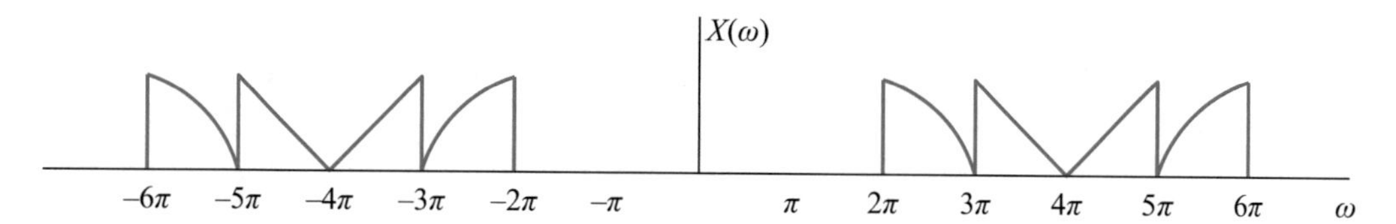

Figure 16.10: Transform of the modulated signal.

16.5 ANSWERS

1. Match the figures with the signal (see Table 16.2).
2. (a) 1855 kHz

   (b) 2310 kHz
3. (a) 1555 kHz

   (b) 2010 kHz
4. $A = 2$
5. Figure 16.10.

MODULE 17

Discrete Fourier Transform

Nowadays most of the time when you take the Fourier transform it will be using a digital processor—either on the computer, on your phone, or some other device. You are actually evaluating the discrete Fourier transform. Much of what we have already discussed about the continuous Fourier transform carries over to the Discrete Fourier transform so this module will be relatively short. We will begin by developing the discrete time Fourier transform (DTFT) which maps a discrete sequence from the time domain to a continuous transform in the frequency domain. From this we will develop the discrete Fourier transform (DFT) which takes a discrete sequence in the time domain to a discrete sequence in the frequency domain.

17.1 DISCRETE TIME FOURIER TRANSFORM (DTFT)

We have actually already encountered the discrete time Fourier transform in a slightly different guise when we looked at sampling. The sampling process maps a continuous time signal into a discrete time signal. The Fourier transform of that must then be the discrete time Fourier transform. Let's work through the math. Recall that the sampled function can be modeled by a product of the continuous time function $x(t)$ and an impulse train. The Fourier transform of the sampled function is

$$
\begin{aligned}
\mathcal{F}\left[x(t)\sum_{n=-\infty}^{\infty}\delta(t-nT)\right] &= \int_{-\infty}^{\infty} x(t)\sum_{n=-\infty}^{\infty}\delta(t-nT)e^{-j\omega t}\,dt \\
&= \sum_{n=-\infty}^{\infty}\int_{-\infty}^{\infty} x(t)\delta(t-nT)e^{-j\omega t}\,dt \\
&= \sum_{n=-\infty}^{\infty} x(nT)e^{-j\omega nT}
\end{aligned}
$$

where we have used the sifting property of the delta function. If we set $T = 1$ we get

$$
\sum_{n=-\infty}^{\infty} x(nT)e^{-j\omega nT} = \sum_{n=-\infty}^{\infty} x[n]e^{-jn\omega}
$$

where we have replaced the parentheses with brackets to emphasize the fact that we are now dealing with a discrete sequence. This is the discrete time Fourier transform of the sequence $\{x[n]\}$. The convention is to represent the Fourier transform as $X(e^{j\omega})$. Sticking with the convention we get

$$X(e^{j\omega}) = \sum_{n=-\infty}^{\infty} x[n]e^{-jn\omega}$$

So what about the inverse Fourier transform? Remember that the Fourier transform of the sampled function was periodic with period $2\pi/T$. Given that we set $T = 1$ the DTFT is periodic with period 2π.

Take another look at the previous equation. On the right hand side you have an expansion in terms of $e^{jn\omega}$. It looks suspiciously like the Fourier expansion of a periodic function so very likely that the same tricks we used in the case of the Fourier series to come up with an expression for the Fourier series coefficients a_n will work to find the expression for $x[n]$. In order to do this we need only to notice that

$$\frac{1}{2\pi}\int_{2\pi} e^{-jn\omega}e^{jk\omega}d\omega = \begin{cases} 1 & k = n \\ 0 & k \neq n \end{cases}$$

$$x[n] = \frac{1}{2\pi}\int_{2\pi} X(e^{j\omega})e^{jn\omega}d\omega$$

Example 17.1 Find the discrete time Fourier transform of

$$x[n] = \begin{cases} 1 & n = 0, 1, 2, 3 \\ 0 & \text{otherwise} \end{cases}$$

$$\begin{aligned}
X(e^{j\omega}) &= \sum_{n=-\infty}^{\infty} x[n]e^{-jn\omega} \\
&= \sum_{n=0}^{3} e^{-jn\omega} \\
&= 1 + e^{-j\omega} + e^{-j2\omega} + e^{-j3\omega} \\
&= e^{-j\omega/2}\left(e^{j\omega/2} + e^{-j\omega/2}\right) + e^{-j5\omega/2}\left(e^{j\omega/2} + e^{-j\omega/2}\right) \\
&= \left(e^{j\omega/2} + e^{-j\omega/2}\right)\left((e^{-j\omega/2} + e^{-j5\omega/2}\right) \\
&= 2\cos(\omega/2)\left(e^{-j\omega/2} + e^{-j5\omega/2}\right)
\end{aligned}$$

17.2 PROPERTIES OF THE DTFT

Most of the properties of the discrete time Fourier transform are very much like those of the continuous time Fourier transform. We will simply list them here.

1. **Linearity**

$$\mathcal{F}\left[\alpha x_1[n] + \beta x_2[n]\right] = \alpha X_1(e^{j\omega}) + \beta X_2(e^{j\omega})$$

2. **Convolution property**

$$\mathcal{F}\left[x_1[n] \circledast x_2[n]\right] = X_1(e^{j\omega}) \cdot X_2(e^{j\omega})$$

3. **Time shift property**

$$\mathcal{F}\left[x[n-n_o]\right] = e^{-j\omega n_o} X(e^{j\omega})$$

4. **Multiplication property**

$$\mathcal{F}\left[x_1[n] \cdot x_2[n]\right] = X_1(e^{j\omega}) \circledast X_2(e^{j\omega})$$

There is one that is a bit different.

17.2.1 DIFFERENTIATION IN THE FREQUENCY DOMAIN

By taking the derivative of the discrete time Fourier transform in terms of ω we can find the discrete time Fourier transform of $nx[n]$ in terms of the discrete time Fourier transform of $x[n]$. This is very easy to show

$$\begin{aligned}
\frac{dX(e^{j\omega})}{d\omega} &= \frac{d}{d\omega} \sum_{n=-\infty}^{\infty} x[n]e^{-jn\omega} \\
&= \sum_{n=-\infty}^{\infty} \frac{d}{d\omega} x[n]e^{-jn\omega} \\
&= \sum_{n=-\infty}^{\infty} x[n] \frac{d}{d\omega} e^{-jn\omega} \\
&= \sum_{n=-\infty}^{\infty} -jnx[n]e^{-jn\omega}
\end{aligned}$$

Multiplying both sides by j

$$j\frac{dX(e^{j\omega})}{d\omega} = \sum_{n=-\infty}^{\infty} nx[n]e^{-jn\omega}$$

Remembering *Bill* and *Bob* this means that

$$\mathcal{F}\left[nx[n]\right] = j\frac{dX(e^{j\omega})}{d\omega}$$

17.3 DISCRETE FOURIER TRANSFORM

The reason we are interested in this discrete representation is that much of the processing nowadays is conducted in the discrete domain. This means that not only is the time domain representation of the signal discrete—so are the frequency representations. The discrete time Fourier transform is a transform for discrete time functions—the representation in the time domain is discrete. However, the transform itself is a continuous periodic function of ω. To obtain a transform which will give us discrete representations in both the time and frequency domain we turn to the Discrete Fourier Transform (DFT).

We begin with the DTFT and discretize it by sampling the transform $X(e^{j\omega n})$ N times over a period. Remembering that the period is 2π we get

$$X(e^{jk(2\pi/N)}) = \sum_{n=-\infty}^{\infty} x[n]e^{-jnk(2\pi/N)}$$

Having reduced the number of frequency components from infinity to N can we still recover $x[n]$ from $X(e^{jk(2\pi/N)})$? The answer is yes, but only under specific conditions. Let's first show the condition under which we can recover $x[n]$ from $X(e^{jk(2\pi/N)})$. We will then see why this condition is necessary.

We can recover $x[n]$ from $X(e^{jk(2\pi/N)})$ if $x[n]$ is nonzero only for L consecutive values of n where $L \leq N$. To show this we will use the fact that

$$\sum_{n=0}^{N-1} e^{jk(2\pi/N)n} = \begin{cases} N & k = 0, \pm N, \pm 2N, \ldots \\ 0 & \text{otherwise} \end{cases}$$

Showing this is simple—we just use the geometric sum formula for $k \neq 0, \pm N, \ldots$.

$$\begin{aligned} \sum_{n=0}^{N-1} e^{jk(2\pi/N)n} &= \sum_{n=0}^{N-1} \left(e^{jk(2\pi/N)}\right)^n \\ &= \frac{e^0 - \left(e^{jk(2\pi/N)}\right)^{(N-1+1)}}{1 - \left(e^{jk(2\pi/N)}\right)} \\ &= \frac{1 - e^{jk2\pi}}{1 - \left(e^{jk(2\pi/N)}\right)} \\ &= 0 \end{aligned}$$

where we have used the fact that $e^{jk2\pi} = 1$.

If $x[n] = 0$ for $n \geq N$ we can rewrite the limits of summation in our expression for $X(e^{jk(2\pi/N)})$ as

$$X(e^{jk(2\pi/N)}) = \sum_{n=0}^{N-1} x[n]e^{-jnk(2\pi/N)}$$

Multiply both sides both sides of the equation for $X(e^{jk(2\pi/N)})$ with $e^{jlk(2\pi/N)}$ and sum from $k = 0$ to $N - 1$.

$$\sum_{k=0}^{N-1} X(e^{jk(2\pi/N)})e^{jlk(2\pi/N)} = \sum_{k=0}^{N-1}\sum_{n=0}^{N-1} x[n]e^{-jk(2\pi/N)n}e^{jlk(2\pi/N)}$$

Combining the two exponentials of the right hand side we get

$$\begin{aligned}\sum_{k=0}^{N-1} X(e^{jk(2\pi/N)})e^{jlk(2\pi/N)} &= \sum_{k=0}^{N-1}\sum_{n=0}^{N-1} x[n]e^{j(l-n)(2\pi/N)k} \\ &= \sum_{n=0}^{N-1} x[n]\sum_{k=0}^{N-1} e^{j(l-n)(2\pi/N)k}\end{aligned}$$

The inner sum on the right hand side is either equal to N or to 0. It is equal to N when $n - l = 0$ or $n = l$, and 0 for all other values of n. Therefore,

$$\sum_{k=0}^{N-1} X(e^{jk(2\pi/N)})e^{jlk(2\pi/N)} = x[l]N$$

or

$$x[l] = \frac{1}{N}\sum_{k=0}^{N-1} X(e^{jk(2\pi/N)})e^{jlk(2\pi/N)}$$

For convenience let's define

$$X[k] = X(e^{jk(2\pi/N)})$$

Then we get the DFT equations—the forward DFT

$$\boxed{X[k] = \sum_{n=0}^{N-1} x[n]e^{-jnk(2\pi/N)}}$$

and the inverse DFT

$$\boxed{x[n] = \frac{1}{N}\sum_{k=0}^{N-1} X[k]e^{jnk(2\pi/N)}}$$

Therefore, given a discrete time sequence $\{x[n]\}$ which is zero for $n < 0$ and $n \geq N$ we can find the spectral representation at discrete frequency values. And given the discrete spectral representation $\{X[k]\}$ we can recover the discrete time sequence $\{x[n]\}$.

What is the effect of sampling the frequency representation in the general case when we do not have a restriction on the extent of the nonzero values of $\{x[n]\}$? To answer this we begin with the sampled DTFT equation

$$X[k] = \sum_{n=-\infty}^{\infty} x[n]e^{-jnk(2\pi/N)}$$

where we have used $X[k] = X(e^{jk(2\pi/N)})$ and rewrite the infinite sum as a sum of finite sums.

$$X[k] = \cdots + \sum_{n=-2N}^{-N-1} x[n]e^{-jnk(2\pi/N)} + \sum_{n=-N}^{-1} x[n]e^{-jnk(2\pi/N)} + \sum_{n=0}^{N-1} x[n]e^{-jnk(2\pi/N)} + \sum_{n=N}^{2N-1} x[n]e^{-jnk(2\pi/N)} + \cdots$$

Each of the finite sums are of the form

$$\sum_{n=mN}^{mN+N-1} x[n]e^{-jnk(2\pi/N)}$$

where m varies from negative infinity to infinity. So we can rewrite the expression for $X[k]$ as

$$X[k] = \sum_{m=-\infty}^{\infty} \sum_{n=mN}^{mN+N-1} x[n]e^{-jnk(2\pi/N)}$$

This in turn can be rewritten as

$$X[k] = \sum_{m=-\infty}^{\infty} \sum_{n=0}^{N-1} x[n+mN]e^{-j(n+mN)k(2\pi/N)}$$

Let's examine the complex exponential in this equation.

$$e^{-j(n+mN)k(2\pi/N)} = e^{-jnk(2\pi/N)} \cdot e^{-jmNk(2\pi/N)} = e^{-jnk(2\pi/N)} \cdot e^{-jmk2\pi}$$

Noting that

$$e^{-jmk2\pi} = 1$$

we have

$$e^{-j(n+mN)k(2\pi/N)} = e^{-jnk(2\pi/N)}$$

Substituting this back into the expression for $X[k]$

$$X[k] = \sum_{m=-\infty}^{\infty} \sum_{n=0}^{N-1} x[n+mN]e^{-jnk(2\pi/N)}$$

Switching the order of summation we get

$$X[k] = \sum_{n=0}^{N-1} \left[\sum_{m=-\infty}^{\infty} x[n+mN] \right] e^{-jnk(2\pi/N)}$$

Define

$$x_p[n] = \sum_{m=-\infty}^{\infty} x[n+mN]$$

The expression for $X[k]$ now becomes

$$X[k] = \sum_{n=0}^{N-1} x_p[n] e^{-jnk(2\pi/N)}$$

Comparing this to the earlier boxed expression for $X[k]$ with $x_p[n]$ in place of $x[n]$. From that earlier work we know that given $\{X[k]\}$ we can recover $x_p[n]$. Let's take a look at the relationship between $x_p[n]$ and $x[n]$. The expression for $x_p[n]$ should look somewhat familiar. We used something similar to this when we were developing the Fourier transform of continuous time signals. We called it the periodic extension of $x(t)$. We can see that $x_p[n]$ is clearly periodic. To see the relationship between $x_p[n]$ and $x[n]$ let's write out $x_p[n]$ for a few values of n.

$$\begin{aligned}
x_p[0] &= \cdots + x[-2N] + x[-N] + x[0] + x[N] + x[2N] + \cdots \\
x_p[1] &= \cdots + x[1-2N] + x[1-N] + x[1] + x[1+N] + x[1+2N] + \cdots \\
\vdots \quad & \quad \vdots \\
x_p[N-1] &= \cdots + x[-N-1] + x[-1] + x[N-1] + x[2N-1] + x[3N-1] + \cdots
\end{aligned}$$

So if $x[n] = 0$ outside the range $0 \leq n \leq N-1$ then $x_p[n] = x[n]$. If not, $x_p[n]$ is an aliased version of $x[n]$. Notice the similarity with our study of sampling. Just as sampling in the time domain resulted in a periodic extension in the frequency domain, sampling in the frequency domain results in a periodic extension in the time domain. Just as increasing the number of samples in the time domain spread the replicas in the frequency domain, increasing the number of samples in the frequency domain spreads out the replicas in the time domain. And just as an insufficient number of samples (undersampling) in the time domain results in aliasing in the frequency domain, an insufficient sampling in the frequency domain results in aliasing in the time domain. The bottom line though is that if we have a time limited sampled signal we can get a unique spectral profile of it in the discrete domain. This allows us to perform operations in the spectral domain using digital processors that we otherwise would not be able to do.

In the DFT literature there is an additional notational shortcut that is often used. Define

$$W_N = e^{-j\frac{2\pi}{N}}$$

Table 17.1: Periodic extension of $x[n]$

$x[0]$	$x[1]$	$x[2]$	$x[3]$	$x[4]$	$x[5]$	$x[6]$	$x[7]$	$x[8]$	$x[9]$	$x[10]$	$x[11]$
1	a	a^2	a^3	1	a	a^2	a^3	1	a	a^2	a^3

then The forward DFT becomes

$$\boxed{X[k] = \sum_{n=0}^{N-1} x[n]W_N^{nk}}$$

and the inverse DFT becomes

$$\boxed{x[n] = \frac{1}{N}\sum_{k=0}^{N-1} X[k]W_N^{-nk}}$$

17.4 PROPERTIES OF THE DFT

Unlike the case of the Fourier transform, the discrete time Fourier transform, and even the Fourier series, we are dealing with signals that are implicitly periodic in both domains. This means that while the sequence is only defined for the values of the indices in the interval $0 \leq n \leq N-1$ the operations we perform on them using the DFT assumes that this is only one period of a periodic function. Therefore, there is an inherent assumption that the sequence has nonzero values outside of the range $0 \leq n \leq N-1$. In order to deal with this assumption we need to take a closer look at operations which need the value of the sequence for indices outside of this range. Let's examine shifts in the time domain.

17.4.1 LINEAR AND CIRCULAR SHIFTS

The values of the time sequence $x[n]$ is only available for $n = 0, 1, \ldots N-1$ and the frequency sequence $X[k]$ for $k = 0, 1, \ldots, N-1$. If we need to express $x[n]$ for n outside of the range $0, 1, \ldots, N-1$ we can do so by evaluating $x[n \bmod N]$. Consider a simple example where $N = 4$. Let's suppose $x[n] = a^n$ for $n = 0, 1, 2, 3$ then we can write the periodic extension of $x[n]$ as in Table 17.1.

We can see that $x[4] = x[0] = 1$ and $x[5] = x[1] = a$ and so on. In other words $x[n] = x[n \bmod N]$.

So what happens if we shift the signal? Do we have to consider the samples of the periodic extension of the signal in neighboring periods? Or can we still operate on the values of $x[n]$ in the interval $0 \leq n \leq N-1$? We will try and answer these questions using a simple example. Consider the graphical representation of $x[n]$ in Figure 17.1 where we have indicated the range between 0 and $N-1$ (where $N = 4$) with the dashed box.

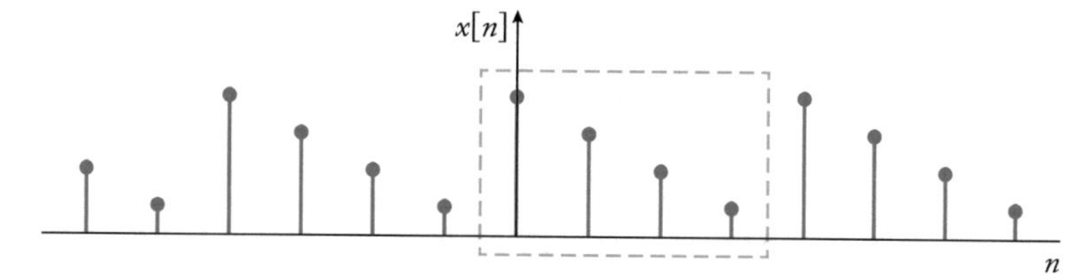

Figure 17.1: A discrete function.

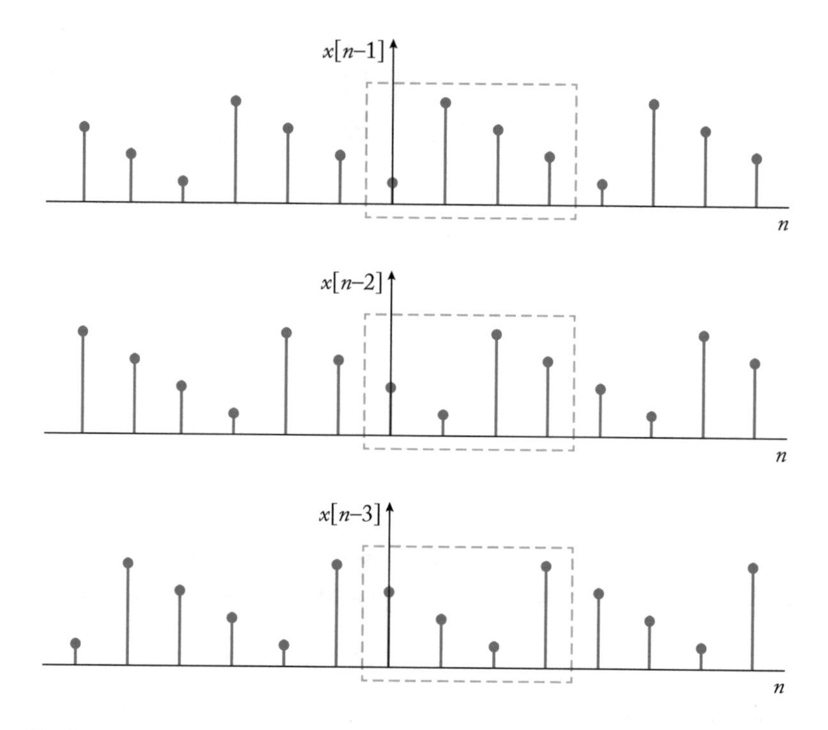

Figure 17.2: Shifts of the discrete function.

Shifting the function one time unit at a time we get the sequences shown in Figure 17.2. As before we have indicated the range of n between 0 and $N-1$ with a dashed box. Notice that while the exact values for $n = 0, 1, 2,$ and 3 is different in each shift, all the different values of $x[n]$ are within the window $n = 0, 1, 2, 3$.

If we remove all the values outside the range $0 \leq n \leq N-1$ as shown in Figure 17.3 we can see what each linear shift looks like if we only look at the values in the range $0 \leq n \leq N-1$. If we only look within the window each time we linearly shift the periodic extension of the sequence the sequence in the window is *circularly shifted.* After a shift the last value in the window prior to the shift becomes the first value in the window after the shift. Therefore, by using circular

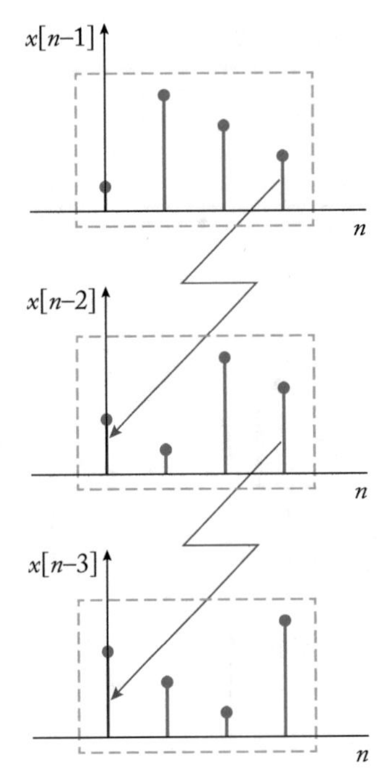

Figure 17.3: Circular shifts.

shifts when linear shifts are required we can satisfy both the implicit periodicity required by the development of the DFT and the fact that the sequence is only defined for N values.

Example 17.2 Find the discrete time Fourier transform of

$$x[n] = \begin{cases} 1 & n = 0, 1, 2, 3 \\ 0 & \text{otherwise} \end{cases}$$

This is exactly the function for which we found the discrete time Fourier transform. Let's pick $N = 4$ and see what the DFT looks like. Taking $N = 4$ the forward DFT equation is

$$X[k] = \sum_{n=0}^{N-1} x[n]e^{-jnk(2\pi/4)} = \sum_{n=0}^{N-1} x[n]e^{-jnk(\pi/2)}$$

Plugging in different values of k we get.

$$\begin{aligned}
X[0] &= 1+1+1+1 \\
&= 4 \\
X[1] &= 1+e^{-j\frac{\pi}{2}}+e^{-jj\pi}+e^{-j\frac{3\pi}{2}} \\
&= 1-j-1+j=0 \\
X[2] &= 1+e^{-j\pi}+e^{-j2\pi}+e^{-j3\pi} \\
&= 1-1+1-1=0 \\
X[3] &= 1+e^{-j\frac{3\pi}{2}}+e^{-j3\pi}+e^{-j\frac{9\pi}{2}} \\
&= 1+j-1-j=0
\end{aligned}$$

This looks like a very different result than what we obtained using the DTFT! We can check by sampling the discrete time Fourier transform at multiples of $\pi/2$. Recall that the DTFT was

$$X(e^{j\omega}) = 2\cos(\omega/2)\left(e^{-j\omega/2}+e^{-j5\omega/2}\right)$$

For $k=0$ we have

$$X[0] = X(e^{j0}) = 2\cos(0)\left(e^{-j0}+e^{-j0}\right) = 2(1+1) = 4$$

For $k=1$ we have

$$\begin{aligned}
X[1] = X(e^{j\pi/2}) &= 2\cos(\pi/4)\left(e^{-j\pi/4}+e^{-j5\pi/4}\right) \\
&= \frac{2}{\sqrt{2}}\left(\frac{1}{\sqrt{2}}-j\frac{1}{\sqrt{2}}-\frac{1}{\sqrt{2}}+j\frac{1}{\sqrt{2}}\right) \\
&= 0
\end{aligned}$$

For $k=2$ we have

$$\begin{aligned}
X[2] = X(e^{j\pi}) &= 2\cos(\pi/2)\left(e^{-j\pi/2}+e^{-j5\pi/2}\right) \\
&= 0(-j-j) \\
&= 0
\end{aligned} \tag{17.1}$$

And finally for $k=3$ we have

$$\begin{aligned}
X[3] = X(e^{j3\pi/2}) &= 2\cos(3\pi/4)\left(e^{-j3\pi/4}+e^{-j15\pi/4}\right) \\
&= -\frac{2}{\sqrt{2}}\left(-\frac{1}{\sqrt{2}}-j\frac{1}{\sqrt{2}}+\frac{1}{\sqrt{2}}+j\frac{1}{\sqrt{2}}\right) \\
&= 0
\end{aligned}$$

We get the same result. So the DFT is a sampling of the DTFT. But why did we get this particular result? If we take these values of $x[n]$ to be the values of a periodic function with period 4, then the periodic extension of this is a function that is 1 for all values of n! The spectral profile of a constant function will only have a value for a frequency of zero, which is what $X[0]$ is. All other values will be zero. What can we do with this? Well we could pick N to be larger so that this effect goes away. If we pick $N = 8$ we get $X[0] = 4$, $X[1] = 1 - j2.414$, $X[2] = 0$, $X[3] = 1 - j0.414$, $X[4] = 0$, $X[5] = 1 + j0.414$, $X(6) = 0$, and $X[7] = 1 + j2.414$—a somewhat more interesting result. This process of adding zeros is called *zero padding* and it allows us to sample the DTFT more finely.

With this under our belt we can begin our look at the properties of the discrete Fourier transform. All the properties of the continuous time Fourier transform carry over to the discrete time with appropriate modifications. Let's begin with the most important and easiest to show.

17.4.2 LINEARITY

The discrete Fourier transform, like it's continuous time and discrete time counterparts, is linear—keeping in mind the implied period N

$$DFT\,[\alpha x[n] + \beta y[n]] = \alpha X[k] + \beta Y[k]$$

where $X[k]$ is the discrete Fourier transform of $x[n]$ and $Y[k]$ is the discrete Fourier transform of $y[n]$. We can easily show this using the linearity of the summation operation.

17.4.3 TIME SHIFT

If the DFT of the time sequence $x[n]$ is the frequency sequence $X[k]$ let's find the DFT of the time shifted sequence $x[n - n_o]$.

$$DFT\,[x[n - n_o]] = \sum_{n=0}^{N-1} x[n - n_o]W^{nk}$$

multiplying and dividing the right hand side by $W^{-n_o k}$.

$$\begin{aligned} DFT\,[x[n - n_o]] &= W^{n_o k} \sum_{n=0}^{N-1} x[n - n_o]W^{nk}W^{-n_o k} \\ &= W^{n_o k} \sum_{n=0}^{N-1} x[n - n_o]W^{(n-n_o)k} \end{aligned}$$

We would like to only use the values of $x[n]$ in the interval $0 \ldots N - 1$ which we can if we replace $n - n_o$ by $(n - n_o) \bmod N$. To keep the correspondence we also need to do the same to $W^{(n-n_o)k}$ but can we? Let's take a look at $W^{k(n \bmod N)}$.

If $0 \leq n < N$ them $n \bmod N = n$ and

$$W^{k(n \bmod N)} = W^{kn}$$

if $n \geq N$ then $n \bmod N = n - mN$ for some integer m and

$$\begin{aligned} W^{k(n \bmod N)} &= e^{-k(n \bmod N)j\frac{2\pi}{N}} \\ &= e^{-k(n-mN)\frac{2\pi}{N}} \\ &= e^{-kn\frac{2\pi}{N}} e^{kmN\frac{2\pi}{N}} \\ &= e^{-kn\frac{2\pi}{N}} e^{km(2\pi)} \\ &= e^{-kn\frac{2\pi}{N}} \\ &= W^{kn} \end{aligned}$$

If $n < 0$ then $n \bmod N = n + mN$ for some m and by the same process as above we can show that $W^{k(n \bmod N)} = W^{kn}$.

Using this we can write

$$\begin{aligned} DFT\left[x[n - n_o]\right] &= W^{n_o k} \sum_{n=0}^{N-1} x[(n - n_o) \bmod N] W^{((n-n_o) \bmod N)k} \\ &= W^{n_o k} X[k] \end{aligned}$$

where we have used the fact that the order in which we sum elements within a finite sum does not change the sum.

17.4.4 CONVOLUTION

Just as in the case of the continuous time and discrete time Fourier transforms, the DFT of the convolution of two sequences is the product of the DFT of the individual sequences. However, in order to incorporate the implicit periodicity of the sequences we need to map any index that falls outside of the range between 0 and $N - 1$ back into this range. As shown earlier we can do this by using the modulo operation. The convolution property thus becomes

$$DFT\left[\sum_{k=0}^{N-1} x[k]h[(n-k) \bmod N]\right] = X(k)H(k)$$

17.5 THE FAST FOURIER TRANSFORM

In practice directly computing the DFT can be very resource intensive with the number of computations growing with N^2. Because of this, a number of fast algorithms have been developed with the most popular being the Fast Fourier Transform (FFT) algorithm developed by James W. Cooley and John W. Tukey in 1965. A description of the algorithm is beyond the scope of this course but if you are interested in using it MATLAB has an *fft* function you can play with.

17.6 SUMMARY

In this module we developed the discrete Fourier transform (DFT) using the discrete time Fourier transform (DTFT) as a stepping stone. We pointed out some of the possible issues that may arise when using a DFT but we didn't really go into much detail. There is much more to learn about this very useful transform and its implementation. You can get a much more detailed introduction in digital signal processing courses. These courses are both a great deal of fun and immensely useful. Please do take one.

17.7 EXERCISES

(Answers on the following page)

1. Find the discrete time Fourier transform of the sequence

$$x[n] = \left(\frac{1}{2}\right)^n u[n]$$

2. Find the discrete time Fourier transform of the sequence

$$x[n] = \left(\frac{1}{2}\right)^{|n|}$$

3. Find the discrete time Fourier transform of

$$x[n] = \begin{cases} \left(\frac{1}{2}\right)^n & n = 0, 1, 2, 3 \\ 0 & \text{otherwise} \end{cases}$$

4. Find the discrete Fourier transform of

$$x[n] = \left(\frac{1}{2}\right)^n \quad n = 0, 1, 2, 3$$

with $N = 4$. Check your answer by taking samples of the answer to the previous question every $\pi/2$.

5. Find the discrete Fourier transform of

$$x[n] = \left(\frac{1}{2}\right)^n \quad n = 0, 1, 2, 3$$

with $N = 8$. Check your answer by taking samples of the answer to the question 3 every $\pi/4$.

17.8 ANSWERS

1.
$$X(e^{j\omega}) = \frac{1}{1 - 0.5e^{-j\omega}}$$

2.
$$X(e^{j\omega}) = \frac{0.75}{1 - \cos(\omega) + 0.25}$$

3.
$$X(e^{j\omega}) = \left(1 + \frac{1}{4}e^{-j2\omega}\right)\left(1 + \frac{1}{2}e^{-j\omega}\right)$$

4. $X[0] = 1.875$, $X[1] = 0.75 - j0.375$, $X[2] = 0.625$, $X[3] = 0.75 + j0.375$

5. $X[0] = 1.875$, $X[1] = 1.265 - j0.692$, $X[2] = 0.75 - j0.375$, $X[3] = 0.735 - j0.192$, $X[4] = 0.625$, $X[5] = 0.735 + j0.192$, $X[6] = 0.75 + j0.375$, $X[7] = 1.265 + j0.692$

MODULE 18

The Laplace Transform – Introduction

The Fourier transform is an immensely useful tool for understanding signals. However, when it comes to analyzing systems there is a limitation to the use of the Fourier transform. Recall that the Fourier transform exists for all signals satisfying the Dirichlet conditions. Namely:

1. The function $x(t)$ should be absolutely integrable.

$$\int |x(t)| < \infty$$

2. The function has bounded variation which means that it has only a finite number of zero crossings in any finite interval.

3. The function has only a finite number of finite discontinuities.

The last two conditions usually are not a hindrance to analyzing systems as it would be unusual to have a system that has an impulse response that does not satisfy these conditions. However, when it comes to analyzing systems, the first condition can sometimes be problematic. A linear time invariant system is completely described by its impulse response $h(t)$. Recall that for a system to be stable we need

$$\int_{-\infty}^{\infty} |h(t)|dt < \infty$$

which is the first Dirichlet condition. This means that a system has to be stable in order for the Fourier transform of its impulse response—the transfer function—to exist. What if we want to work with unstable systems (and we do)? The Fourier transform for such a system does not exist, however, the idea of giving up the convenience of using the transfer function to relate the input and output is not a pleasant one to contemplate. Luckily we have a workaround which will allow us to retain the idea of a transfer function, allow us to relate the input and output of a linear time-invariant system using multiplication, and analyze the stability of systems under different conditions. To understand this workaround which is at the heart of our view of the Laplace transform let's consider a toy example.

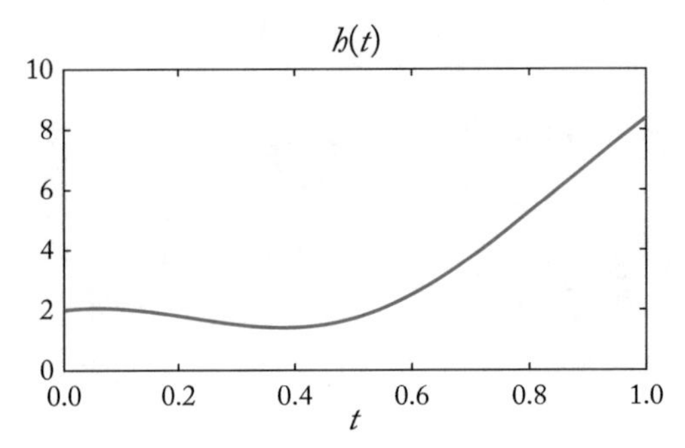

Figure 18.1: An example of the impulse response of an unstable system.

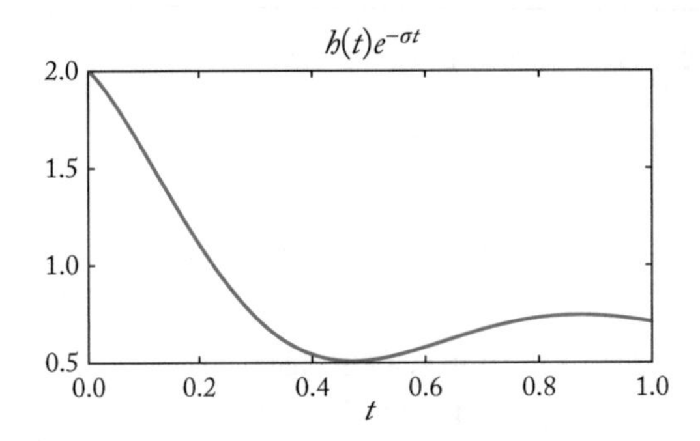

Figure 18.2: Weighting the impulse response of an unstable system with a decaying exponential.

Example 18.1 Suppose we have a system with an impulse response shown in Figure 18.1.[1] Based on what we can see the first Dirichlet condition is not satisfied and, therefore, we cannot find the Fourier transform of this impulse response.

So here's our little dodge. What if we multiply the impulse response with a decaying exponential $e^{-\sigma t}$ to get the signal shown in Figure 18.2. This weighted impulse response looks like it would satisfy the Dirichlet conditions. Given that this weighted impulse response satisfies the Dirichlet conditions we can find it's Fourier transform. True, but

1. So what? What is the use of finding the Fourier transform of $h(t)e^{-\sigma t}$ when we are interested in the Fourier transform of $h(t)$.

2. Does the result depend on the value of σ used.

The answer to the second question is yes, and we will spend some time looking at that. In Figure 18.3 we show the effect of different values of σ with the impulse response of our toy example. You can clearly see that certain values of σ actually exacerbate the problem while other

[1]For the curious this is a plot of $h(t) = \cos(2\pi t) + e^{2t}$.

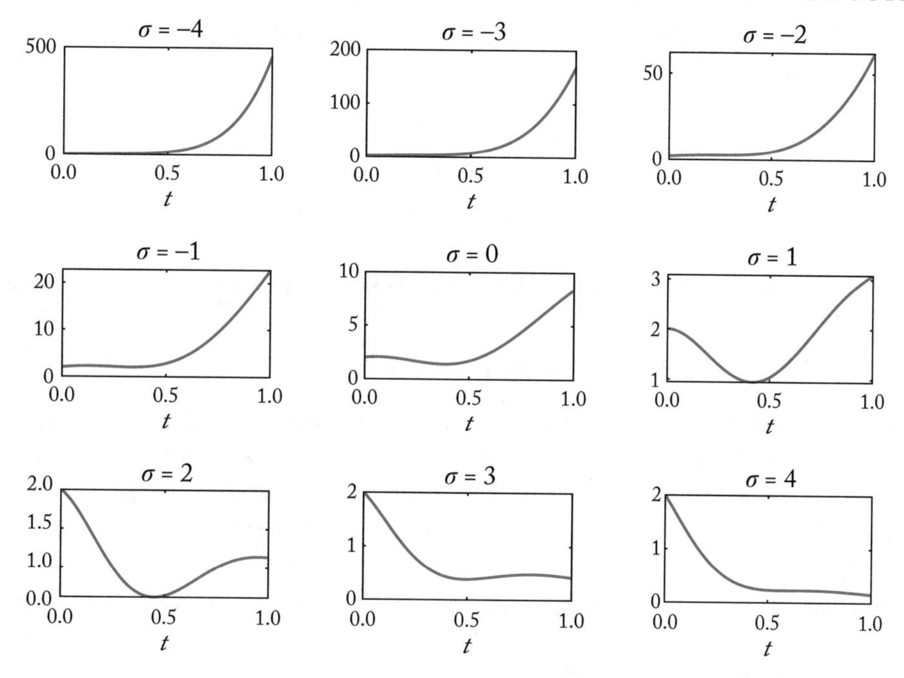

Figure 18.3: Weighting the impulse response of an unstable system with a decaying exponential $e^{-\sigma t}$ for different values of σ.

values of σ result in a product that satisfies the Dirichlet conditions. The range of values for which the Dirichlet conditions are satisfied is going to be important in what follows.

The answer to the first question is that this method gives us a way of analyzing systems and understanding the issues related to their stability as well as providing a backdoor to exploring their spectral properties. And it gives us a transfer function for simply relating the input and output of a linear time-invariant system. Which is what we will do in this and the following modules.

Let's go through this development again, but this time with a general impulse response $h(t)$. As with out toy problem we can weight $h(t)$ with an exponential $e^{-\sigma t}$ such that for some values of σ

$$\int_{-\infty}^{\infty} |h(t)e^{-\sigma t}|dt < \infty$$

Then as $h(t)e^{-\sigma t}$ satisfies the Dirichlet condition we can find the Fourier transform of this weighted function

$$\begin{aligned}\mathcal{F}\left[h(t)e^{-\sigma t}\right] &= \int_{-\infty}^{\infty}[h(t)e^{-\sigma t}]e^{-j\omega t}dt \\ &= \int_{-\infty}^{\infty} h(t)e^{-(\sigma+j\omega)}dt\end{aligned}$$

If we rename $\sigma + j\omega$ such that $s = \sigma + j\omega$, then the transform shown will become a function of s

$$H(s) = \int_{-\infty}^{\infty} h(t)e^{-st}dt$$

$H(s)$ is called the *Laplace transform* of $h(t)$. Unlike the Fourier transform $H(\omega)$ which when it exists exists for all ω the Laplace transform $H(s)$ may only exist for a restricted set of values of σ. To see why this is so consider the exponential functions $e^{-2t}u(t)$ and $e^{2t}u(t)$. The first is an exponentially decreasing function which clearly satisfies the Dirichlet conditions and hence for which the Fourier transform exists. The second is an exponentially increasing function which just as clearly violates the first Dirichlet condition and, hence, for which the Fourier transform does not exist.

Let's begin with the Laplace transform of $x(t) = e^{-2t}u(t)$.

$$\begin{aligned}X(s) &= \int_{-\infty}^{\infty} e^{-2t}u(t)e^{-st}dt \\ &= \int_{0}^{\infty} e^{-2t}e^{-st}dt \\ &= \int_{0}^{\infty} e^{-(s+2)t}dt \\ &= -\frac{1}{s+2}e^{-(s+2)t}\,|_0^{\infty}\end{aligned}$$

Things have proceeded pretty smoothly up to this point. But now we have to figure out what happens when we evaluate $e^{-(s+2)t}$ for $t \to \infty$ and $t = 0$. The latter is not an issue. When we set $t = 0$ we get e^0 which is 1. The first is a bit more problematic. To see how we resolve this, let's write s in terms of its component parts σ and ω. If we set $s = \sigma + j\omega$ in the equation above we get

$$X(s) = -\frac{1}{s+2}e^{-(\sigma+j\omega+2)t}\,|_0^{\infty}$$

Splitting apart the exponent into the real and imaginary parts we get

$$X(s) = -\frac{1}{s+2}e^{-(\sigma+2)t}e^{-j\omega t}\,|_0^{\infty}$$

Lets see what happens to each of the two factors $e^{-(\sigma+2)t}$ and $e^{-j\omega t}$ as $t \to \infty$. Let's start with $e^{-j\omega t}$.

Truth be told, it is not at all clear what value $e^{-j\omega t}$ will take as $t \to \infty$. However, thanks to our old friend Euler we do know that

$$e^{-j\omega t} = \cos(\omega t) - j\sin(\omega t)$$

The magnitude of this complex number is given by

$$\begin{aligned} \left|e^{-j\omega t}\right| &= \sqrt{\cos^2(\omega t) + \sin^2(\omega t)} \\ &= = \sqrt{1} \\ &= 1 \end{aligned}$$

So, while we do not know the value of $e^{-j\omega t}$ when t goes to infinity we do know that whatever it is, its magnitude will be equal to one. Hopefully this much information will suffice.

Let's now look at the other factor, $e^{-(\sigma+2)t}$. First let us look at what the exponent does as $t \to \infty$.

$$\lim_{t\to\infty} -(\sigma+2)t = ?$$

The limit is either $-\infty$ or $+\infty$ depending on whether $\sigma + 2$ is positive or negative. For $\sigma + 2 > 0$, which happens when $\sigma > -2$, the limit is $-\infty$. When $\sigma + 2$ is negative or $\sigma + 2 < 0$, which happens when $\sigma < -2$, the limit is $+\infty$. Therefore,

$$\lim_{t\to\infty} e^{-(\sigma+2)t} = \begin{cases} 0 & \sigma > -2 \\ +\infty & \sigma < -2 \end{cases}$$

We have enough information now to finish evaluating the Laplace transform integral.

$$\begin{aligned} X(s) &= -\frac{1}{s+2}(0-1) \quad \sigma > -2 \\ &= \frac{1}{s+2} \quad \sigma > -2 \end{aligned}$$

Note that the $1/(s+2)$ is the Laplace transform only for $\sigma > -2$. The set of values in the s plane for which $\sigma > -2$, shown in Figure 18.4, is called the *region of convergence* (also denoted by ROC) of the Laplace transform and is an essential part of the Laplace transform.

This bears repeating:

The region of convergence is an integral part of the Laplace transform.

A couple of things to notice before we leave this example. Notice that the function, $x(t) = e^{-2t}u(t)$, satisfies the Dirichlet conditions and, therefore, has a Fourier transform. Looking at

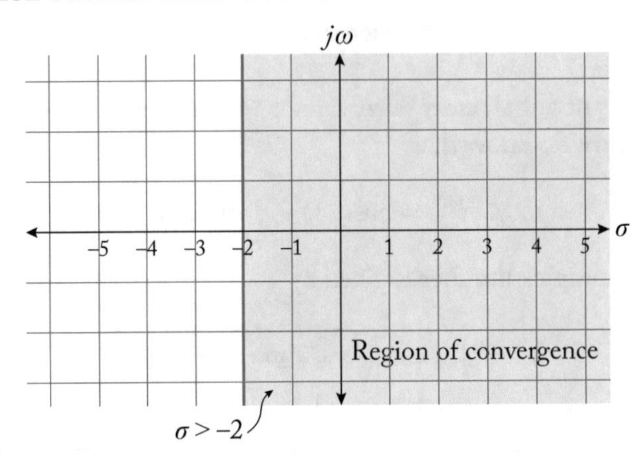

Figure 18.4: The region of convergence for the Laplace transform of $e^{-2t}u(t)$.

the integral equation for the Laplace transform

$$\begin{aligned} X(s) &= \int_{-\infty}^{\infty} x(t)e^{-st}\,dt \\ &= \int_{-\infty}^{\infty} x(t)e^{-(\sigma+j\omega)t}\,dt \end{aligned}$$

if we set $\sigma = 0$ in this equation we get the Fourier transform. But, we can only do this if the $\sigma = 0$ line is contained in the region of convergence. Notice in Figure 18.4 the line $\sigma = 0$ is contained in the region of convergence. If we set $\sigma = 0$ in this Laplace transform we get

$$X(\omega) = \frac{1}{s+2}|_{\sigma=0} = \frac{1}{\sigma + j\omega + 2}|_{\sigma=0} = \frac{1}{j\omega + 2}$$

Computing the Fourier transform of $e^{-2t}u(t)$ directly we see that this is indeed the case. Finally, if $x(t)$ was the impulse response of a linear time invariant system just the fact that the Fourier transform exists means that the impulse response is absolutely integrable. Which in turn means that the system is stable in the bounded-input-bounded-output (BIBO) sense. So, for a system with impulse response $h(t)$ the following three statements are equivalent.

1. The system is BIBO stable.
2. The Fourier transform of the impulse response exists.
3. The line $\sigma = 0$ in the s-plane is part of the region of convergence for the Laplace transform of $h(t)$.

This also suggests that for an unstable system if we could somehow move the boundary of the region of convergence so that the $\sigma = 0$ line is within the region of convergence the unstable system could be made stable. We will pursue this idea later on when we talk about feedback.

Let's go through the same process and find the Laplace transform for $x(t) = e^{2t}u(t)$. Once again we start with the integral definition of the Laplace transform

$$X(s) = \int_{-\infty}^{\infty} e^{2t}u(t)e^{-st}\,dt$$

and proceed as we did for $x(t) = e^{-2t}u(t)$.

$$\begin{aligned}
X(s) &= \int_{-\infty}^{\infty} e^{2t}u(t)e^{-st}\,dt \\
&= \int_{0}^{\infty} e^{2t}e^{-st}\,dt \\
&= \int_{0}^{\infty} e^{-(s-2)t}\,dt \\
&= -\frac{1}{s-2}e^{-(s-2)t}\,|_0^{\infty} \\
&= -\frac{1}{s-2}e^{-(\sigma+j\omega-2)t}\,|_0^{\infty} \\
&= -\frac{1}{s-2}e^{-(\sigma-2)t}e^{-j\omega t}\,|_0^{\infty} \\
&= -\frac{1}{s-2}(0-1) \qquad \sigma > 2 \\
&= \frac{1}{s-2} \qquad \sigma > 2
\end{aligned}$$

As before the region of convergence, $\sigma > 2$ in this case, consists of those values of σ for which the integral converges, i.e., does not go to infinity. A graphical representation of the region of convergence is shown in Figure 18.4. In this case the function, $x(t) = e^{2t}u(t)$, does not satisfy the first Dirichlet condition and, therefore, does not have a Fourier transform. Looking at the region of convergence we can see that the line $\sigma = 0$ is not included in the region of convergence so we could not use the trick of setting $\sigma = 0$ in the Laplace transform to get the Fourier transform. Also if $x(t)$ were the impulse response of a linear time invariant system then we can use the fact that the Fourier transform does not exist to infer that the impulse response is not absolutely integrable. Which, in turn, means that the system is not BIBO stable.

18.1 SUMMARY

In this module (Figure 18.5)

- We have introduced the Laplace transform

$$X(s) = \int_{-\infty}^{\infty} x(t)e^{-st}\,dt$$

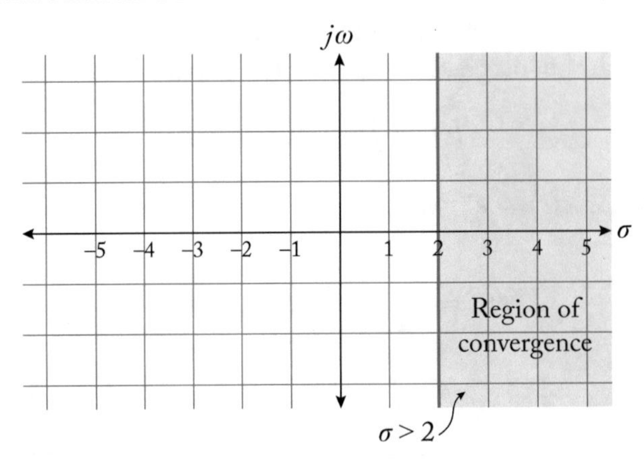

Figure 18.5

- The transform exists for a range of values of σ. This range is called the Region of Convergence (ROC).
- If the region of convergence includes the $j\omega$ axis the Fourier transform exists.
- If the Fourier transform exists and x(t) is an impulse response then the system, for which x(t) is the impulse response, is stable.

18.2 EXERCISES

(Answers on the following page)

What is the region of convergence for the Laplace transform of

1.
$$x(t) = e^{-4t}u(t)$$

2.
$$x(t) = e^{4t}u(t)$$

3.
$$x(t) = tu(t)$$

4.
$$x(t) = t^2$$

5.
$$x(t) = t^2u(t)$$

6.
$$x(t) = \cos(4t)u(t)$$

7.
$$x(t) = \sin(4t)u(t)$$

8.
$$x(t) = e^{-4t}u(t) + e^{-2t}u(t)$$

9.
$$x(t) = e^{4t}u(t) + e^{-2t}u(t)$$

18.3 ANSWERS

1. ROC: $\sigma > -4$
2. ROC: $\sigma > 4$
3. ROC: $\sigma > 0$
4. ROC: Does not converge
5. ROC: $\sigma > 0$
6. ROC: $\sigma > 0$
7. ROC: $\sigma > 0$
8. ROC: $\sigma > -2$
9. ROC: $\sigma > 4$

MODULE 19

Uniqueness and Linearity of the Laplace Transform

In this module we look at two properties of the Laplace transform that make it very useful—its uniqueness and its linearity. The Laplace transform, like the Fourier transform, affords us an alternative view of a signal. However, remember that both a function of time and its Laplace transform are different representations of the **same** physical process. Therefore, in order to accord with reality the function of time and its Laplace transform have to form a unique pair—each implying the other. This being the case it is important that we establish the conditions under which the Laplace transform is unique—the function of time and its Laplace transform form a pair.

The linearity property makes the Laplace transform convenient to use. Without this property we probably would not have been talking about it. So let's take a look at these properties.

19.1 THE LAPLACE TRANSFORM IS UNIQUE – AS LONG AS WE INCLUDE THE ROC IN THE TRANSFORM

A rather long sub-heading, but we really want to emphasize the fact that the region of convergence is an integral part of the Laplace transform. Let's see why the emphasis needs to be there with an example. Recall in the last module we obtained the Laplace transform of

$$x(t) = e^{-2t}u(t)$$

as

$$X(s) = \frac{1}{s+2} \quad \sigma > -2$$

To see the reason why we need to include the region of convergence let's find the Laplace transform of

$$x(t) = -e^{-2t}u(-t)$$

$$\begin{aligned}
X(s) &= -\int_{-\infty}^{\infty} e^{-2t}u(-t)e^{-st}\,dt \\
&= -\int_{-\infty}^{0} e^{-2t}e^{-st}\,dt \\
&= -\int_{-\infty}^{0} e^{-(s+2)t}\,dt \\
&= \frac{1}{s+2}e^{-(s+2)t}\,\Big|_{-\infty}^{0} \\
&= \frac{1}{s+2}e^{-(\sigma+j\omega+2)t}\,\Big|_{-\infty}^{0} \\
&= \frac{1}{s+2}e^{-(\sigma+2)t}e^{-j\omega t}\,\Big|_{-\infty}^{0}
\end{aligned}$$

As in the case of the computation of the Laplace transform of $e^{-2t}u(t)$ we need to consider what happens at the lower limit of negative infinity. The magnitude of $e^{-j\omega t}$ is always bounded by one so, as before we need to look at the behavior of the term $e^{-(\sigma+2)t}$. This term will either go to infinity or to zero, in other words diverge or converge, depending on whether the exponent is positive or negative. The exponent is negative—the integral converges—when $\sigma + 2 < 0$ or $\sigma < -2$. Therefore,

$$\begin{aligned}
X(s) &= \frac{1}{s+2}(1-0) \qquad \sigma < -2 \\
&= \frac{1}{s+2} \qquad \sigma < -2
\end{aligned}$$

Notice we ended up with same algebraic expression for $X(s)$, $1/(s+2)$, for both $e^{-2t}u(t)$ and for $-e^{-2t}u(-t)$ even though as shown in Figure 19.1 the two signals are completely different. Therefore, if we simply used the algebraic expression of $X(s)$ we would be unable to uniquely describe the particular function of time for which $X(s)$ is the Laplace transform. We have to include the region of convergence for $X(s)$ and $x(t)$ to be a unique pair. Notice we say *region* of convergence. As you can see from Figure 19.2, $X(s)$ is valid over a *region* in the s-plane.

Notice the difference in the regions for the two functions. For the signal $e^{-2t}u(t)$ which only takes on non-zero values to the right of a point (in this case $t = 0$)—a right-sided function—the region of convergence is to the right of the boundary $\sigma = -2$. For the signal $-e^{-2t}u(-t)$ which only takes on non-zero values to the left of a point ($t = 0$)—a left-sided function—the region of convergence is to the left of the boundary $\sigma = -2$. Finally, notice that in one case the region of convergence includes the $j\omega$ axis while in the other case it does not. Before we discuss why this is important let us generalize our example.

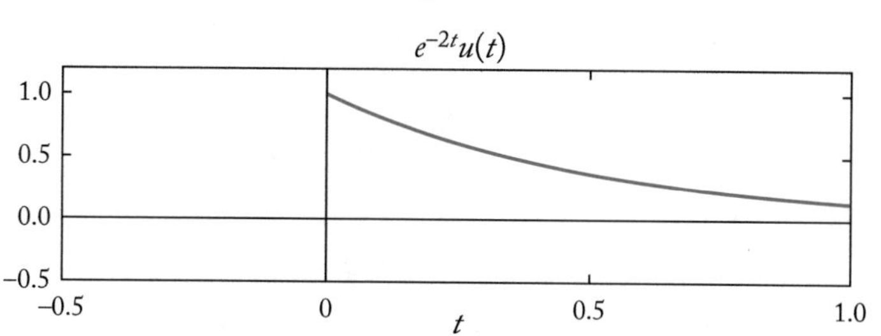

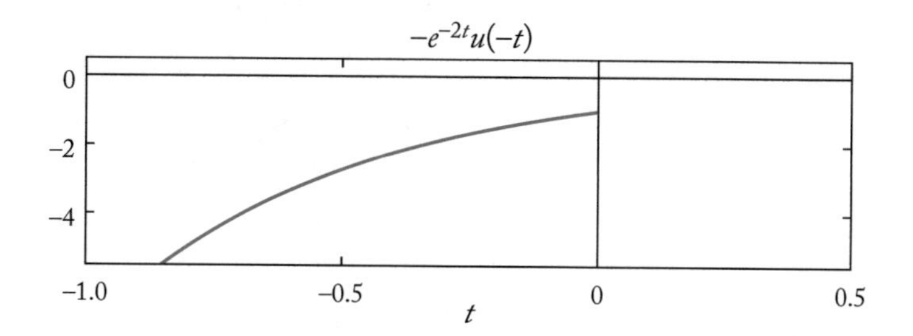

Figure 19.1: The signals $e^{-2t}u(t)$ and $-e^{-2t}u(-t)$.

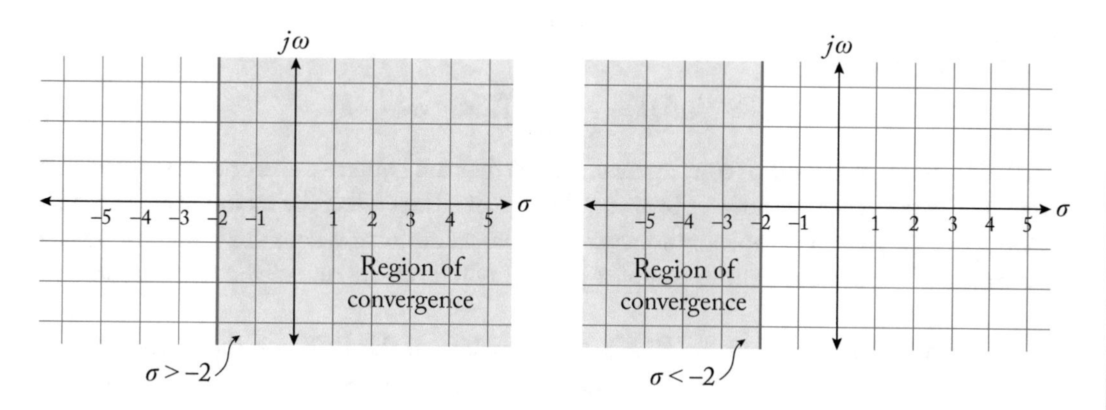

Figure 19.2: The region of convergence for the Laplace transform of $e^{-2t}u(t)$ and $-e^{-2t}u(-t)$.

Let's consider the signal

$$x(t) = e^{-at}u(t)$$

For $a > 0$ this is a decaying exponential with absolute integral

$$\begin{aligned}\int_{-\infty}^{\infty} e^{-at}u(t)dt &= \int_{0}^{\infty} e^{-at}dt \\ &= -\frac{1}{a}e^{-at}|_0^{\infty}\end{aligned}$$

When we evaluate e^{-at} for $t \to \infty$, because $a > 0$, $\lim_{t\to\infty} e^{-at} \to 0$. Therefore,

$$\mathcal{F}\left[e^{-at}u(t)\right] = \frac{1}{a + j\omega}$$

However, for $a < 0$

$$\lim_{t\to\infty} e^{-at} \to \infty$$

Therefore, the Fourier transform of $e^{-at}u(t)$ exists for $a > 0$ but not for $a < 0$. As you might surmise the Laplace transform of $e^{-at}u(t)$ exists for both all values of a. The Laplace transform is given by

$$\begin{aligned} X(s) &= \int_{-\infty}^{\infty} e^{-at}u(t)e^{-st}\,dt \\ &= \int_{0}^{\infty} e^{-(s+a)t}\,dt \\ &= -\frac{1}{s+a}e^{-(s+a)t}\Big|_0^{\infty} \\ &= \frac{1}{s+a} \quad \text{for } \sigma > -a \end{aligned}$$

We write this as

$$\mathcal{L}\left[e^{-at}u(t)\right] = \frac{1}{s+a} \quad \text{ROC: } \sigma > -a$$

where ROC stands for region of convergence. Notice that the Laplace transform exists regardless of whether a is positive or negative. However, the sign of a does affect the region of convergence. Because $s = \sigma + j\omega$ is a complex number it takes its values over a complex plane. If we look at the region of convergence in the complex plane it will look like one of the regions shown in Figure 19.3.

Notice a few things about the ROC. The boundary of the ROC on the left is a vertical line. The reason for this is that the ROC depends only on the values of σ. These values are represented on the horizontal axis. The values of σ for which $\sigma > -a$ will be all values in the semi-infinite plane to the left of the line $\sigma = -a$. This line is the vertical line that intersects the σ axis at $\sigma = -a$. We can see from this that the boundaries for the region of convergence will always be either a vertical line or infinity. Also notice that the ROC in the case where $a > 0$ includes the $j\omega$ axis while the ROC for the case $a < 0$ does not include the $j\omega$ axis. The $j\omega$ axis is the line where $\sigma = 0$. If we plug $\sigma = 0$ into the Laplace transform equation we get the Fourier transform. This means that when the ROC includes the $j\omega$ axis the equation for the Fourier transform converges and the Fourier transform exists. If we look back we see this is precisely true in this case—the Fourier transform exists if $a > 0$ and does not exist when $a < 0$. If $x(t)$ is the impulse response of a linear time invariant system we can draw a further conclusion from this. Recall that the Fourier transform only exists if the Dirichlet conditions are satisfied. This

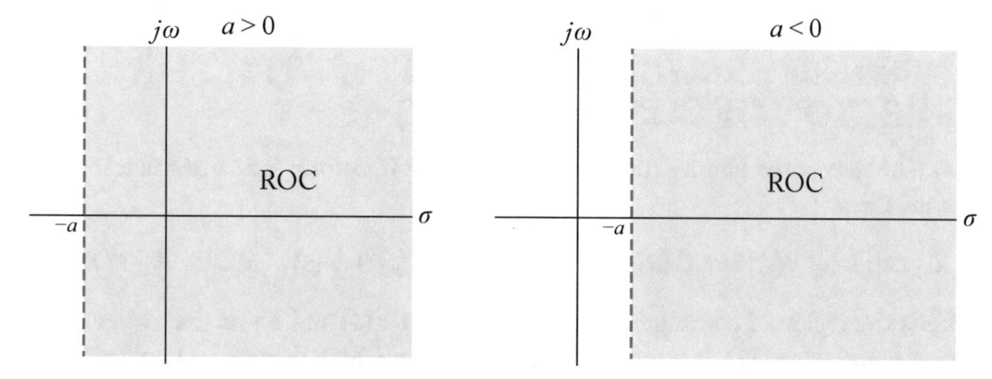

Figure 19.3: The region of convergence for the Laplace transform of $e^{-at}u(t)$ for positive and negative values of a.

means that if the $j\omega$ axis is included in the region of convergence the Dirichlet conditions are satisfied. The first Dirichlet condition was that of absolute integrability

$$\int_{-\infty}^{\infty} |x(t)| < \infty$$

But if $x(t)$ is an impulse response satisfying this condition means that the system is stable. Therefore, if $x(t)$ is an impulse response and the region of convergence of its Laplace transform includes the $j\omega$ axis, the system for which $x(t)$ is the impulse response is stable.

Finally, let us look at the limiting condition when $a = 0$. In this case

$$x(t) = e^{-at}u(t) = e^{0}u(t) = u(t)$$

and

$$X(s) = \frac{1}{s} \quad \text{ROC: } \sigma > 0$$

To further understand the importance of stating the region of convergence consider the function $x(t) = -e^{-at}u(-t)$. For this signal we get the Laplace transform

$$X(s) = \frac{1}{s+a} \qquad \sigma < -a$$

Notice that the ratio $\frac{1}{s+a}$ is the same for both $e^{-at}u(t)$ and $-e^{at}u(-t)$. What is different between the Laplace transforms is the region of convergence. Without the region of convergence we can't differentiate between them. Which brings us back to the heading for this section **the Laplace transform is unique—as long as we include the ROC in the transform**.

19.2 THE LAPLACE TRANSFORM IS A LINEAR TRANSFORM (BUT BE CAREFUL WITH THE REGION OF CONVERGENCE)

Just as was the case of the Fourier transform, the Laplace transform is an integral transform and is, therefore, linear.

$$\mathcal{L}\left[\alpha x(t) + \beta y(t)\right] = \alpha \mathcal{L}\left[x(t)\right] + \beta \mathcal{L}\left[y(t)\right] = X(s) + Y(s) \qquad \text{ROC: } R_X \cap R_Y$$

where R_X is the region of convergence corresponding to $X(s)$ and R_Y is the region of convergence corresponding to $Y(s)$. In the equation above we use $\mathcal{L}\left[\cdot\right]$ to denote the Laplace transform, α and β are constants, and we use lower case letters to represent the time domain functions and upper case letters to denote their Laplace transforms. Because the Laplace transform of a function $x(t)$ can be viewed as the Fourier transform of $x(t)e^{-\sigma t}$, many of the properties of the Fourier transform translate over for the Laplace transform. We will go through and list these properties later. For now, let's just go through a few examples.

Example 19.1 Consider the sum of two exponentials

$$x(t) = e^{-2t}u(t) + 2e^{-3t}u(t)$$

Because of the linearity of the Laplace transform this is simply the sum of the Laplace transforms of the individual exponentials which we can easily compute as:

$$\begin{aligned} \mathcal{L}\left[e^{-2t}u(t)\right] &= \frac{1}{s+2} \qquad \sigma > -2 \\ \mathcal{L}\left[e^{-3t}u(t)\right] &= \frac{1}{s+3} \qquad \sigma > -3 \end{aligned}$$

We can just take the linear combination of these two to find $X(s)$, however, we have to be a bit careful with the region of convergence. As you can see the region of convergence for each term in the sum is different. As the region of convergence for the sum we pick the intersection of the regions of convergence for each terms as the region of convergence for the sum should be those values of σ for which each term of the sum converges. Thus,

$$X(s) = \frac{1}{s+2} + \frac{2}{s+3} = \frac{3s+7}{(s+2)(s+3)} \qquad \sigma > -2$$

Example 19.2 Let's bring in the sinusoids.

$$x(t) = e^{-2t}\cos(6t)u(t)$$

But sinusoids are just exponentials who have met Euler.

$$
\begin{aligned}
x(t) &= e^{-2t}\cos(6t)u(t) \\
&= e^{-2t}\left(\frac{e^{j6t}+e^{-j6t}}{2}\right)u(t) \\
&= \frac{1}{2}e^{-(2-j6)t}u(t)+\frac{1}{2}e^{-(2+j6)t}u(t)
\end{aligned}
$$

We know that

$$
\mathcal{L}\left[e^{-at}u(t)\right] = \frac{1}{s+a} \qquad \sigma > -a
$$

We could just replace a with $2 - j6$ and $2 + j6$, and for the most part we will. The only place we have to be a bit careful is in the definition of the region of convergence. Recall that σ is the real part of s. So for the region of convergence we only compare σ with the real part of the exponents. The real part of the exponent for both terms is 2, so,

$$
X(s) = \frac{1}{2}\left(\frac{1}{s+2-j6}+\frac{1}{s+2+j6}\right) \qquad \sigma > -2
$$

or

$$
X(s) = \frac{s+2}{s^2+4s+40} \qquad \sigma > -2
$$

Example 19.3 Let's mix it up a bit

$$
x(t) = 2e^{-3t}u(t) + e^{-2t}\cos(6t)u(t)
$$

Again using the linearity of the Laplace transform

$$
\begin{aligned}
X(s) &= \frac{2}{s+3}+\frac{1}{2}\left(\frac{1}{s+2-j6}+\frac{1}{s+2+j6}\right) \qquad \text{ROC: } (\sigma > -3) \cap (\sigma > -2) \\
&= \frac{2}{s+3}+\frac{1}{2}\left(\frac{2s+4}{s^2+4s+40}\right) \qquad \sigma > -2 \\
&= \frac{2s^2+8s+80+s^2+5s+6}{(s+3)(s^2+4s+40)} \qquad \sigma > -2 \\
&= \frac{3s^2+13s+86}{(s+3)(s^2+4s+40)} \qquad \sigma > -2
\end{aligned}
$$

where we have used the fact that

$$
(\sigma > -3) \cap (\sigma > -2) = \sigma > -2
$$

to get the region of convergence.

Example 19.4 Let's do one last signal

$$x(t) = e^{-2|t|}$$

We can write $x(t)$ as

$$x(t) = e^{-2t}u(t) + e^{2t}u(-t)$$

We know that

$$\mathcal{L}\left[e^{-2t}u(t)\right] = \frac{1}{s+2} \quad \sigma > -2$$

and

$$\mathcal{L}\left[e^{2t}u(-t)\right] = \frac{1}{s-2} \quad \sigma < 2$$

Combining the two

$$\mathcal{L}\left[e^{-2|t|}\right] = \frac{1}{s+2} + \frac{1}{s-2} \qquad -2 < \sigma < 2$$

or

$$X(s) = \frac{2s}{(s+2)(s-2)} \qquad -2 < \sigma < 2$$

19.3 SUMMARY

In this module we have looked at two properties of the Laplace transform.

1. We have seen the importance of the region of convergence in making the Laplace transform unique. If we do not specify the region of convergence the Laplace transform is not necessarily unique.

2. We have seen the usefulness of the linearity property in the computation of the Laplace transform.

19.4 EXERCISES

(Answers on the following page)

Find the Laplace transform including the regions of convergence for the following signals.

1.
$$x(t) = e^{-4t}\cos(2t)u(t)$$

2.
$$x(t) = e^{-4t}\cos(6t)u$$

3.
$$x(t) = e^{-2t}\sin(10t)$$

4.
$$x(t) = u(t)$$

5.
$$x(t) = (1 + e^{-t})u(t)$$

6.
$$x(t) = e^{-t}u(t) + e^{2t}u(-t)$$

7.
$$x(t) = \delta(t)$$

8.
$$x(t) = \delta(t - t_o)$$

9.
$$x(t) = \sum_{n=0}^{\infty} \delta(t - nt_o)$$

10.
$$x(t) = u\left(t + \frac{1}{2}\right) - u\left(t - \frac{1}{2}\right)$$

19.5 ANSWERS

1.
$$X(s) = \frac{s+4}{s^2+8s+20} \quad \sigma > -4$$

2.
$$X(s) = \frac{s+4}{s^2+8s+52} \quad \sigma > -4$$

3.
$$X(s) = \frac{10}{s^2+4s+104}$$

4.
$$X(s) = \frac{1}{s} \quad \sigma > 0$$

5.
$$X(s) = \frac{2s+1}{s^2+s} \quad \sigma > 0$$

6.
$$X(s) = \frac{-3}{s^2-s-2} \quad -1 < \sigma < 2$$

7.
$$X(s) = 1 \quad \forall\sigma$$

8.
$$X(s) = e^{-st_o} \quad \forall\sigma$$

9.
$$X(s) = \frac{1}{1-e^{-st_o}} \quad \sigma > 0$$

10.
$$X(s) = \frac{e^{\frac{s}{2}} - e^{-\frac{s}{2}}}{s}$$

MODULE 20

Laplace Transform – Poles and Zeros

In this module we introduce the concept of poles and zeros which will significantly increase the utility of the Laplace transform for us. We will immediately make use of this concept to show how we can characterize the transfer function of a system simply by using the location of the poles and zeros. This will not by any means be the only place we will make use of poles and zeros. But we are getting ahead of ourselves. Let's first define poles and zeros.

Notice that in almost all of the cases we have looked at the Laplace transform is a ratio of polynomials in s

$$X(s) = \frac{N(s)}{D(s)}$$

Almost the only exception we shall see in our work is the Laplace transform whose region of convergence was the entire s-plane. This does not mean that all Laplace transforms are ratios of polynomials. It's just that most of the Laplace transforms we are interested in are ratios of polynomials.

We can write the polynomials $N(s)$ and $D(s)$ as a product of factors and a gain term in the following manner

$$X(s) = \frac{N(s)}{D(s)} = A\frac{(s - z_1)(s - z_2)\ldots(s - z_m)}{(s - p_1)(s - p_2)\ldots(s - p_n)}$$

Assuming that $X(s)$ is the transfer function of a linear time invariant system the roots of the numerator polynomial $\{z_i\}$ are called the *zeros* of the system. The name is very appropriate because whenever $s = z_i$ for any i, the transfer function $X(s)$ goes to zero. Zeros are represented in the s-plane with circles. The roots of the denominator polynomial $\{p_i\}$ are called the poles of the system and these are the values at which the value of the transfer function goes to infinity. Poles are represented in the s-plane by crosses.

Let's look at the pole-zero diagrams of the various examples we have introduced previously.

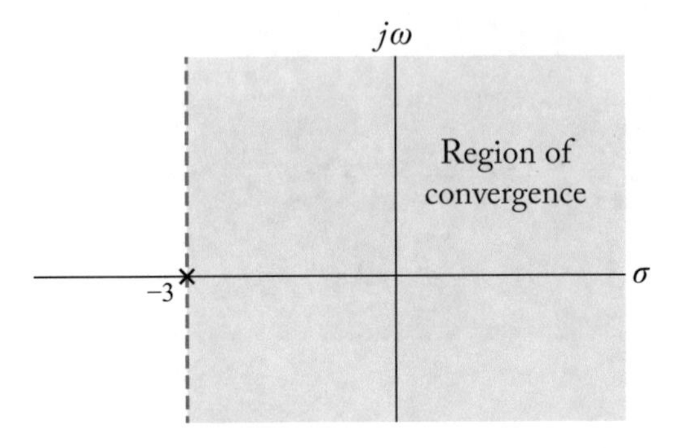

Figure 20.1: Pole zero pattern for $e^{-3t}u(t)$.

Example 20.1 Let's begin with the simplest example

$$x(t) = e^{-3t}u(t)$$

The Laplace transform is given by

$$X(s) = \frac{1}{s+3} \qquad \sigma > -3$$

There is only one pole and no zeros as can be seen from the pole-zero diagram shown in Figure 20.1. The pole is at -3.

Notice that the function of time is nonzero to the right of a point—the origin in this case—and the region of convergence is right-sided—to the right of a boundary. Furthermore the boundary is defined by the pole.

Example 20.2 If we now consider the left-sided version of this

$$x(t) = -e^{-3t}u(-t)$$

The Laplace transform is given by

$$X(s) = \frac{1}{s+3} \qquad \sigma < -3$$

We obtain the same pole-zero pattern but the region of convergence is left sided with the boundary being defined by the pole (Figure 20.2).

The correspondence between the "sidedness" of the time function and the region of convergence is something we will see over and over again.

Example 20.3 Consider the case where we had the sum of two exponentials

$$x(t) = e^{-2t}u(t) + 2e^{-3t}u(t)$$

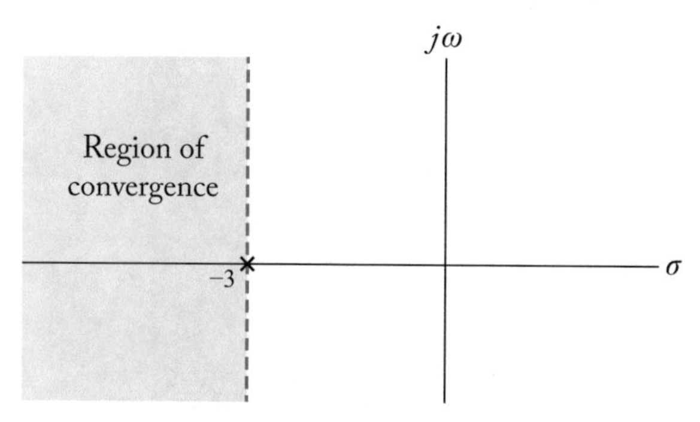

Figure 20.2: Pole zero pattern for $-e^{-3t}u(-t)$.

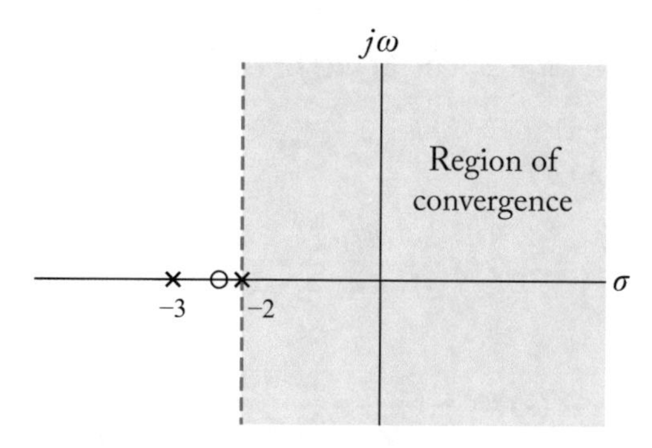

Figure 20.3: Pole zero pattern for $e^{-2t}u(t) + 2e^{-3t}u(t)$.

The Laplace transform is given by

$$X(s) = \frac{1}{s+2} + \frac{2}{s+3} = \frac{3s+7}{(s+2)(s+3)} \quad \sigma > -2$$

There are two poles, at -2 and -3, and a zero at $-7/3$. We plot the poles and zero as shown in Figure 20.3.

Notice that as expected the region of convergence is right-sided. Furthermore, the rightmost pole defines the boundary of the region of convergence. This makes sense. If the boundary was to the left of the rightmost pole the region of convergence would include the pole where we know that the Laplace transform does not converge. Finally, notice that the poles are both on the real axis. As the poles match up with the exponents of the two component exponentials and the exponents are both real it makes sense that the poles would be on the real axis.

Example 20.4 Let's look at the next example which was a damped sinusoid.

$$x(t) = e^{-2t}\cos(6t)u(t)$$

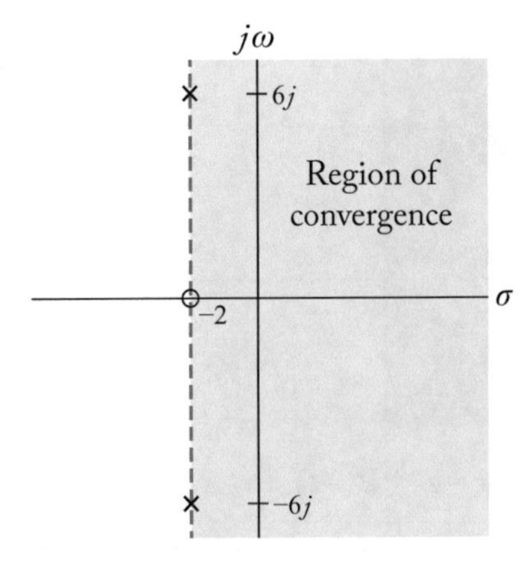

Figure 20.4: Pole zero pattern for $e^{-2t}\cos(6t)u(t)$.

The Laplace transform is given by

$$X(s) = \frac{s+2}{s^2+4s+40} \qquad \sigma > -2$$

which can be written as

$$X(s) = \frac{s+2}{(s+2-j6)(s+2+j6)}$$

to make explicit the fact that there is a zero at $s = -2$ and poles at $s = -2 - j6$ and $s = -2 + j6$. The pole-zero pattern is shown in Figure 20.4.

Notice again that the region of convergence is right-sided and is bounded by the rightmost poles. Unlike the previous case the poles are complex. We will see that this is always the case when we have oscillations in the impulse response. The oscillations correspond to sinusoids which are represented by complex exponentials—hence the complex poles.

Example 20.5 The next example contains a decaying exponential and a damped sinusoid.

$$x(t) = 2e^{-3t}u(t) + e^{-2t}\cos(6t)u(t)$$

The Laplace transform is given by

$$X(s) = \frac{3s^2+13s+86}{(s+3)(s^2+4s+40)} \qquad \sigma > -2$$

The pole-zero pattern is shown in Figure 20.5. As before a right-sided function results in a right-sided region of convergence, and the region of convergence is bounded on the left by the rightmost poles.

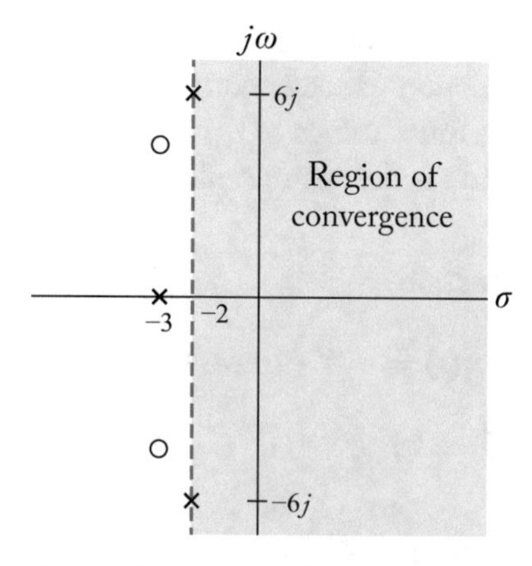

Figure 20.5: Pole zero pattern for $2e^{-3t}u(t) + e^{-2t}\cos(6t)u(t)$.

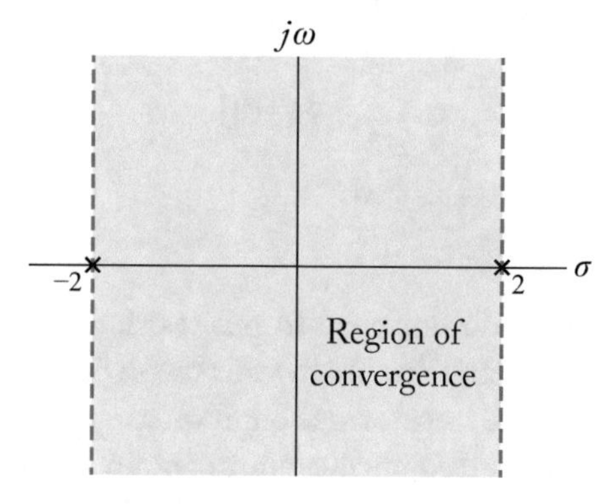

Figure 20.6: Pole zero pattern for $e^{-2|t|}$.

Example 20.6 Our final example is the Laplace transform of the signal

$$x(t) = e^{-2|t|}$$

which is

$$X(s) = \frac{2s}{(s+2)(s-2)} \qquad -2 < \sigma < 2$$

The pole-zero pattern is shown in Figure 20.6.

In this case unlike the previous cases the time domain signal is two-sided. The region of convergence becomes the intersection of a left-sided and a right-sided region of convergence resulting in the bounded region of convergence shown in Figure 20.6.

In all of these examples, the region of convergence is some subset of the s-plane. Are there functions with regions of convergence that are the entire s-plane? To see the answer recall that the region of convergence consists of those values of σ for which the integral of $|x(t)e^{-\sigma t}|$ is finite. If $x(t)$ is a finite valued function which is nonzero only over a finite interval, the integral of $|x(t)e^{-\sigma t}|$ will be finite for all finite values of σ; and, therefore, the Laplace transform will have a region of convergence which is the entire s-plane.

Example 20.7 Consider the function

$$x(t) = e^{-2t}\,(u(t) - u(t-1))$$

The Laplace transform $X(s)$ is given by

$$\begin{aligned} X(s) &= \int_{-\infty}^{\infty} e^{-2t}(u(t) - u(t-1))e^{-st}\,dt \\ &= \int_{0}^{1} e^{-(s+2)t}\,dt \\ &= -\frac{1}{s+2}e^{-(s+2)t}|_0^1 \\ &= \frac{1-e^{-(s+2)}}{s+2} \end{aligned}$$

There are no infinite limits so we do not need to put restrictions on σ which means that the transform exists over the entire s-plane. In other words the region of convergence is the entire s-plane. When we look at the Laplace transform though there appears to be a fly in the ointment. We have said that the region of convergence cannot contain any poles and it looks like there is a pole at $s = -2$. But is there really a pole at $s = -2$? We have been dealing with rational polynomials in which the roots of the denominator are the places where the function goes to infinity, and hence the roots of the denominator are the poles. The $X(s)$ obtained here, however, is not a rational polynomial, so to check if $s = -2$ is indeed a pole we need to evaluate $X(s)$ at $s = -2$

$$X(s)|s = -2 = \frac{1-e^{-(-2+2)}}{-2+2} = \frac{0}{0}$$

Applying L'Hopitals's rule we get

$$\frac{e^{-(s+2)}}{1}|_{s=-2} = 1$$

Therefore, $s = -2$ is not a pole; and the region of convergence is indeed the entire s-plane.

So to recap, here are the rules about the regions of convergence we can deduce from our examples:

1. If $x(t)$ is right sided, the region of convergence is to the right of a boundary. The boundary is determined by the right-most pole.
2. If $x(t)$ is left sided, the region of convergence is to the left of a boundary. The boundary is determined by the left-most pole.
3. if $x(t)$ is two sided, the region of convergence is a strip bounded on both sides by boundaries determined by poles.
4. The region of convergence cannot contain any poles.
5. If $x(t)$ is of finite duration and is absolutely integrable then the region of convergence is the entire plane.

20.1 FREQUENCY RESPONSE

The Laplace transform is a generalization of the Fourier transform. If the region of convergence of the Laplace transform includes the $j\omega$ axis we can obtain the Fourier transform by evaluating the Laplace transform along the $j\omega$ axis. So if we have a simple linear system with impulse response $h(t) = e^{-at}u(t)$. The corresponding Laplace function is

$$H(s) = \frac{1}{s+a}; \quad \sigma > -a$$

If $a > 0$ then the $j\omega$ axis is included in the region of convergence. So we can obtain the Fourier transform by simply setting $\sigma = 0$ in the Laplace transform.

$$\begin{aligned} H(s) &= \frac{1}{s+a}; \quad \sigma > -a \\ &= \frac{1}{\sigma + j\omega + a}; \quad \sigma > -a \\ H(\omega) &= H(s)\,|_{\sigma=0} = \frac{1}{j\omega + a} \end{aligned}$$

which, you may recall, is the Fourier transform of $e^{-at}u(t)$. Notice that this only works if $\sigma = 0$ is in the region of convergence which is true for $-a < 0$.

The nice thing about the Laplace transform is that simply by looking at the location of the poles and zeros we can get a fair idea of what the transfer function looks like. Here we will just look at the magnitude of the transfer function but we can get a fair idea of the phase from the pole zero locations as well. As our example, let's use the one pole system above with the pole at $a = -0.5$. Given that

$$|H(s)| = \frac{1}{|s - 0.5|}$$

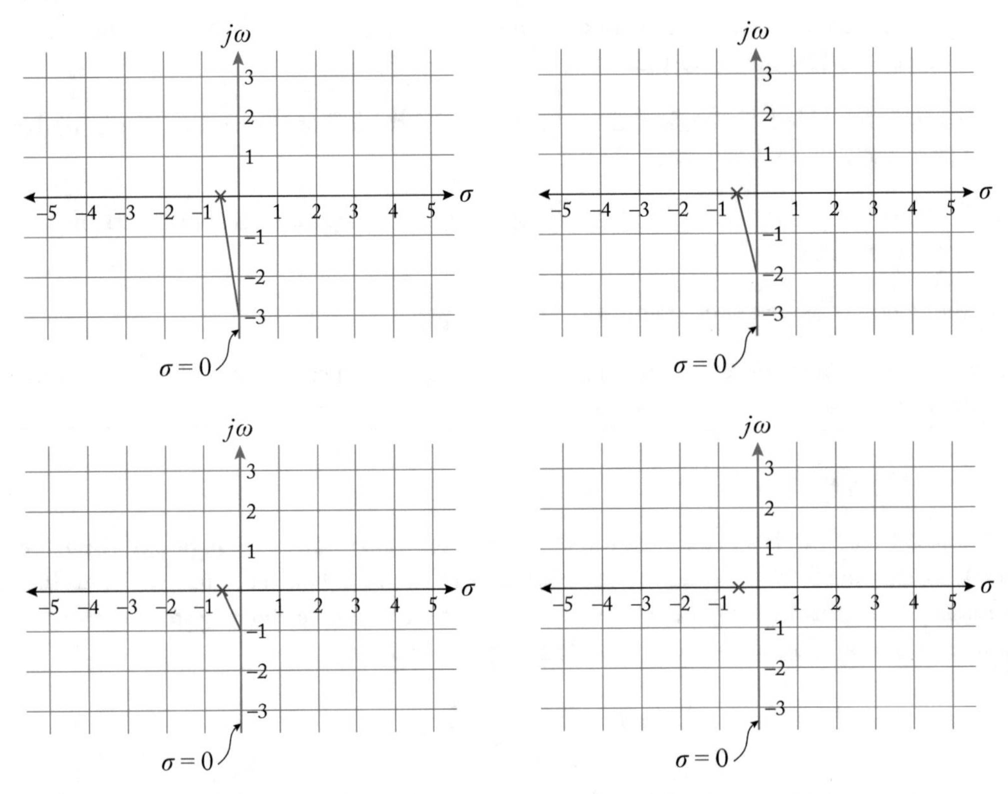

Figure 20.7: Time response and frequency response of a system with a single pole at $s = 0.5$.

To evaluate the transfer function we have to stay on the $j\omega$ axis. As we move up and down the $j\omega$ axis the magnitude of the transfer function will be 1 divided by the distance between where we are on the axis and the pole at $s = 0.5$. The situation for four conditions $\omega = -3$, $\omega = -2$, $\omega = -1$, and $\omega = 0$ is shown in Figure 20.7.

When $\omega = -3$ the distance to the pole is the length of the line connecting the point $\omega = -3$ to the pole at $s = -0.5$. We can calculate this using Pythogarus' theorem to be about 3.04. The reciprocal of that is about 0.33. Therefore, $|H(-3)| = 0.33$. Similarly we can calculate $|H(-2)| = 1/\sqrt{(4 + 0.25)} = 0.48$, $|H(-1)| = 1/\sqrt{(1 + 0.25)} = 0.89$ and $|H(0)| = 2$. If we now proceed up the $j\omega$ axis we will encounter these values in reverse—$|H(1)| = 1/\sqrt{(1 + 0.25)} = 0.89$, $|H(2)| = 1/\sqrt{(4 + 0.25)} = 0.48$, and $|H(3)| = 0.33$. Notice that the magnitude is even as we would expect for the Fourier transform of a real valued function.

The time response and the magnitude of the transfer function for this system are shown in Figure 20.8.

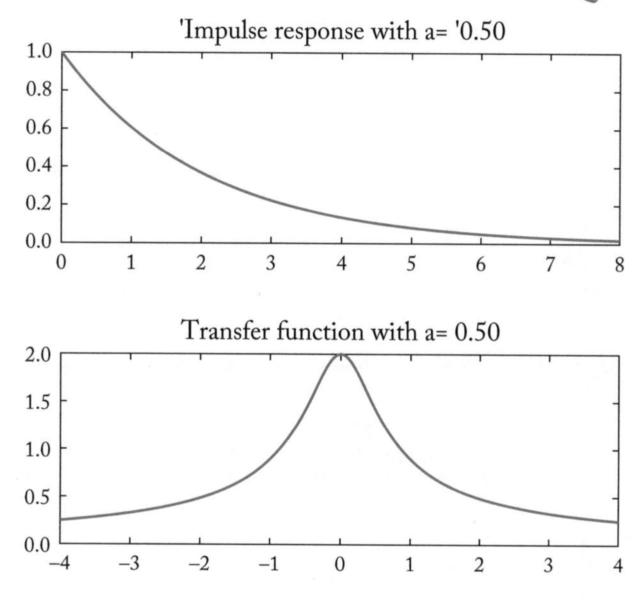

Figure 20.8: Time response and frequency response of a system with a single pole at $s = 0.5$.

By moving the pole around we can change the time and frequency response of this system. Pick $a = 2$ that is, move the pole from -0.5 to -2, and we get the responses shown in Figure 20.9.

The reason we can evaluate the Laplace transform on the $j\omega$ axis is that in this case the $j\omega$ axis is included in the region of convergence. As we move up and down the $j\omega$ axis we move closer and farther away from the pole. The closest we get to the pole is at $\omega = 0$. As the pole is a root of the denominator the closer we approach it the smaller the denominator becomes and hence the greater the magnitude. We can see in this for both values of the a. If we bring the pole closer to the $j\omega$ axis we enhance this behavior and the increase in the amplitude of the magnitude of the transfer function is greater for $a = 0.5$ than it is for $a = 2$.

Consider a different example with the pole-zero pattern shown in Figure 20.10.

There is a zero at 0 and poles at $-2 \pm j12$. Let's move along the $j\omega$ axis in the positive direction. At $\omega = 0$ we have a zero, so the transfer function goes to zero at $\omega = 0$. As we move further in the positive direction we come closer to the pole located at $-2 + j12$. We would never actually reach the pole as it is off the $j\omega$ axis. Instead the transfer function peaks as ω approaches 12 and decays again as we move away from the pole. The magnitude of the transfer function is plotted in Figure 20.11.

You can see how by placing poles and zeros we can obtain a desired response. Once we have the desired response we can use the pole-zero patterns to obtain the transfer function as a function of s. For example, for the case we have been looking at it is easy to see that the transfer

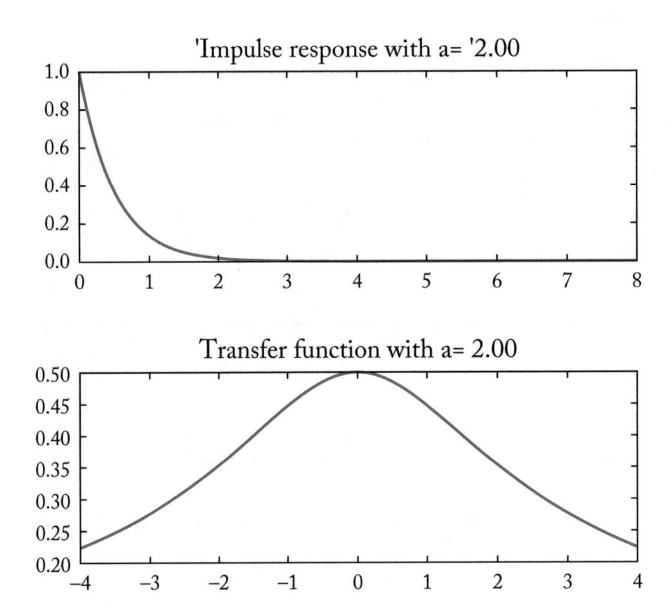

Figure 20.9: Time response and frequency response of a system with a single pole at $s = 2.0$.

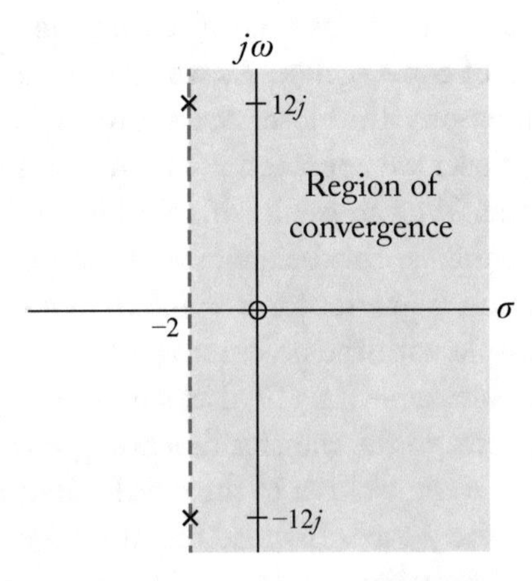

Figure 20.10: Different example with the pole-zero pattern.

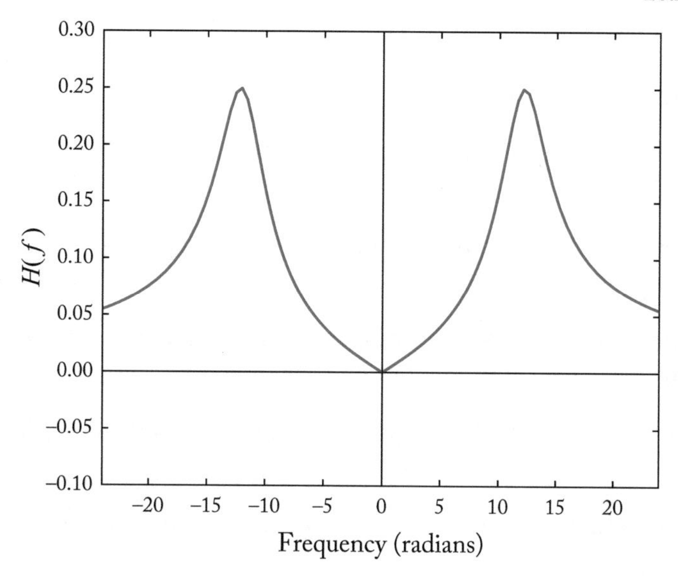

Figure 20.11: The magnitude of the transfer function.

function is

$$H(s) = \frac{s}{(s+2+j12)(s+2-j12)} = \frac{s}{s^2+4s+148}$$

From the transfer function we can obtain different hardware realizations using some well known algorithmic approaches.

20.2 SUMMARY

In this module we learned that:

- The Laplace transform of most of the functions we are interested in are ratios of polynomials in s.
- The roots of the numerator polynomial are called the zeros of the system.
- The roots of the denominator polynomials are called the poles of the system.
- The region of convergence is bounded by vertical lines through the poles and possibly infinity.
- The poles and zeros can tell us a lot about the frequency response of those systems that have a Fourier transform.
- We can design simple filters using pole zero placement.

20.3 EXERCISES

(Answers on the following page)

1. Given a system with a right-sided impulse response and poles at $-2, -1$, what is the region of convergence.
2. Given a system with a right-sided impulse response and poles at $-2, 2$, what is the region of convergence.
3. Given a system with a left-sided impulse response and poles at $-2, -1$, what is the region of convergence.
4. Given a system with a left-sided impulse response and poles at $-2, 2$, what is the region of convergence.
5. Given a system with a two-sided impulse response and poles at $-2, -1$, what is the region of convergence.
6. You have a system with a two-sided impulse response and poles at $-2, -1$. Is this system stable?
7. You have a system with a two-sided impulse response and poles at $-2, 2$. Is this system stable?
8. You have a system with a right-sided impulse response and poles at $-2, 2$. Is this system stable?
9. By inspection what are the pole locations of the function $x(t) = e^{-5t}u(t)$?
10. By inspection what are the pole locations of the function $x(t) = e^{5t}u(t)$?
11. By inspection what are the pole locations of the function $x(t) = e^{-5t}u(-t)$?
12. By inspection what are the pole locations of the function $x(t) = e^{5t}u(-t)$?
13. By inspection what are the pole locations of the function $x(t) = e^{-5t}\cos(4t)u(t)$?

20.4 ANSWERS

1. $\sigma > -1$
2. $\sigma > 2$
3. $\sigma < -2$
4. $\sigma < -2$
5. $-2 < \sigma < -1$
6. No
7. Yes
8. No
9. -5
10. 5
11. -5
12. 5
13. $-5 \pm j4$

MODULE 21

The Inverse Laplace Transform

Unlike our discussion of the Fourier transform, for the Laplace transform you might have noticed that we have focused exclusively on the forward transform and made no mention of the inverse transform. Is that because the inverse transform is difficult or because it is easy? The answer is Yes. The integral equation to find the inverse Laplace transform is given by

$$x(t) = \frac{1}{2\pi j} \int_{\sigma - j\infty}^{\sigma + j\infty} X(s) e^{st} \, ds$$

where σ is in the region of convergence and $s = \sigma + j\omega$. Clearly not a trivial task for most cases. However, recall that in practice the kinds of Laplace transforms we are most interested in are rational polynomials.

$$X(s) = \frac{N(s)}{D(s)} = A \frac{(s - z_1)(s - z_2)\ldots(s - z_m)}{(s - p_1)(s - p_2)\ldots(s - p_n)}$$

We will see that it is really easy to find the inverse Laplace transform of rational polynomials in s using the uniqueness property of the Laplace transform. So let's begin.

If for a given $X(s)$ the degree of the numerator polynomial $N(s)$ is less than the degree of the denominator polynomial $D(s)$ we can always expand $X(s)$ in terms of the factors of $D(s)$ using partial fraction expansion. If the degree of the numerator is greater than or equal to the degree of the denominator then we can simply divide the numerator with the denominator and we will end up with a polynomial in s and a ratio of polynomials in which the degree of the denominator is greater than the degree of the numerator. Let's assume for the moment that the degree of the denominator is greater than the degree of the numerator. We will deal later with the case where this is not true. In this module we will deal with the case where the denominator polynomial $D(s)$ has distinct roots which may be real or complex. The case where the roots are repeated is also easy to deal with but we need an additional property of the Laplace transform so we will hold of on that for the moment. If the denominator polynomial only has distinct roots p_i then we can write $X(s)$ as

$$X(s) = \sum_{i=1}^{n} \frac{A_i}{s - p_i}$$

As in the case of the forward Laplace transform, the inverse Laplace transform is a linear transform so the inverse transform of a sum is the sum of the inverse transforms and

$$\mathcal{L}^{-1}[X(s)] = \mathcal{L}^{-1}\left[\sum_{i=1}^{n} \frac{A_i}{s-p_i}\right] = \sum_{i=1}^{n} A_i \mathcal{L}^{-1}\left[\frac{1}{s-p_i}\right]$$

We have previously seen the following Laplace transforms:

$$\begin{aligned} \mathcal{L}\left[e^{-at}u(t)\right] &= \frac{1}{s+a} \quad \sigma > -a \\ \mathcal{L}\left[-e^{-at}u(-t)\right] &= \frac{1}{s+a} \quad \sigma < -a \end{aligned}$$

So we can take the inverse Laplace transform of the individual terms and then use the linearity and uniqueness of the Laplace transform (being very careful with the regions of convergence) to add together the individual sequences and determine the overall inverse Laplace transform.

Let's take a look at some examples.

Example 21.1 Find the inverse Laplace transform of

$$X(s) = \frac{3s+5}{s^2+3s+2} \quad \sigma > -1$$

The first thing we do is to write the denominator polynomial in terms of its factors.

$$\begin{aligned} X(s) &= \frac{3s+5}{s^2+3s+2} \\ &= \frac{3s+5}{(s+2)(s+1)} \end{aligned}$$

Then we use partial fraction expansion

$$\frac{3s+5}{(s+2)(s+1)} = \frac{A_1}{s+2} + \frac{A_2}{s+1}$$

Let me show you how I do partial fraction expansion. There are a number of techniques that people use. Whichever one you use is fine.

To find A_1 multiply both sides by $(s+2)$ and evaluate at $s=-2$

$$\frac{3s+5}{(s+2)(s+1)}(s+2)|_{s=-2} = A_1 + \frac{A_2}{(s+1)}(s+2)|_{s=-2}$$

then

$$A_1 = \frac{3(-2)+5}{-2+1} = 1$$

To find A_2 multiply both sides by $(s+1)$ and evaluate at $s=-1$

$$\frac{3s+5}{(s+2)(s+1)}(s+1)|_{s=-1} = \frac{A_1}{s+2}(s+1)|_{s=-1} + A_2$$

or

$$A_2 = \frac{3(-1)+5}{-1+2} = 2$$

So now our problem becomes one of finding the inverse Laplace transform of

$$\frac{1}{s+2} + \frac{2}{s+1} \quad \sigma > -1$$

Based on our previous experience the inverse Laplace transform of

$$\frac{1}{s+2} \quad \sigma > -1$$

is $e^{-2t}u(t)$ and the inverse Laplace transform of

$$\frac{2}{s+1} \quad \sigma > -1$$

is $2e^{-t}u(t)$. Therefore,

$$x(t) = e^{-2t}u(t) + 2e^{-t}u(t)$$

Let's try a different example with the same rational polynomial but a different region of convergence.

Example 21.2 Find the inverse Laplace transform of

$$X(s) = \frac{3s+5}{s^2+3s+2} \quad \sigma < -2$$

We start out in exactly the same way writing the denominator polynomial in terms of its factors and then expanding using partial fraction expansion. Then we need to find the inverse Laplace transform of

$$\frac{1}{s+2} + \frac{2}{s+1} \quad \sigma < -2$$

The inverse Laplace transform of

$$\frac{1}{s+2} \quad \sigma < -2$$

is $-e^{-2t}u(-t)$. What about the inverse Laplace transform of

$$\frac{2}{s+1} \quad \sigma < -2$$

We know what the inverse Laplace transform of $1/(s+1)$ is when the region of convergence is $\sigma < -1$, but here the region of convergence is $\sigma < -2$. If we look at these two regions we can see that a transform that converges for $\sigma < -1$ will also converge for $\sigma < -2$ as -2 is less than -1. So, in this case, the inverse transform for the region of convergence $\sigma < -2$ is the same as the inverse transform for the region of convergence $\sigma < -1$ which is $-2e^{-t}u(-t)$. Therefore, the inverse Laplace transform if the ROC is $\sigma < -2$

$$x(t) = -e^{-2t}u(-t) - 2e^{-t}u(-t)$$

But what if we have a two sided region of convergence?

Example 21.3 Find the inverse Laplace transform of

$$X(s) = \frac{3s+5}{s^2+3s+2} \quad -2 < \sigma < -1$$

Again we start out in exactly the same way writing the denominator polynomial in terms of its factors and then expanding using partial fraction expansion. Then we need to find the inverse Laplace transform of

$$\frac{1}{s+2} + \frac{2}{s+1} \quad -2 < \sigma < -1$$

We again find the inverse transform of each term separately. First lets find the inverse transform of

$$\frac{1}{s+2} \quad -2 < \sigma < -1$$

But the only pole we have here is at $s = -2$. We only have two regions of convergence $\sigma > -2$ and $\sigma < -2$. The inverse we get for the region $\sigma < -2$ would not be appropriate as this transform would not converge in the region $-2 < \sigma < -1$. However the inverse we get for the region $\sigma > -2$ would converge in the region $-2 < \sigma < -1$. So, the inverse of this term will be $e^{-2t}u(t)$. For the second term

$$\frac{2}{s+1} \quad -2 < \sigma < -1$$

we use a similar argument. In this case the only pole is at $s = -1$ and, therefore, the two regions of convergence are $\sigma < -1$ and $\sigma > -1$. Anything that converges for $\sigma < -1$ will also converge for $-2 < \sigma < -1$, so we use the region of convergence $\sigma < -1$ and the inverse transform for this term is the left sided function $-2e^{-t}u(-t)$. And the inverse Laplace transform for this case is

$$x(t) = e^{-2t}u(t) - 2e^{-t}u(-t)$$

If the roots were complex instead of real nothing really changes.

Example 21.4 Find the inverse Laplace transform of

$$X(s) = \frac{s+3}{s^2+6s+45} \quad \sigma > -3$$

The first step is to find the roots of the denominator polynomial in order to find the factors. The roots of the denominator polynomial are

$$p_{1,2} = \frac{-6 \pm \sqrt{36-180}}{2} = \frac{-6 \pm j12}{2} = -3 \pm j6$$

or

$$X(s) = \frac{s+3}{(s+3-j6)(s+3+j6)} \quad \sigma > -3$$

We can expand this using partial fraction expansion to obtain

$$X(s) = \frac{1/2}{(s+3-j6)} + \frac{1/2}{(s+3+j6)} \quad \sigma > -3$$

Again these are familiar forms. The inverse transform of

$$\frac{1/2}{(s+3-j6)} \quad \sigma > -3$$

is $(1/2)e^{-(3+j6)t}u(t)$ and the inverse transform of

$$\frac{1/2}{(s+3+j6)} \quad \sigma > -3$$

is $(1/2)e^{-(3-j6)t}u(t)$. Putting them together we get

$$\begin{aligned} x(t) &= \frac{1}{2}e^{-(3+j6)t}u(t) + \frac{1}{2}e^{-(3-j6)t}u(t) \\ &= e^{-3t}u(t)\left[\frac{1}{2}e^{-j6t} + \frac{1}{2}e^{j6t}\right] \\ &= e^{-3t}\cos(6t)u(t) \end{aligned}$$

We still haven't done anything with repeated roots or with exponential terms. Let's wait until after we have seen some properties of the Laplace transform for that.

21.1 SUMMARY

In order to find the inverse Laplace transform of a rational polynomial $X(s)$ with distinct roots of the denominator polynomial and a region of convergence $\mathcal{R}$.

1. Find the partial fraction expansion of $X(s)$

$$X(s) = \sum_{i=1}^{n} \frac{A_i}{s - p_i}$$

2. For each term determine if the region of convergence $\mathcal{R}$ is to the right or left of p_i.

   (a) If $\mathcal{R}$ is to the right of pi then

$$\mathcal{L}^{-1}\left[\frac{A_i}{s - p_i}\right] = A_i e^{p_i t} u(t)$$

   (b) If $\mathcal{R}$ is to the left of p_i then

$$\mathcal{L}^{-1}\left[\frac{A_i}{s - p_i}\right] = -A_i e^{p_i t} u(-t)$$

3. The inverse transform $x(t)$ is the sum of the inverse transform of each term in the summation.

21.2 EXERCISES

(Answers on the following page)

1. Find the inverse Laplace transform of

$$X(s) = \frac{2s}{s^2 - 1} \quad \sigma > 1$$

2. Find the inverse Laplace transform of

$$X(s) = \frac{2s}{s^2 + 1} \quad \sigma > 0$$

3. Find the inverse Laplace transform of

$$X(s) = \frac{2s}{s^2 - 1} \quad \sigma < -1$$

4. Find the inverse Laplace transform of

$$X(s) = \frac{s + 5}{s^2 + 10s + 29} \quad \sigma > -5$$

5. Find the inverse Laplace transform of

$$X(s) = \frac{2s + 6}{s^2 + 6s + 10} \quad \sigma > -3$$

6. Find the inverse Laplace transform of

$$X(s) = \frac{3s + 8}{s^2 + 5s + 6} \quad \sigma > -2$$

7. Find the inverse Laplace transform of

$$X(s) = \frac{3s + 8}{s^2 + 5s + 6} \quad -3 < \sigma < -2$$

8. Find the inverse Laplace transform of

$$X(s) = \frac{2s}{s^2 - 1} \quad -1 < \sigma < 1$$

9. Find the inverse Laplace transform of

$$X(s) = \frac{3s + 6}{s^2 + 4s + 13} \quad \sigma > -2$$

10. Find the inverse Laplace transform of

$$X(s) = \frac{-2}{s^2 - 1} \quad -1 < \sigma < 1$$

21.3 ANSWERS

1.

$$x(t) = e^{-t}u(t) + e^{t}u(t)$$

2.

$$x(t) = 2\cos(t)u(t)$$

3.

$$x(t) = -e^{-t}u(-t) - e^{t}u(-t)$$

4.

$$x(t) = e^{-5t}\cos(2t)u(t)$$

5.

$$x(t) = 2e^{-3t}\cos(t)u(t)$$

6.

$$x(t) = e^{-3t}u(t) + 2e^{-2t}u(t)$$

7.

$$x(t) = e^{-3t}u(t) - 2e^{-2t}u(-t)$$

8.

$$x(t) = e^{-t}u(t) - e^{t}u(-t)$$

9.

$$x(t) = 3e^{-2t}\cos(3t)u(t)$$

10.

$$x(t) = e^{-|t|}$$

MODULE 22

Properties of Laplace Transforms

Now that we have some feel for the Laplace transform and the importance of the regions of convergence let us go back as we promised and look at some of the properties of the Laplace transform. As we had said earlier these are essentially the same properties as those of the Fourier transform with the addition of regions of convergence.

22.1 CONVOLUTION PROPERTY

As in the case of the Fourier transform, the Laplace transform of the convolution of two signals in the time domain is the product of the Laplace transform of each signal. For the product to converge, each of the transforms has to converge and, therefore, the region of convergence of the product is a region which contains the intersection of the region of convergence of the Laplace transform of each signal.

$$\mathcal{L}\{x_1(t) \circledast x_2(t)\} = X_1(s)X_2(s) \quad \text{ROC: containing } R_1 \cap R_2$$

where R_1 is the region of convergence for $X_1(s)$ and R_2 is the region of convergence for $X_2(s)$.

The proof of the convolution property for Laplace transform follows the same steps as the proof of the convolution property for Fourier transforms.

$$\begin{aligned} x_1(t) \circledast x_2(t) &= \int_{-\infty}^{\infty} x_1(\tau)x_2(t-\tau)d\tau \\ \mathcal{L}\{x_1(t) \circledast x_2(t)\} &= \int_{-\infty}^{\infty}\int_{-\infty}^{\infty} x_1(\tau)x_2(t-\tau)d\tau e^{-st}dt \end{aligned}$$

substituting $u = t - \tau$ we get

$$\begin{aligned} \mathcal{L}\{x_1(t) \circledast x_2(t)\} &= \int_{-\infty}^{\infty}\int_{-\infty}^{\infty} x_1(\tau)x_2(u)e^{-s(\tau+u)}d\tau du \\ &= \int_{-\infty}^{\infty}\left[\int_{-\infty}^{\infty} x_1(\tau)e^{-s\tau}d\tau\right]x_2(u)e^{-su}du \\ &= X_1(s)X_2(s) \end{aligned}$$

Notice that the region of convergence is one *containing* $R_1 \cap R_2$. Why "containing?" Consider the following example:

Example 22.1

$$\begin{aligned} x_1(t) &= e^{-2t}u(t) - e^{-3t}u(t) \\ x_2(t) &= 3e^{-5t}u(t) - 2e^{-4t}u(t) \end{aligned}$$

The Laplace transforms are given by

$$\begin{aligned} X_1(s) &= \frac{1}{s+2} - \frac{1}{s+3} \quad \text{ROC}: \ \sigma > -2 \\ &= \frac{1}{(s+2)(s+3)} \quad \text{ROC}: \ \sigma > -2 \\ X_2(s) &= \frac{3}{s+5} - \frac{2}{s+4} \quad \text{ROC}: \ \sigma > -4 \\ &= \frac{s+2}{(s+4)(s+5)} \quad \text{ROC}: \ \sigma > -4 \end{aligned}$$

where the regions of convergence are determined by the rightmost poles. The intersection of the two regions would be the region > -2. However, when we multiply $X_1(s)$ and $X_2(s)$, the $(s+2)$ factor cancels out; and the rightmost pole is the pole at $s = -3$.

$$X_1(s)X_2(s) = \frac{1}{(s+2)(s+3)} \cdot \frac{s+2}{(s+4)(s+5)} = \frac{1}{(s+3)(s+4)(s+5)} \quad \text{ROC}: \ \sigma > -3$$

This pole zero cancellation is what causes the "containing" formulation—a pole that is restricting the region of convergence in one of the Laplace transforms may get cancelled out by a zero in the other Laplace transform.

As in the case of Fourier transforms, the convolution property allows us to relate the input and output of linear time-invariant systems through the transfer function which, in this case, is the Laplace transform of the impulse response. Given a linear time invariant system with impulse response $h(t)$, the output $y(t)$ is given by the convolution of the input $x(t)$ with the impulse response.

$$y(t) = x(t) \circledast h(t)$$

Hence,

$$Y(s) = X(s)H(s)$$

where $Y(s)$ is the Laplace transform of the output, $X(s)$ is the Laplace transform of the input and $H(s)$ is the Laplace transform of the impulse response.

22.2 TIME SHIFTING

A shift in the time domain, as in the case of the Fourier transform, results in the Laplace transform of the shifted signal being a product of the Laplace transform of the original signal and a complex exponential. The difference is that the complex exponential instead of being $e^{-j\omega t_o}$ is e^{-st_o}.

$$\mathcal{L}\{x(t-t_o)\} = e^{-st_o}X(s) \quad \text{ROC} : R$$

where R is the region of convergence for $X(s)$. Notice that the region of convergence did not change. The reason for this is because the multiplication with e^{-st_o} does not introduce any new poles or change the location of the poles in the original transform. As poles determine the boundaries of the region of convergence no change in the pole locations means no change in the boundaries of the region of convergence.

The proof is quite simple.

$$\mathcal{L}[x(t-t_o)] = \int_{-\infty}^{\infty} x(t-t_o)e^{-st}\,dt$$

Substitute $\tau = t - t_o$ which also means that $t = \tau + t_o$ into this equation and we get

$$\begin{aligned}
\mathcal{L}[x(t-t_o)] &= \int_{-\infty}^{\infty} x(\tau)e^{-s(\tau+t_o)}\,d\tau \\
&= \int_{-\infty}^{\infty} x(\tau)e^{-s\tau}e^{-st_o}\,d\tau \\
&= e^{-st_o}\int_{-\infty}^{\infty} x(\tau)e^{-s\tau}\,d\tau \\
&= e^{-st_o}X(s)
\end{aligned}$$

In the last module we had looked at how to find the inverse Laplace transform when $X(s)$ is a rational polynomial. Using this property we can extend our ability to find the Laplace transform to situations where we have complex exponents as well.

Example 22.2 Let's find the inverse Laplace transform of

$$X(s) = \frac{2se^{-s} + 4e^{-s}}{s^2 + 4s + 40}; \quad \sigma > -2$$

We can rewrite this as

$$X(s) = X_1(s)e^{-s}; \quad \sigma > -2$$

where

$$X_1(s) = \frac{2s + 4}{s^2 + 4s + 40}; \quad \sigma > -2$$

We can go through our standard routine to find the inverse Laplace transform of $X_1(s)$ and then use the time shifting property to find the inverse Laplace transform of $X(s)$. First we find the roots of the denominator.

$$X_1(s) = \frac{2s+4}{(s+2-j6)(s+2+j6)}; \quad \sigma > -2$$

Then we expand $X_1(s)$ using partial fraction expansion.

$$X_1(s) = \frac{1}{s+2-j6} + \frac{1}{s+2+j6}; \quad \sigma > -2$$

Taking the inverse transform of each term and then adding them together we get

$$x_1(t) = 2e^{-2t}\cos(6t)u(t)$$

Because $X(s) = X_1(s)e^{-s}$ we can use the time shifting property to find $x(t)$ as

$$x(t) = x_1(t-1) = 2e^{-2(t-1)}\cos(6(t-1))u(t-1)$$

22.3 SHIFTING IN THE S-DOMAIN

Just as multiplication with a complex exponential in Laplace domain results in a shift in the time domain, multiplication with a complex exponential in the time domain results in a shift in the Laplace domain.

$$\mathcal{L}\left\{e^{s_o t}x(t)\right\} = X(s - s_o) \quad \text{ROC}: R + \sigma_o$$

where σ_o is the real part of s_o. To show this let's evaluate the Laplace transform integral

$$\begin{aligned}\mathcal{L}\left\{e^{s_o t}x(t)\right\} &= \int_{-\infty}^{\infty} e^{s_o t}x(t)e^{-st}\,dt \\ &= \int_{-\infty}^{\infty} x(t)e^{-(s-s_o)t}\,dt\end{aligned}$$

We can see that if this integral converges it will result in $X(s - s_o)$. Let's see the conditions under which it converges. The Laplace transform of $x(t)$ converges for $s \in R$, therefore this integral will converge for $s - s_o \in R$. We know the region of convergence is going to be one of four possible regions

All of the s-plane	
$\sigma > \alpha$	for some real α
$\sigma < \alpha$	for some real α
$\alpha < \sigma < \beta$	for some real α and β

To find the region of convergence for the Laplace transform of $e^{s_o t}x(t)$ all we need to do is replace σ with $\sigma - \sigma_o$. This will change the region of convergence

$$\begin{aligned}
\text{All of the } s\text{-plane} &\Rightarrow \text{All of the } s\text{-plane} \\
\sigma > \alpha \quad \text{for some real } \alpha &\Rightarrow \sigma > \alpha + \sigma_o \quad \text{for some real } \alpha \\
\sigma < \alpha \quad \text{for some real } \alpha &\Rightarrow \sigma < \alpha + \sigma_o \quad \text{for some real } \alpha \\
\alpha < \sigma < \beta \quad \text{for some real } \alpha \text{ and } \beta &\Rightarrow \alpha + \sigma_o < \sigma < \beta + \sigma_o \quad \text{for some real } \alpha \text{ and } \beta
\end{aligned}$$

To write it more concisely, the new region of convergence will be $R + \sigma_o$.

Multiplication with an exponential causes a shift in the s-domain corresponding to the real part of the exponent. This results in a movement of the poles and zeros in the s-plane. The movement can take a stable system and make it unstable or vice versa.

Example 22.3 Consider the time function

$$x(t) = e^{-2t}u(t)$$

with Laplace transform

$$X(s) = \frac{1}{s+2} \qquad \sigma > -2$$

If $x(t)$ is the impulse response of a linear time invariant system you can clearly see that the system is stable by looking at the time domain response (a decaying exponential) or the s-domain function (the region of convergence includes the $j\omega$ axis). If we multiply $x(t)$ by e^{3t} by this property the Laplace transform will be

$$X(s-3) = \frac{1}{s-3+2} = \frac{1}{s-1} \qquad \sigma > 1$$

The pole shifts from -2 to 1 as does the boundary of the region of convergence. The region of convergence no longer includes the $j\omega$ axis so the system is no longer stable. We could also see this in the time domain

$$e^{3t}x(t) = e^{3t}e^{-2t}u(t) = e^{t}u(t)$$

which is a growing exponential.

We can also go the other way. If we have a system which is not stable—say a causal system whose transfer function contains poles in the right half s-plane, we can use this property to shift the pole over the $j\omega$ axis into the left half s-plane and thus make the system stable. We will come back to this idea of shifting in the s-domain when we discuss feedback systems.

22.4 TIME SCALING

Scaling in the case of the Laplace transform works exactly the same as scaling in the case of the Fourier transform with the additional complication of the region of convergence. Let's start with the simple case of the Laplace transform of $x(\alpha t)$ where $\alpha > 0$.

$$\mathcal{L}[x(\alpha t)] = \int_{-\infty}^{\infty} x(\alpha t)e^{-st}dt$$

Let's define $\tau = \alpha t$. As $\alpha > 0$ this won't effect the limits—$\alpha \times \infty = \infty$—and $d\tau = \alpha dt$. Substituting these into the integral we get

$$\begin{aligned}\mathcal{L}[x(\alpha t)] &= \int_{-\infty}^{\infty} x(\tau)e^{-s\frac{\tau}{\alpha}}\frac{1}{\alpha}d\tau \\ &= \frac{1}{\alpha}\int_{-\infty}^{\infty} x(\tau)e^{-\frac{s}{\alpha}\tau}d\tau\end{aligned}$$

If this equation converges the result will be $X(\frac{s}{\alpha})$ where $X(s)$ is the Laplace transform of $x(t)$. If the region of convergence for $X(s)$ was $\sigma > a$ the region of convergence for $X(\frac{s}{\alpha})$ will be $\frac{s}{\alpha} > a$ or $\sigma > \alpha a$. Similarly if the region of convergence for $X(s)$ was $\sigma < a$ the region of convergence for $X(\frac{s}{\alpha})$ will be $\sigma < \alpha a$. Finally, if $b < \sigma < a$ is the region of convergence for $X(s)$, the region of convergence for $X(\frac{s}{\alpha})$ is $\alpha b < \sigma < \alpha a$. But this was all for $\alpha > 0$. What happens if $\alpha < 0$?

First our variable substitution doesn't go as smoothly as before. When we substitute $\tau = \alpha t$ this effects the limits. The lower limit for the integral with respect to t was $-\infty$. With respect to τ, because t is being multiplied by a negative number the lower limit will be $+\infty$. Similarly the upper limit will go from being $+\infty$ to $-\infty$. We can flip the limits back to the original form by multiplying the integral with negative one. So the transform will now be

$$\begin{aligned}\mathcal{L}[x(\alpha t)] &= \int_{\infty}^{-\infty} x(\tau)e^{-s\frac{\tau}{\alpha}}\frac{1}{\alpha}d\tau \\ &= -\frac{1}{\alpha}\int_{-\infty}^{\infty} x(\tau)e^{-\frac{s}{\alpha}\tau}d\tau\end{aligned}$$

Noting that

$$\frac{1}{|\alpha|} = \begin{cases} \frac{1}{\alpha} & \text{for } \alpha > 0 \\ -\frac{1}{\alpha} & \text{for } \alpha < 0 \end{cases}$$

We can combine the case for $\alpha > 0$ and $\alpha < 0$ into

$$\mathcal{L}[x(\alpha t)] = \frac{1}{|\alpha|}\int_{-\infty}^{\infty} x(\tau)e^{-\frac{s}{\alpha}\tau}d\tau$$

which is equal to $\frac{1}{|\alpha|}X(\frac{s}{\alpha})$ when the integral converges.

The region of convergence when $\alpha < 0$ will also change from the case when $\alpha > 0$. Where $X(s)$ converged for $\sigma > a$, $X(\frac{s}{\alpha})$ will converge for $\sigma < \alpha a$ and where $X(s)$ converged for $\sigma < a$, $X(\frac{s}{\alpha})$ will converge for $\sigma > \alpha a$. We put all this together and simply say that

$$\mathcal{L}\{x(\alpha t)\} = \frac{1}{|\alpha|} X\left(\frac{s}{\alpha}\right) \quad \text{ROC} : \alpha R$$

where R is the region of convergence for $X(s)$.

Scaling will also effect pole zero locations. If $s = a$ was a pole or zero for $X(s)$, $s = \alpha a$ will be the pole or zero for $X(\frac{s}{\alpha})$. This effect is particularly drastic when α is negative because it flips the poles and zeros from the right half plane to the left half plane and vice versa.

Example 22.4 Suppose $x(t)$ is a function with Laplace transform

$$X(s) = \frac{s+1}{(s+2)(s+4)}; \quad \sigma > -2$$

Clearly if $x(t)$ is the impulse response of a system the system is both causal and stable. Let us take a look at what happens when we scale $x(t)$ with $1/2$ and $-1/2$.

$$\begin{aligned}
\mathcal{L}\{x(t/2)\} &= 2X(2s); \quad \sigma > -1 \\
&= 2\frac{2s+1}{(2s+2)(2s+4)}; \quad \sigma > -1 \\
&= \frac{s+1/2}{(s+1)(s+2)}; \quad \sigma > -1
\end{aligned}$$

The system with impulse response $x(t/2)$ would still be a causal stable system albeit with poles and zero closer to the $j\omega$ axis. When we scale by $-1/2$ the situation changes

$$\begin{aligned}
\mathcal{L}\{x(-t/2)\} &= 2X(-2s); \quad \sigma < 1 \\
&= 2\frac{-2s+1}{(-2s+2)(-2s+4)}; \quad \sigma < 1 \\
&= -\frac{s-1/2}{(s-1)(s-2)}; \quad \sigma < 1
\end{aligned}$$

The system with impulse response $x(-t/2)$ is still stable as the $j\omega$ axis is still in the region of convergence but it is no longer causal.

22.5 DIFFERENTIATION IN THE TIME DOMAIN

To find the Laplace transform of the derivative of a function $x(t)$ with Laplace transform $X(s)$ with region of convergence R is relatively simple (assuming all derivatives and integrals involved exist). We begin with the formal definition of the inverse Laplace transform.

$$x(t) = \frac{1}{2\pi j}\int_{\sigma-j\infty}^{\sigma+j\infty} X(s)e^{st}\,ds$$

Taking the derivative with respect to time of both sides

$$\begin{aligned}\frac{d}{dt}x(t) &= \frac{d}{dt}\left[\frac{1}{2\pi j}\int_{\sigma-j\infty}^{\sigma+j\infty} X(s)e^{st}\,ds\right]\\ &= \frac{1}{2\pi j}\int_{\sigma-j\infty}^{\sigma+j\infty} X(s)\frac{d}{dt}e^{st}\,ds\\ &= \frac{1}{2\pi j}\int_{\sigma-j\infty}^{\sigma+j\infty} sX(s)e^{st}\,ds\end{aligned}$$

or in other words,

$$\mathcal{L}\left\{\frac{dx(t)}{dt}\right\} = sX(s) \quad \text{ROC : containing } R$$

but why "containing" R?

This property is widely used when solving differential equations. Differential equations have been a common way of representing the dynamics of systems. The reason for that word is that if $X(s)$ happened to have a pole at 0 which was a boundary of the region of convergence the additional factor of s in the numerator might cancel that pole out resulting in a bigger region of convergence which contains the original region of convergence.

Example 22.5 Consider the continuous-time linear time-invariant system with impulse response $h(t)$ and transfer function $H(s)$ for which the input $x(t)$ and the output $y(t)$ are related by the differential equation

$$\frac{d^2y(t)}{dt^2} + 2\frac{dy(t)}{dt} - 8y(t) = 2\frac{dx(t)}{dt} + 5x(t)$$

Let's assume that the initial conditions are all zero. We can find the transfer function for this system by taking the Laplace transform of both sides of the differential equation and using the differentiation property.

$$s^2Y(s) + 2sY(s) - 8Y(s) = 2sX(s) + 5X(s)$$

or

$$Y(s)\left(s^2+2s-8\right)=X(s)(2s+5)$$

Therefore,

$$H(s)=\frac{Y(s)}{X(s)}=\frac{2s+5}{(s-4)(s+2)}$$

22.6 DIFFERENTIATION IN THE S-DOMAIN

$$\mathcal{L}\{-tx(t)\}=\frac{dX(s)}{ds}\quad \text{ROC}: R$$

To show this start with the definition of the Laplace transform

$$X(s)=\int_{-\infty}^{\infty}x(t)e^{-st}dt$$

Taking the derivative with respect to s on both sides and switching the order of integration and differentiation we get

$$\begin{aligned}\frac{d}{ds}X(s) &= \frac{d}{ds}\int_{-\infty}^{\infty}x(t)e^{-st}dt\\ &= \int_{-\infty}^{\infty}\frac{d}{ds}\left(x(t)e^{-st}\right)dt\\ &= \int_{-\infty}^{\infty}\left(-tx(t)\right)e^{-st}dt\end{aligned}$$

or

$$\mathcal{L}\{-tx(t)\}=\frac{dX(s)}{ds}$$

Taking the derivative of a rational polynomial will not change the location of the roots in the denominator so the region of convergence does not change. This property comes in handy when we have factors involving t^n in $x(t)$

Example 22.6 Find the Laplace transform of

$$x(t)=te^{-2t}u(t)$$

We know that

$$\mathcal{L}\left[e^{-2t}u(t)\right]=\frac{1}{s+2};\quad \sigma>-2$$

and

$$\frac{d}{ds}\left[\frac{1}{s+2}\right]=\frac{-1}{(s+2)^2}$$

Therefore, using the differentiation property

$$\mathcal{L}\left[te^{-2t}u(t)\right] = \frac{1}{(s+2)^2}; \quad \sigma > -2$$

22.7 INTEGRATION IN THE TIME DOMAIN

To find out what happens in the Laplace domain when we integrate in the time domain we take a slightly circuitous route. What we want to find is

$$\mathcal{L}\left[\int_{-\infty}^{t} x(\tau)d\tau\right] = ?$$

Instead what we will do is find the Laplace transform of $x(t) \circledast u(t)$. We know that the Laplace transform of a convolution is the product of the individual Laplace transforms.

$$\mathcal{L}\left[x(t) \circledast u(t)\right] = X(s)\mathcal{L}\left[u(t)\right]$$

So what is the Laplace transform of $u(t)$? We know that

$$\mathcal{L}\left[e^{-at}u(t)\right] = \frac{1}{s+a}; \quad \sigma > -a$$

If we pick a to be zero then e^{-at} is equal to 1 and we find the Laplace transform of $u(t)$. That is

$$\mathcal{L}\left[u(t)\right] = \frac{1}{s}; \quad \sigma > 0$$

The final piece of the puzzle is the convolution integral

$$x(t) \circledast u(t) = \int_{-\infty}^{\infty} x(\tau)u(t-\tau)d\tau = \int_{-\infty}^{t} x(\tau)d\tau$$

where we have used the fact that $u(t-\tau)$ is 1 for $\tau < t$ and 0 for $\tau > t$.

Putting all of this together we get

$$\mathcal{L}\left[\int_{-\infty}^{t} x(\tau)d\tau\right] = \frac{1}{s}X(s) \quad \text{ROC} : R \cap \{\sigma > 0\}$$

Notice that in this case we do increase the number of poles. The $1/s$ term introduces a pole at $s = 0$. Hence, the region of convergence shrinks to either the right or left of the line $\sigma = 0$ depending on whether we have a left sided or right sided function.

22.8 SUMMARY

In this module we looked at the following properties of the Laplace transform:

1. **Convolution property:** The Laplace transform of the convolution of two signals is the product of the Laplace transforms of the individual signals with a region of convergence which contains the intersection of the regions of convergence of the individual Laplace transforms.

$$\mathcal{L}\{x_1(t) \circledast x_2(t)\} = X_1(s)X_2(s) \quad ROC: \text{containing } R_1 \cap R_2$$

2. **Time shifting:** If we introduce a shift in the time domain representation of a signal this shows up as a complex exponential factor in the Laplace domain. The region of convergence is not effected.

$$\mathcal{L}\{x(t-t_o)\} = e^{-st_o}X(s) \quad ROC: R$$

3. **Frequency shifting:** If we multiply the time signal by a complex exponential this results in a shift in the Laplace domain corresponding to the real part of the exponent. This shift is also reflected in the poles and zeros and hence in the region of convergence.

$$\mathcal{L}\left\{e^{s_o t}x(t)\right\} = X(s-s_o) \quad ROC: R+\sigma_o$$

4. **Time scaling:** Scaling a function in the time domain results in a movement in the pole zero locations and hence the region of convergence.

$$\mathcal{L}\{x(\alpha t)\} = \frac{1}{|\alpha|}X(\frac{s}{\alpha}) \quad ROC: \alpha R$$

5. **Time differentiation:** Differentiation of a signal in the time domain results in a multiplication of the Laplace transform of the signal with s.

$$\mathcal{L}\left\{\frac{dx(t)}{dt}\right\} = sX(s) \quad ROC: \text{containing } R$$

6. **Frequency differentiation:** Multiplication of a signal in time with $-t$ corresponds to the differentiation of its Laplace transform with respect to s.

$$\mathcal{L}\{-tx(t)\} = \frac{dX(s)}{ds} \quad ROC: R$$

7. **Integration in time:** Integration in the time domain results in the division of the Laplace transform of the signal by s.

$$\mathcal{L}\left[\int_{-\infty}^{t} x(\tau)d\tau\right] = \frac{1}{s}X(s) \quad ROC: R \cap \{\sigma > 0\}$$

22.9 EXERCISES

(Answers on the following page)

1. Given the input $x(t) = e^{-2t}u(t)$ and a linear time-invariant system with transfer function

$$H(s) = \frac{1}{s+1}$$

   Find the output $y(t)$.

2. The input $x(t)$ and the output $y(t)$ of a linear time invariant system are given by

$$\begin{aligned} x(t) &= e^{-2t}u(t) \\ y(t) &= e^{-2t}u(t) + e^{-3t}u(t) \end{aligned}$$

   Find the transfer function $H(s)$.

3. Using the differentiation in time property find the Laplace transform of

$$x(t) = te^{-2t}u(t)$$

4. Use the fact that

$$\mathcal{L}\left[e^{-at}u(t)\right] = \frac{1}{s+a}$$

   and the time shift property to find the Laplace transform of

$$x(t) = u(t) - u(t-4)$$

   Confirm your result by direct evaluation of the Laplace transform.

5. The poles of the Laplace transform of $x(t)$ are at -2 and -4. Where are the poles of the Laplace transform of $x(-t)$?

6. If $x(t)$ is an even function of time what can you say about the poles of $x(t)$ and $x(-t)$?

7. The input output relationship for a linear time invariant system is given by

$$\frac{d^2y(t)}{dt^2} + 3\frac{dy(t)}{dt} + 2y(t) = \frac{dx(t)}{dt} + 5x(t)$$

   Find the transfer function for this system.

8. Given a linear time invariant system described by the differential equation

$$\frac{d^2y(t)}{dt^2} + 3\frac{dy(t)}{dt} + 2y(t) = \frac{dx(t)}{dt} + 5x(t)$$

   and an input $e^{-5t}u(t)$ what is the output $y(t)$?

9. Use the integration property to find the Laplace transform of $tu(t)$.

10. A linear time invariant system has an impulse response $e^{-2t}u(t)$. What is the output of this system if the input is the unit step function $u(t)$?

22.10 ANSWERS

1.
$$y(t) = e^{-t}u(t) - e^{-2t}u(t)$$

2.
$$H(s) = \frac{2s + 5}{s + 3}$$

3.
$$X(s) = \frac{1}{(s + 2)^2}$$

4.
$$X(s) = \frac{1}{s}\left[1 - e^{-4s}\right]$$

5. 2, 4

6. They are the same.

7.
$$H(s) = \frac{s + 5}{s^2 + 3s + 2}$$

8.
$$y(t) = e^{-t}u(t) - e^{-2t}u(t)$$

9.
$$\frac{1}{s^2} \quad \sigma > 0$$

10.
$$\frac{1}{2}\left(1 - e^{-2t}\right)$$

MODULE 23

Analysis of Systems with Feedback

If we examine any complex system we will find that the behavior of the system depends on the existence of feedback mechanisms. A feedback is a measure of some or all of the outputs of a system which is then used to affect the input to the system. This feedback can be positive or negative. The system can be as simple as the heating system in your house, or more complicated like the cruise control on your car or much more complicated like the economy or the political system of a country or probably the most complicated system we deal with regularly—our own body. We can use feedback to make an unstable system stable, to reduce the sensitivity of the system to parameter variations, to make the system more responsive, or to construct the inverse of a system. We can also make use of feedback to make sure that the system under consideration is producing a desired output. To study all of this in any detail requires a course of its own but we can briefly look at some of these uses using some simple examples

23.1 STABILIZING A SYSTEM

Consider a really simple unstable system, a system with impulse response $e^t u(t)$. The output of this system can increase exponentially. We could control this exponential growth by measuring the output and when the output increased we could reduce the input by a commensurate amount. A block diagram of this conceptual system is shown in Figure 23.1. The system in this form is a feedback system with a negative feedback. We have represented the system we are attempting to stabilize—the system with impulse response $e^t u(t)$—with the transfer function where the transfer function of this system is

$$H(s) = \frac{1}{s-1}; \quad \sigma > 1$$

How do we pick what values of K will allow the system to be stable? Here is where we can use the Laplace domain view of the system. We know the response is right-sided, therefore, the region of convergence will be to the right of a boundary. To be stable this region of convergence has to include the $j\omega$ axis. So we need to pick the gain K in order to make sure that the poles of the overall system are in the left half plane. In order to do that we have to first find the transfer

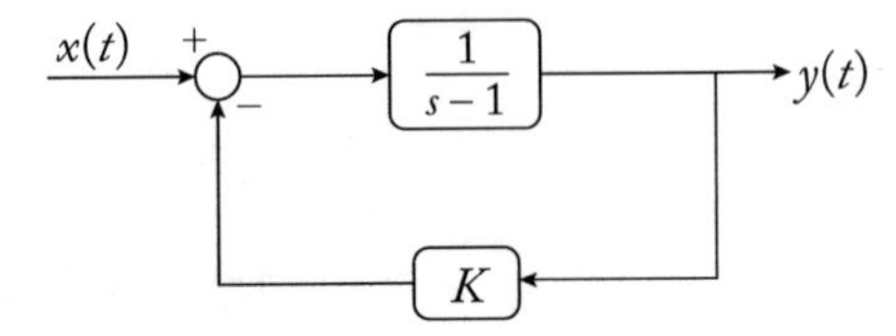

Figure 23.1: A simple feedback system.

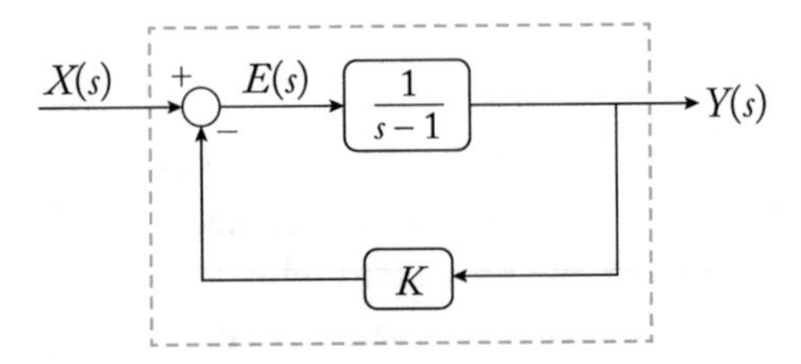

Figure 23.2: A simple feedback system redrawn.

function of the overall system. Let's use the labeled version of the system, shown in Figure 23.2, to compute the transfer function.

In the figure we have replaced the time domain expression of the input and output with their Laplace transform and we have added another label $E(s)$ in the figure. This strategy of adding labels, or defining additional variables, in complicated systems makes these systems much easier to analyze. At first this seems a bit counterintuitive as we are increasing the number of variables but as we will see in the end it simplifies the analysis problem considerably.

So let's find the transfer function of the overall system—the system outlined by the dashed box. We can see that $E(s)$ is the difference between the input $X(s)$ and the feedback. The feedback in turn is the gain K times the output $Y(s)$. Or

$$E(s) = X(s) - KY(s)$$

The output is given by

$$Y(s) = E(s) \cdot \frac{1}{s-1}$$

Substituting for $E(s)$ we get

$$\begin{aligned} Y(s) &= (X(s) - KY(s)) \cdot \frac{1}{s-1} \\ &= \frac{X(s)}{s-1} - \frac{KY(s)}{s-1} \end{aligned}$$

Collecting all the $Y(s)$ terms on the lefthand side we get

$$Y(s) + \frac{KY(s)}{s-1} = \frac{X(s)}{s-1}$$

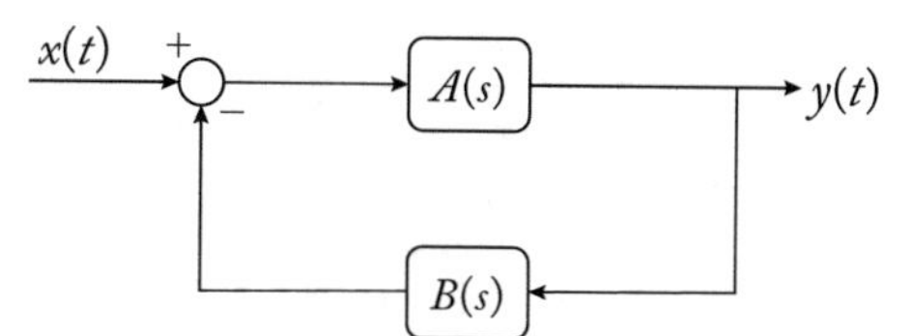

Figure 23.3: A general feedback system.

or

$$Y(s)\left[1+\frac{K}{s-1}\right]=\frac{X(s)}{s-1}$$

Combining the terms on the left hand side

$$Y(s)\cdot\frac{s-1+K}{s-1}=\frac{X(s)}{s-1}$$

and

$$H(s)=\frac{Y(s)}{X(s)}=\frac{1}{s-1+K}$$

This system has a pole at $1-K$. If we now pick $K>1$ then that pole will lie in the left half plane and the system will be stable.

Notice that we haven't changed the system in any fundamental way apart from making it stable. We initially had a system with an exponential response. We still have a system with an exponential response. Except now, if we pick $K>1$, the response is a decaying exponential.

Let's repeat this with more general components. Suppose the transfer function of the system we want to control is $A(s)$ and when we feed the output back we not only amplify or attenuate it we also shape it by passing it through a system with transfer function $B(s)$. This more general system is shown in Figure 23.3. Let's repeat the steps above to calculate the overall transfer function. As before, it is convenient in calculating the overall transfer function $H(s)$ to define a variable $E(s)$ at the output of the summer.

$$E(s)=X(s)-B(s)Y(s)$$

Then

$$Y(s)=E(s)A(s)$$

Substituting for $E(s)$

$$\begin{aligned}Y(s)&=(X(s)-B(s)Y(s))\,A(s)\\&=A(s)X(s)-A(s)B(s)Y(s)\end{aligned}$$

and ...

$$H(s)=\frac{A(s)}{1+A(s)B(s)}$$

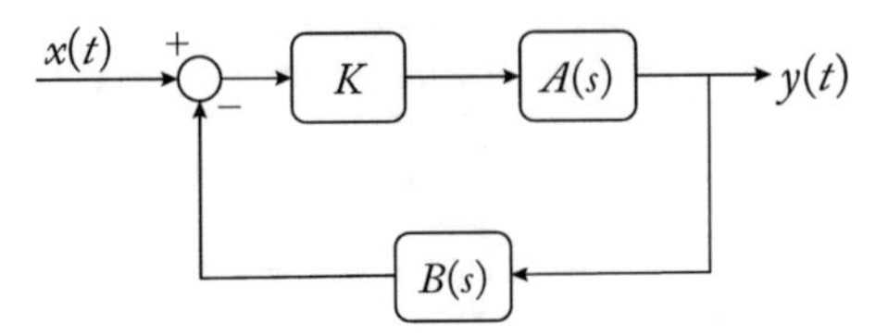

Figure 23.4: A general feedback system.

This is a much more complicated result compared to our initial system. Depending on the poles and zeros of $B(s)$ the new system may have a frequency response which is much different from the original system. We can pick $B(s)$ such that the overall system response is what we want.

Another commonly used variation of a general system is shown in Figure 23.4. If we go through our previous analysis to find the overall transfer function we get

$$H(s) = \frac{KA(s)}{1 + KA(s)B(s)}$$

In our first system increasing the value of the gain K moved the poles further to the right. In this system depending on the poles and zeros of $KA(s)B(s)$, increasing the gain can have a number of different consequences as the poles of the overall system move around. This movement of the poles is summarized in a diagram called the *root locus* diagram. We defer further discussion of this to your control system's class.

23.2 MAKING A SYSTEM MORE RESPONSIVE

This is almost a cheat as it is simply a variation of the previous idea. Because we can move the pole around using feedback we can make the system more responsive. Consider a single pole system with impulse response

$$a(t) = e^{-2t}u(t)$$

and suppose we use a simple gain in the feedback. So in terms of the system of Figure 23.3 we have

$$\begin{aligned} A(s) &= \frac{1}{s+2}; \quad \sigma > -2 \\ B(s) &= K \end{aligned}$$

The overall transfer function for this system would be

$$\begin{aligned} H(s) &= \frac{\frac{1}{s+2}}{1 + K\frac{1}{s+2}} \\ &= \frac{1}{s+2+K}; \quad \sigma > -(2+K) \end{aligned}$$

The response for this system is

$$h() = e^{-(K+2)t}u(t)$$

Depending on the value of K this exponential decays faster than the original response.

23.3 IMPLEMENTING THE INVERSE OF A SYSTEM

Suppose we want to cancel out the effect of a system. Finding the inverse system is not always easy but we can use the properties of feedback to obtain a close approximation to the inverse. Let's suppose the system whose inverse we want to obtain has a transfer function $G(s)$. In the system of Figure 23.3 set

$$\begin{aligned} A(s) &= K \\ B(s) &= G(s) \end{aligned}$$

The transfer function of the overall system then becomes

$$H(s) = \frac{K}{1 + KG(s)}$$

For K sufficiently large such that $KG(s) >> 1$ we can approximate the denominator $1 + KG(s)$ by $KG(s)$ and

$$H(s) \approx \frac{1}{G(s)}$$

23.4 PRODUCING A DESIRED OUTPUT

Often what we want from a system is to control a system to generate a prescribed output. There are many such systems you use in your daily life. Consider your heating system. You set the temperature you want and your heater or air conditioner cycles on and off to achieve that temperature. Or consider cruise control. Here you set a desired speed. If the measured speed is less than the desired speed the error signal (the difference between the desired and measured output) is positive and that might result in an action similar to increasing the pressure on the accelerator. If the measured output is larger than the desired output the error is negative and this might lead to actions to slow the car down. This is the realm of control systems. To make what we are trying to do more clear in the control system literature $B(s)$ is generally placed in-line with $A(s)$ as shown in Figure 23.5. In the control system literature the system with transfer function $A(s)$ is often called the *plant* that needs to be controlled, while the system with transfer function $B(s)$ is called the *controller*. Notice that what is driving the controller is the difference between the desired signal $x(t)$ and the measured signal $y(t)$. In the cruise control example the plant is your car and the output is the measured speed of the car.

Let's derive the overall transfer function $H(s)$. For this configuration

$$E(s) = X(s) - Y(s)$$

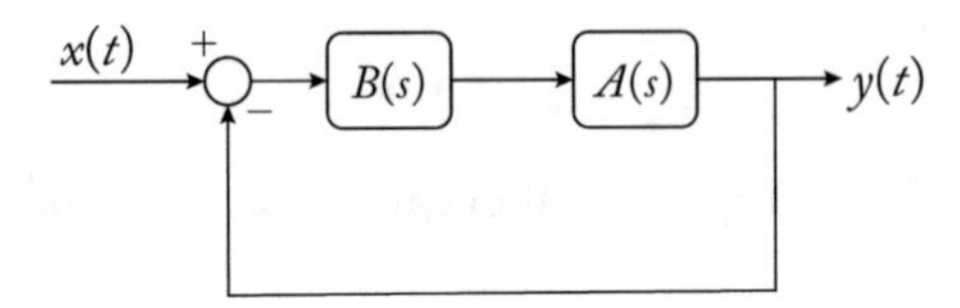

Figure 23.5

and

$$Y(s) = A(s)B(s)E(s)$$

Plugging in the expression for $E(s)$.

$$\begin{aligned} Y(s) &= A(s)B(s)\left(X(s) - Y(s)\right) \\ &= A(s)B(s)X(s) - A(s)B(s)Y(s) \end{aligned}$$

Moving $A(s)B(s)Y(s)$ over to the left hand side we get

$$Y(s) + A(s)B(s)Y(s) = A(s)B(s)X(s)$$

or

$$Y(s)(1 + A(s)B(s)) = A(s)B(s)X(s)$$

and

$$H(s) = \frac{A(s)B(s)}{1 + A(s)B(s)}$$

If we used our previous examples of $A(s)$ and $B(s)$ we get

$$H(s) = \frac{K\frac{1}{s-1}}{1 + K\frac{1}{s-1}}$$

Multiplying through by $s - 1$ we get

$$H(s) = \frac{K}{s - 1 + K}$$

As in the previous configuration the pole is now at $1 - K$ and by setting $K > 1$ we can make this system stable.

In this particular case we used the controller $B(s) = K$. This meant that the control signal sent to the plant was proportional to the error signal. This form of a controller is called a *proportional controller*. When the output deviates from the set-point, a signal proportional to the amount of deviation controls the plant.

In some systems the (proportional) amount of error may be too small to effect the plant and the error can achieve a nonzero steady state value. If you want the error between the desired signal and the measured signal to go to zero one way you can do that is to add an integrator to the controller. The integrator will integrate the error signal and even when the error signal is small over a period of time the cumulative effect will be large. Recall that integration of a function in the time domain corresponds to multiplication with $1/s$ in the Laplace domain. So we set the controller to be

$$B(s) = K_p + K_i \frac{1}{s}$$

where K_p is the proportional gain and K_i is the gain for the integrator. For obvious reasons this kind of controller is called a *PI* controller.

Both the proportional controller and the PI controller react to the error signal. If we want some level of anticipatory control we can add a derivative term as well. The derivative measures the slope of the error signal. If the slope is high this would point to an imminent sharp change in the error signal and the controller can begin to compensate for that sharp change. A derivative in the time domain corresponds to multiplication in the frequency domain so the controller becomes

$$B(s) = K_p + \frac{1}{s} K_i + s K_d$$

This kind of controller is called a PID controller.

Example 23.1 Cruise Control Let's simulate a simplified version of the cruise control to see the effect of the different gain parameters. There are a lot of factors that go into modeling a car including the mass of the car, drag on the car, the rolling friction, lift, etc. We will use a very simplified model where the force exerted by the engine is used to overcome friction which will be proportional to the speed of the car and move the car. The force exerted by the engine is the input $f(t)$ and the velocity of the car $v(t)$ is the output. This output is measured and subtracted from the set point of the cruise control to form the input of the PID controller. The output of the controller is the force exerted on the car.

The equation relating the force exerted on the car and its speed is

$$f(t) - cv(t) = ma(t)$$

where c is the damping constant. We can rewrite the acceleration as a derivative of the velocity and rewrite this equation as

$$m \frac{dv}{dt} + cv(t) = f(t)$$

Taking the Laplace transform of both sides we can find the transfer function of the car as

$$A(s) = \frac{1}{ms + c}; \quad \sigma > -c/m$$

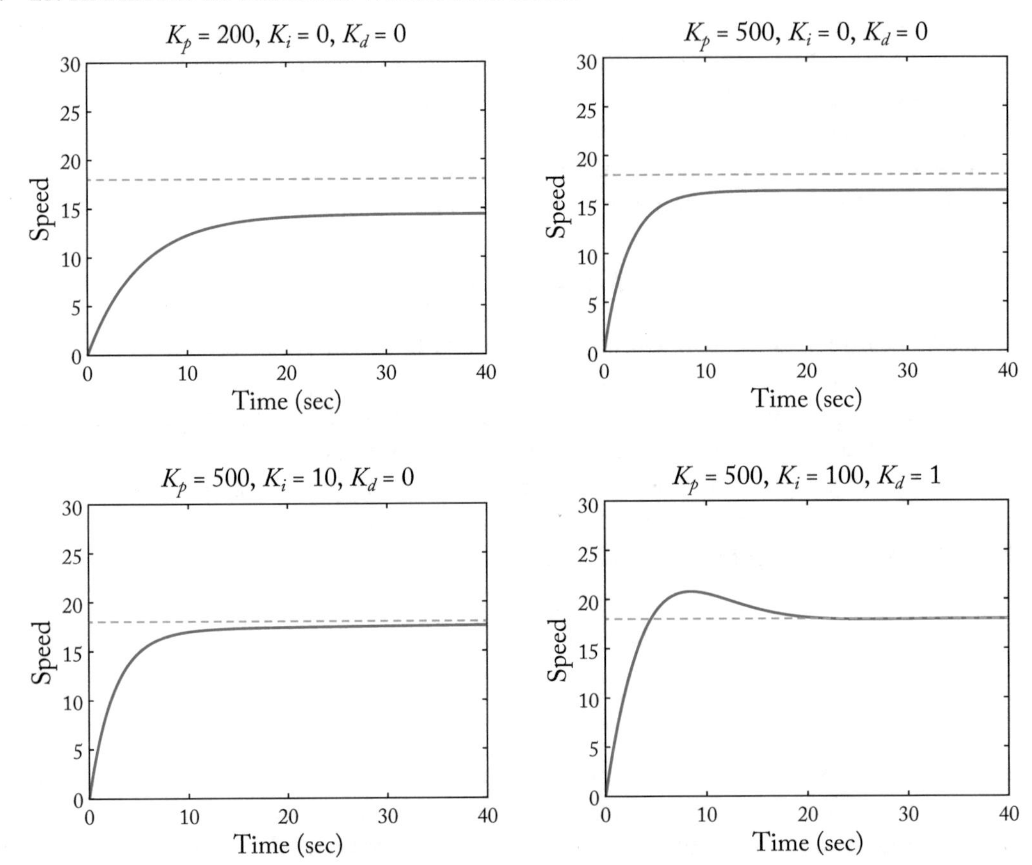

Figure 23.6: Behavior of a simple cruise control system.

The mass of my 2006 Prius is 1310 kg, and a reasonable value for the damping constant would be 50 Nsec/m. Assuming we set the cruise control to a little over 40 mph, which translates to 18 m/sec, we simulated this system with different values of K_p, K_i, and K_d. Let's look at some of the results shown in Figure 23.6. We begin with setting K_i and K_d set to zero so we only have a proportional controller. You can see in the top two panels of the figure that as we increase the value of the proportional gain from 200 to 500 the rise time of the response improves. However, even with a large value for the proportional constant K_p there is a consistent, or steady state, error between the speed achieved and the desired speed. The desired speed is plotted as a dotted line. This is where the integral part of the controller comes in handy. The lower left panel shows the result when we change the coefficient K_i from 0 to 10. You can see a considerable decrease in the steady state error. If we increase K_i further we can pretty much eliminate the steady state error but at the cost of an overshoot.

The MATLAB program used to generate these result is shown below if you are interested in playing with this.

```
mass = 1310;   % mass of the car
fric = 50;  % damping constant
car = tf([1],[mass fric]); % transfer function of the car (plant)
ref = 18; % cruise control set point

figure

Kp = 200; % Proportional gain
Ki = 0; % integral gain
Kd = 0; % derivative gain
control = pid(Kp,Ki,Kd); % Controller
H = feedback(control*car,1); % overall transfer function
t = 0:0.1:40;
[y1] = step(ref*H,t); % step response

Kp = 500; % Proportional gain
Ki = 0; % integral gain
Kd = 0;% derivative gain
control = pid(Kp,Ki,Kd);% Controller
H = feedback(control*car,1); % overall transfer function
t = 0:0.1:40;
[y2] = step(ref*H,t);% step response

Kp = 500;% Proportional gain
Ki = 10;% integral gain
Kd = 0;% derivative gain
control = pid(Kp,Ki,Kd);% Controller
H = feedback(control*car,1); % overall transfer function
t = 0:0.1:40;
[y3] = step(ref*H,t);% step response


Kp = 500;% Proportional gain
Ki = 100;% integral gain
Kd = 0;% derivative gain
```

```
control = pid(Kp,Ki,Kd);% Controller
H = feedback(control*car,1); % overall transfer function
t = 0:0.1:40;
[y4] = step(ref*H,t);% step response

tt = tiledlayout(2,2,'TileSpacing','Compact','Padding','Compact');
nexttile
plot(t,y1,'-k','LineWidth',2)
axis([0 40 0 30])
title('Kp=200, Ki=0, Kd=0')
yline(ref,':','LineWidth',2)
xlabel('time (sec)')
ylabel('speed')

nexttile
plot(t,y2,'-k','LineWidth',2)
axis([0 40 0 30])
title('Kp=500, Ki=0, Kd=0')
yline(ref,':','LineWidth',2)

nexttile
plot(t,y3,'-k','LineWidth',2)
axis([0 40 0 30])
title('Kp=500, Ki=10, Kd=0')
yline(ref,':','LineWidth',2)

nexttile
plot(t,y4,'-k','LineWidth',2)
axis([0 40 0 30])
title('Kp=500, Ki=100, Kd=0')
yline(ref,':','LineWidth',2)
```

23.5 SUMMARY

In this module we introduced the idea of feedback and showed how we can use feedback to stabilize a system, to make it more responsive, to obtain the inverse of a system, and to control the output of a system in order to get a desired response.

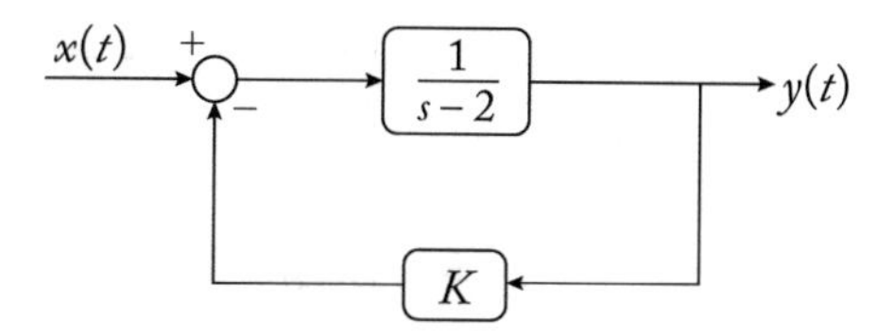

Figure 23.7: A simple feedback system.

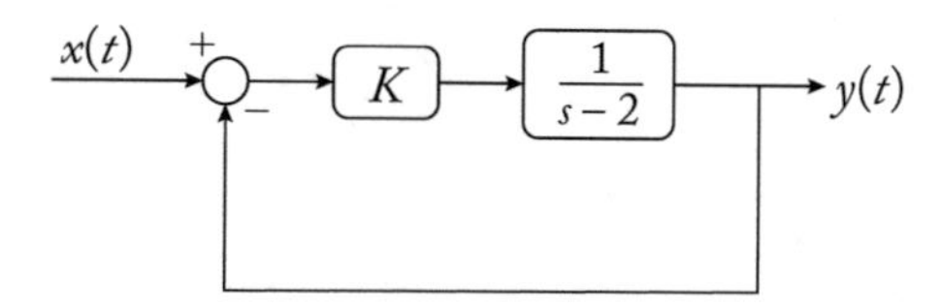

Figure 23.8: Another simple feedback system.

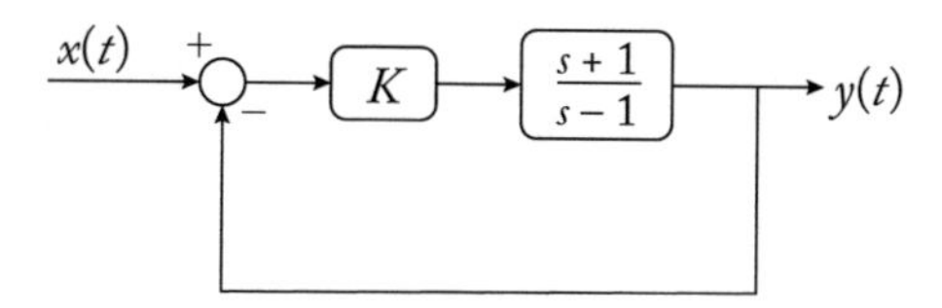

Figure 23.9: A slightly more complicated system.

23.6 EXERCISES

(Answers on the following page)

1. In the system shown in Figure 23.7 what are the poles of the system if $K = 1$? Is this system stable.

2. In the system shown in Figure 23.7 what are the poles of the system if $K = 3$? Is this system stable.

3. In the system shown in Figure 23.8 what are the poles of the system if $K = 1$? Is this system stable.

4. In the system shown in Figure 23.8 what are the poles of the system if $K = 3$? Is this system stable.

5. In the system shown in Figure 23.9 what are the poles and zeros of the system if $K = 3$? Is this system stable.

6. In the system shown in Figure 23.9 what are the poles and zeros of the system if $K = 5$? Is this system stable.

7. For the system shown in Figure 23.7 for what values of K is the system stable?

8. For the system shown in Figure 23.8 for what values of K is the system stable?

9. For the system shown in Figure 23.9 for what values of K is the system stable?

23.7 ANSWERS

1. The pole is at $s = 1$. The system is not stable.
2. The pole is at $s = -1$. The system is stable.
3. The pole is at $s = 1$. The system is not stable.
4. The pole is at $s = -1$. The system is stable.
5. The pole is at $s = -0.5$ and the zero is at $s = -1$. The system is stable.
6. The pole is at $s = -2/3$ and the zero is at $s = -1$. The system is stable.
7. For the system to be stable we need $K > 2$.
8. For the system to be stable we need $K > 2$.
9. for the system to be stable we need $K > 1$ or $K < -1$.

MODULE 24

Z-Transform

Except for our brief foray into the discrete Fourier transform until now we have been talking about continuous systems. Now we move from analyzing continuous-time systems to analyzing discrete-time systems. To analyze continuous-time systems we used the Laplace transform which we obtained as a generalization of Fourier transform. In the same way, we can obtain the Z-transform as a generalization of the discrete-time Fourier transform. We can also develop the Z-transform directly, somewhat akin to how we got into the idea of using $e^{j\omega t}$ as a building block for representing functions of time. Let's try both—you can pick the approach you prefer.

24.1 THE Z-TRANSFORM AS A GENERALIZATION OF THE DTFT

Recall that we introduced the discrete time Fourier transform as a direct consequence of sampling. Given a continuous time signal $x(t)$ we can model the sampling process as a product of the continuous time function and an impulse train. If we normalize the sampling interval to 1 we get the discrete time Fourier transform

$$X(e^{j\omega}) = \sum_{n=-\infty}^{\infty} x[n]e^{-jn\omega}$$

which is periodic with period 2π. We can be certain the discrete time Fourier transform exists if the discrete time sequence $\{x[n]\}$ is absolutely summable, that is

$$\sum_{n=-\infty}^{\infty} |x[n]| < \infty$$

Note that if $\{x[n]\}$ is the impulse response of a linear time-invariant discrete time system then this also means that the corresponding system is stable in the bounded input bounded output (BIBO) sense. What if it isn't and we still want to analyze the frequency domain behavior of the system? We can do what we did in the case of the continuous time system. Instead of finding the discrete time Fourier transform of $x[n]$ we can find the transform of the product of $x[n]$ with the exponential r^{-n} given by

$$\sum_{n=-\infty}^{\infty} x[n]r^{-n}e^{-jn\omega}$$

This summation will converge if

$$\sum_{n=-\infty}^{\infty} |x[n]r^{-n}| < \infty$$

The values of r for which this summation will converge would be the region of convergence for this transform. Let's define a new variable z as

$$z = re^{j\omega}$$

z is simply a complex number with a magnitude of r and angle ω. Using this new variable we can rewrite the transform as

$$\sum_{n=-\infty}^{\infty} x[n]z^{-n}$$

As we are summing over n we will be left with a function of z—let's call it $X(z)$. This is the Z-transform of the discrete time sequence $\{x[n]\}$ which converges for certain values of $r = |z|$.

24.2 THE Z-TRANSFORM IN THE CONTEXT OF DISCRETE LTI SYSTEMS

Recall that for a discrete linear time invariant system with an impulse response $h[n]$ with input $x[n]$, the output $y[n]$ is given by the convolution sum

$$y[n] = \sum_{k=-\infty}^{\infty} x[n-k]h[k]$$

For the special case where $x[n] = z^n$, where z is a complex number

$$y[n] = \sum_{k=-\infty}^{\infty} z^{n-k}h[k] = \sum_{k=-\infty}^{\infty} z^n z^{-k}h[k]$$

Pulling z^n out of the summation we get

$$y[n] = z^n \sum_{k=-\infty}^{\infty} h[k]z^{-k}$$

Or, an input of $x[n] = z^n$ in a discrete LTI system results in an output of $y[n] = H(z)z^n$ where

$$H(z) = \sum_{n=-\infty}^{\infty} h[n]z^{-n}$$

Again we have arrived at the Z-transform—$H(z)$ is called the Z-transform of $h[n]$.

In general given a discrete time function $x[n]$ the Z-transform $X(z)$ is given by

$$X(z) = \sum_{n=-\infty}^{\infty} x[n]z^{-n}$$

The values of z for which this summation converges defines the region of convergence for the transform. As we saw earlier the region of convergence is actually defined by the magnitude of z. As in the case for the Laplace transform **the region of convergence is an essential part of the Z-transform**.

24.3 EXPLORING THE Z-TRANSFORM THROUGH EXAMPLES

As in case of the Laplace transform we will explore the Z-transform through examples. Lets begin with two prototypic signals which we will use as our basic patterns for obtaining Z-transforms—one a right sided sequence and the other a left sided sequence. The right sided sequence is

$$x[n] = a^n u[n]$$

The Z-transform is given by

$$\begin{aligned} X(z) &= \mathcal{Z}\left[a^n u[n]\right] \\ &= \sum_{n=-\infty}^{\infty} a^n u[n] z^{-n} \\ &= \sum_{n=0}^{\infty} a^n z^{-n} \\ &= \sum_{n=0}^{\infty} \left(az^{-1}\right)^n \\ &= \frac{\left(az^{-1}\right)^0 - \left(az^{-1}\right)^{\rightarrow\infty}}{1 - az^{-1}} \end{aligned}$$

The second term in the numerator will go to zero if the magnitude of az^{-1} is less than 1 and it will go to infinity if this magnitude of az^{-1} is greater than 1. Therefore,

$$X(z) = \frac{1}{1 - az^{-1}} \quad |az^{-1}| < 1$$

or

$$X(z) = \frac{1}{1 - az^{-1}} \quad |z| > |a|$$

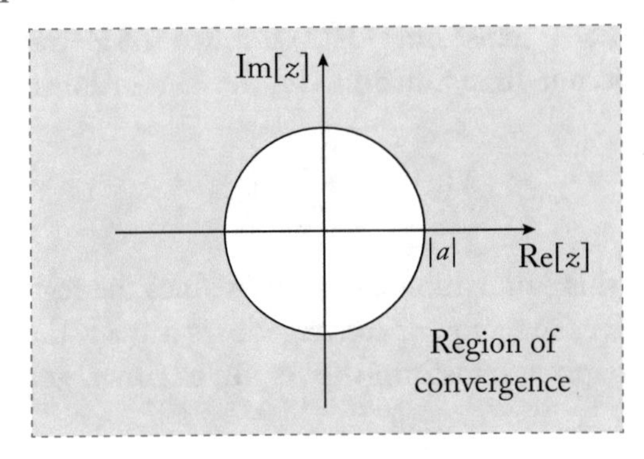

Figure 24.1: Region of convergence $|z| > |a|$ in the complex z-plane.

What does this region of convergence look like? Recall that z is a complex number. We can write this in polar form as

$$z = re^{j\omega}$$

Then

$$|z| = \sqrt{zz^*} = \sqrt{re^{j\omega}re^{-j\omega}} = \sqrt{r^2} = r$$

Therefore, the region $|z| > |a|$ consists of all complex numbers $re^{j\omega}$ with $r > |a|$ as shown in Figure 24.1.

Let's now repeat with a left sided sequence

$$x[n] = a^n u[-n-1]$$

$$\begin{aligned}
X(z) &= \mathcal{Z}\left[a^n u[-n-1]\right] \\
&= \sum_{n=-\infty}^{\infty} a^n u[-n-1] z^{-n} \\
&= \sum_{n=-\infty}^{-1} a^n z^{-n} \\
&= \sum_{n=-\infty}^{-1} \left(az^{-1}\right)^n \\
&= \frac{\left(az^{-1}\right)^{\to -\infty} - \left(az^{-1}\right)^0}{1 - az^{-1}}
\end{aligned}$$

$$\left(az^{-1}\right)^{\to -\infty} \Rightarrow \left(\frac{1}{az^{-1}}\right)^{\to \infty}$$

and

$$\left(\frac{1}{az^{-1}}\right)^{\to\infty} \to 0 \quad \text{when} \quad \left|\frac{1}{az^{-1}}\right| < 1$$

and

$$\left|\frac{1}{az^{-1}}\right| < 1 \Rightarrow \left|\frac{z}{a}\right| < 1 \Rightarrow |z| < |a|$$

Thus,

$$X(z) = -\frac{1}{1 - az^{-1}} \quad |z| < |a|$$

and the region of convergence is the inside of a circle with radius $|a|$. Again the boundary is defined by the pole at $z = a$.

We will use the signals $a^n u[n]$ and $a^n u[-n-1]$ for the Z-transform like we used $e^{-at}u(t)$ and $e^{-at}u(-t)$ for the Laplace transform. As in the case of the Laplace transform where most of the signals we were interested in could be written as some combination of one sided exponentials, in the discrete case most of the signals we are interested in can be fit into the discrete exponential form. Let's see this with a few examples.

Example 24.1 What is the Z-transform of

$$x[n] = (0.9)^n u[n]$$

This is a pretty straightforward application of the development above.

$$\begin{aligned}
X(z) &= \sum_{n=-\infty}^{\infty} (0.9)^n u[n] z^{-n} \\
&= \sum_{n=0}^{\infty} (0.9)^n z^{-n} \quad \text{the lower limit becomes 0 because of } u[n] \\
&= \sum_{n=0}^{\infty} (0.9z^{-1})^n \quad \text{we combine } 0.9^n \text{ and } z^{-n} \\
&= \frac{(0.9z^{-1})^0 - (0.9z^{-1})^{\to\infty}}{1 - (0.9z^{-1})} \quad \text{using the geometric sum formula} \\
&= \frac{1}{1 - 0.9z^{-1}} \quad |0.9z^{-1}| < 1 \\
&= \frac{1}{1 - 0.9z^{-1}} \quad |0.9| < |z| \\
&= \frac{z}{z - 0.9} \quad |z| > |0.9|
\end{aligned}$$

The pole-zero diagram and region of convergence are shown in Figure 24.2.

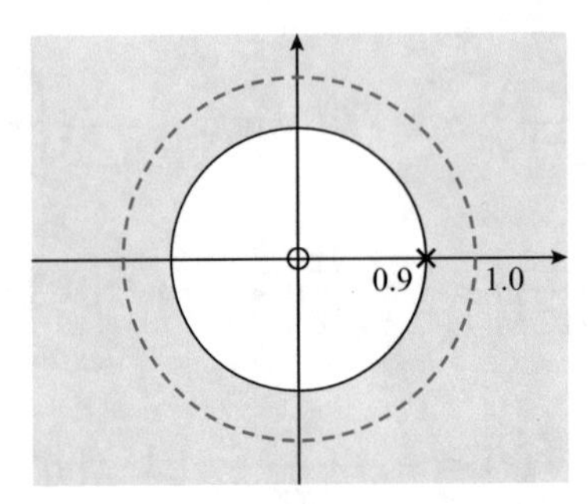

Figure 24.2: The zero is at $z = 0$, the pole is at $z = 0.9$ and the region of convergence is $|z| > 0.9$. The dashed circle is the circle of radius 1.

Let's try a slightly more complicated looking example.

Example 24.2 Find the Z-transform of

$$x[n] = \sin\left(\frac{\pi}{4}n\right) u[n]$$

This looks a bit messier at first but we have Euler on our side. Using Euler's identity we can write the sin function in terms of complex exponentials.

$$\begin{aligned} x[n] &= \left(\frac{1}{2j}e^{j\frac{\pi}{4}n} - \frac{1}{2j}e^{-j\frac{\pi}{4}n}\right) u[n] \\ &= \frac{1}{2j}e^{j\frac{\pi}{4}n}u[n] - \frac{1}{2j}e^{-j\frac{\pi}{4}n}u[n] \end{aligned}$$

Now we will use the linearity of the Z-transform to find the Z-transform of each individual term and then combine the Z-transforms.

$$\begin{aligned} \mathcal{Z}\left[\frac{1}{2j}e^{j\frac{\pi}{4}n}u[n]\right] &= \sum_{n=-\infty}^{\infty} \frac{1}{2j}e^{j\frac{\pi}{4}n}u[n]z^{-n} \\ &= \frac{1}{2j}\sum_{n=0}^{\infty} e^{j\frac{\pi}{4}n}z^{-n} \\ &= \frac{1}{2j}\sum_{n=0}^{\infty}\left(e^{j\frac{\pi}{4}}z^{-1}\right)^n \end{aligned}$$

$$= \frac{1}{2j}\left(\frac{\left(e^{j\frac{\pi}{4}}z^{-1}\right)^{0} - \left(e^{j\frac{\pi}{4}}z^{-1}\right)^{\to\infty}}{1-e^{j\frac{\pi}{4}}z^{-1}}\right)$$

$$= \frac{1}{2j}\left(\frac{1}{1-e^{j\frac{\pi}{4}}z^{-1}}\right) \qquad \left|e^{j\frac{\pi}{4}}z^{-1}\right| < 1$$

$$= \frac{1}{2j}\left(\frac{z}{z-e^{j\frac{\pi}{4}}}\right) \qquad |z| > \left|e^{j\frac{\pi}{4}}\right|$$

What is the magnitude of $e^{j\frac{\pi}{4}}$? You can probably guess but let's slog through the derivation. Using Euler again (this guy Euler is good!).

$$e^{j\frac{\pi}{4}} = \cos\left(\frac{\pi}{4}\right) + j\sin\left(\frac{\pi}{4}\right)$$

and the magnitude of $e^{j\frac{\pi}{4}}$ is simply the square root of the sum of the square of the real and imaginary parts.

$$\left|e^{j\frac{\pi}{4}}\right| = \sqrt{\cos^2\left(\frac{\pi}{4}\right) + \sin^2\left(\frac{\pi}{4}\right)}$$

or

$$\left|e^{j\frac{\pi}{4}}\right| = 1$$

We go through the same process with the second term $\frac{1}{2j}\left(e^{-j\frac{\pi}{4}n}\right)u[n]$ and we get

$$\mathcal{Z}\left[\frac{1}{2j}e^{-j\frac{\pi}{4}n}u[n]\right] = \frac{1}{2j}\left(\frac{z}{z-e^{-j\frac{\pi}{4}}}\right) \qquad |z| > 1$$

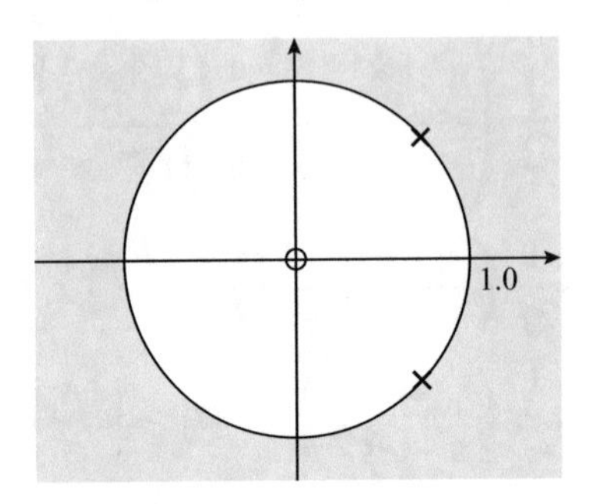

Figure 24.3: The zero is at $z = 0$, the poles are at $z = \pm e^{j\frac{\pi}{4}}$ and the region of convergence is $|z| > 1$.

Combining the two results (notice we have the same region of convergence for both) we get

$$
\begin{aligned}
X(z) &= \frac{1}{2j}\left(\frac{z}{z - e^{j\frac{\pi}{4}}}\right) - \frac{1}{2j}\left(\frac{z}{z - e^{-j\frac{\pi}{4}}}\right) \qquad |z| > 1 \\
&= \frac{1}{2j}\left(\frac{z(z - e^{-j\frac{\pi}{4}}) - z(z - e^{j\frac{\pi}{4}})}{(z - e^{j\frac{\pi}{4}})(z - e^{-j\frac{\pi}{4}})}\right) \qquad |z| > 1 \\
&= \frac{1}{2j}\left(\frac{z^2 - ze^{-j\frac{\pi}{4}} - z^2 + ze^{j\frac{\pi}{4}}}{z^2 - ze^{j\frac{\pi}{4}} - ze^{-j\frac{\pi}{4}} + 1}\right) \qquad |z| > 1 \\
&= \frac{1}{2j}\left(\frac{z(e^{j\frac{\pi}{4}} - e^{-j\frac{\pi}{4}})}{z^2 - z(e^{j\frac{\pi}{4}} + e^{-j\frac{\pi}{4}}) + 1}\right) \qquad |z| > 1 \\
&= \frac{z(e^{j\frac{\pi}{4}} - e^{-j\frac{\pi}{4}})/2j}{z^2 - 2z\cos\left(\frac{\pi}{4}\right) + 1} \qquad |z| > 1 \\
&= \frac{z\sin\left(\frac{\pi}{4}\right)}{z^2 - 2z\cos\left(\frac{\pi}{4}\right) + 1} \qquad |z| > 1
\end{aligned}
$$

That was a lot of algebraic manipulation but that is all it was. The actual machinery for computing the Z-transform was really not much different from the machinery we used to find the Z-transform of $a^n u[n]$. The pole zero diagram and the region of convergence are shown in Figure 24.3.

OK let's take a really simple one now.

Example 24.3 Find the Z-transform of

$$x[n] = u[n]$$

Before we plug this into our Z-transform finding machinery—multiplying by z^{-n} and then summing from $-\infty$ to ∞—let's see if we can use what we already know to find the Z-transform without any of the computation. We already know the Z-transform of $a^n u[n]$. The function $x[n]$ we describe above is simply $a^n u[n]$ with $a = 1$. So in order to find $X(z)$ here all we need to do is substitute $a = 1$ into the Z-transform of $a^n u[n]$. Therefore

$$X(z) = \frac{1}{a - z^{-1}} = \frac{z}{z-1} \quad |z| > 1$$

All the examples we have looked at have been of signals which are non-zero over an infinite interval. What about signals that are non-zero over a finite interval?

Example 24.4 Find the Z-transform of

$$x[n] = u[n] - u[n-4]$$

There are only four non-zero terms here so we can simply write them out.

$$\begin{aligned} X(z) &= \sum_{n=-\infty}^{\infty} (u[n] - u[n-4)z^{-n} \\ &= \sum_{n=0}^{3} z^{-n} \\ &= 1 + z^{-1} + z^{-2} + z^{-3} \end{aligned}$$

easy peasy! Notice that the region of convergence is the entire z-plane except for the point $z = 0$.

OK, but what if the sequence had finite extent but not just non-zero over 3 terms.

Example 24.5 Find the Z-transform of

$$xn = 0.6^n \left(u[n] - u[n-100]\right)$$

Just plug this into the Z-transform equation and turn the crank.

$$\begin{aligned} X(z) &= \sum_{n=-\infty}^{\infty} 0.6^n \left(u[n] - u[n-100]\right) z^{-n} \\ &= \sum_{n=0}^{99} 0.6^n z^{-n} \\ &= \sum_{n=0}^{99} \left(0.6z^{-1}\right)^n \\ &= \frac{\left(0.6z^{-1}\right)^0 - \left(0.6z^{-1}\right)^{100}}{1 - 0.6z^{-1}} \\ &= \frac{1 - \left(0.6z^{-1}\right)^{100}}{1 - 0.6z^{-1}} \end{aligned}$$

Notice that it looks like there is a pole at $z = 0.6$ but if you plug in 0.6 in the expression for $X(z)$ you get a 0/0 form. If you apply L'Hopital's rule you will find that there is no pole at 0.6. The region of convergence is the entire z-plane except for $z = 0$.

24.4 SUMMARY

In this module

1. We defined the Z-transform of a discrete time sequence as

$$X(z) = \sum_{n=-\infty}^{\infty} x[n] z^{-n}$$

   The region of convergence consists of those values of z for which this summation is finite.

2. We derived the Z-transform of two "template" sequences.

$$\begin{aligned} \mathcal{Z}\left[a^n u[n]\right] &= \frac{z}{z-a}; \quad |z| > |a| \\ \mathcal{Z}\left[a^n u[-n-1]\right] &= -\frac{z}{z-a}; \quad |z| < |a| \end{aligned}$$

3. The Z-transform of a sequence that is finite over a finite interval has a region of convergence which is the entire z-plane except possibly for $z = 0$.

24.5 EXERCISES

(Answers on the following page)

Find the Z-transforms of the following sequences.

1.
$$x[n] = 2^n u[n]$$

2.
$$x[n] = 2^{-n} u[n]$$

3.
$$x[n] = 2^n u[-n-1]$$

4.
$$x[n] = 2^{-n} u[-n-1]$$

5.
$$x[n] = 2^n \left(u[n] - u[n-3]\right)$$

6.
$$x[n] = \cos\left(\frac{\pi}{3}n\right) u[n]$$

7.
$$x[n] = a^n \cos\left(\frac{\pi}{3}n\right) u[n]$$

8.
$$x[n] = \delta[n-1]$$

9.
$$x[n] = \left(\frac{1}{2}\right)^n u[n-2]$$

10.
$$x[n] = \left(\frac{1}{2}\right)^n u[2-n]$$

24.6 ANSWERS

1.
$$X(z) = \frac{z}{z-2} \quad |z| > 2$$

2.
$$X(z) = \frac{z}{z-1/2} \quad |z| > \frac{1}{2}$$

3.
$$X(z) = \frac{z}{z-2} \quad |z| < 2$$

4.
$$X(z) = \frac{2z}{1-2z} \quad |z| < \frac{1}{2}$$

5.
$$X(z) = 1 + \frac{2}{z} + \frac{4}{z^2} = \frac{z^2+2z+4}{z^2}$$

6.
$$X(z) = \frac{z^2-0.5}{z^2-z+1} \quad |z| > 1$$

7.
$$X(z) = \frac{z^2-0.5a}{z^2-za+a^2} \quad |z| > |a|$$

8.
$$X(z) = \frac{1}{z}$$

9.
$$X(z) = \frac{\frac{1}{4}}{z^2-\frac{1}{2}z} \quad |z| > \frac{1}{2}$$

10.
$$X(z) = \frac{1}{4z^2(1-2z)} \quad |z| < \frac{1}{2}$$

MODULE 25

Region of Convergence for the Z-Transform

In the previous module we found the Z-transform of a number of sequences and in the process also looked at the region of convergence. In this module we will take a slightly more detailed look at the region of convergence and make an attempt to connect the region of convergence for the Z-transform of a discrete sequence with the Laplace transform of a continuous function.

Let's summarize what we know about the regions of convergence for the Z-transform. We have seen that a right sided signal

$$x[n] = a^n u[n]$$

has a Z-transform

$$X(z) = \frac{1}{1 - az^{-1}} = \frac{z}{z - a}; \quad |z| > a$$

with a region of convergence which is the outside of a circle of radius $|a|$. The boundary of the region of convergence is defined by a pole at $z = a$.

For a left sided signal

$$x[n] = -a^n u[-n - 1]$$

The Z-transform $X(z)$ is given by

$$X(z) = \frac{1}{1 - az^{-1}} = \frac{z}{z - a} \quad |z| < |a|$$

and the region of convergence is the inside of a circle with radius $|a|$ as we would have predicted based on the mapping from the Laplace to the Z-domain. Again the boundary is defined by the pole at $z = a$.

Let's take a look at the Z-transform of a two-sided sequence

$$x[n] = a^{|n|}$$

We can rewrite $x[n]$ as

$$x[n] = a^n u[n] + a^{-n} u[-n - 1]$$

Using the linearity of the Z-transform we can find the Z-transform of each term and then combine them in the intersection of their regions of convergence. For the first term we already have

$$\mathcal{Z}\left[a^n u[n]\right] = \frac{1}{1 - az^{-1}} \quad |z| > |a|$$

For the second term

$$\begin{aligned} \mathcal{Z}\left[a^{-n} u[-n-1]\right] &= \sum_{n=-\infty}^{-1} a^{-n} z^{-n} \\ &= \sum_{n=-\infty}^{-1} \left(a^{-1} z^{-1}\right)^n \\ &= \frac{\left(a^{-1}z^{-1}\right)^{\to -\infty} - \left(a^{-1}z^{-1}\right)^0}{1 - (az)^{-1}} \end{aligned}$$

For the first term in the numerator

$$\left(a^{-1}z^{-1}\right)^{\to -\infty} = (az)^{\to \infty}$$

This will go to zero as long as

$$|az| < 1 \Rightarrow |z| < \frac{1}{|a|}$$

and the Z-transform of the second term becomes

$$\mathcal{Z}\left[a^{-n} u[-n-1]\right] = -\frac{1}{1 - (az)^{-1}} \quad |z| < \frac{1}{|a|}$$

Combining the Z-transform of the two terms

$$\mathcal{Z}\left[a^{|n|}\right] = \frac{1}{1 - az^{-1}} - \frac{1}{1 - (az)^{-1}} \quad |a| < |z| < \frac{1}{|a|}$$

So if $|a| < 1$ the region of convergence is the annular region between circles of radius $|a|$ and $1/|a|$ as shown in Figure 25.1

Note that if $|a| > 1$ then $1/|a| < 1$ and $|z|$ cannot simultaneously be greater than $|a|$ and less than $1/|a|$. Therefore, if $|a| > 1$ the Z-transform of $a^{|n|}$ does not exist.

To summarize what we have seen until now:

1. Each boundary of a region of convergence in the z-plane is a circle centered at the origin. The radius of the circle is the magnitude of a pole of the Z-transform.

2. The region of convergence does not contain any poles.

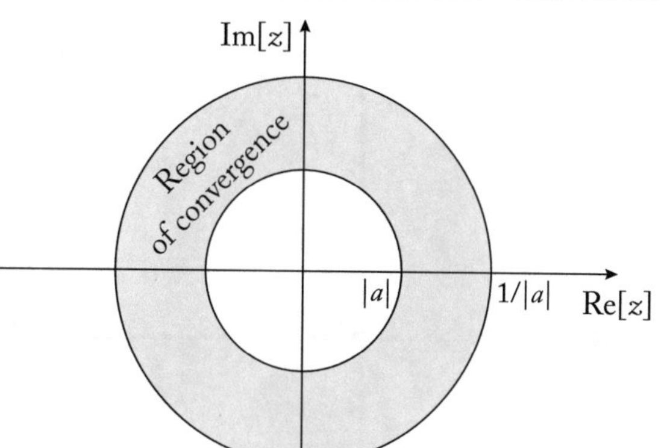

Figure 25.1: An annular region of convergence for a two sided sequence.

3. If $x[n]$ is of finite duration then the region of convergence is the entire plane except perhaps for $z = 0$ or $z = \infty$.
4. If $x[n]$ is a right sided sequence the region of convergence is outside of a circular boundary.
5. If $x[n]$ is a left sided sequence the region of convergence is inside a circular boundary.
6. If $x[n]$ is two sided the region of convergence is between two circular boundaries.

Example 25.1 Consider a system with a transfer function with three poles, two at $0.5 \pm j0.5$ and one at -1.5, one zero at $z = 0$ as shown in Figure 25.2.

The poles define the boundaries of the regions of convergence. Therefore, as shown in Figure 25.3 we end up with three regions of convergence.

The innermost region of convergence will give rise to a left-sided sequence. As the region of convergence does not include the unit circle the corresponding system is not guaranteed to be stable. The outer region of convergence outside of the circle defined by $|z| = 1.5$ corresponds to a right sided sequence. This region of convergence also does not include the unit circle, therefore, the corresponding system is not guaranteed to be stable. Finally, the third region of convergence is the one between the circle of radius 1.5 and the circle of radius $1/\sqrt{2}$. This will correspond to a two sided signal and as the region of convergence includes the unit circle the corresponding system will be stable.

25.1 MAPPING BETWEEN s- AND z-PLANES

How does the magnitude of a affect the discrete time function? For the moment let's assume that a is positive. If the magnitude of a is less than one, then as n increases, the value of $|a|^n$ will

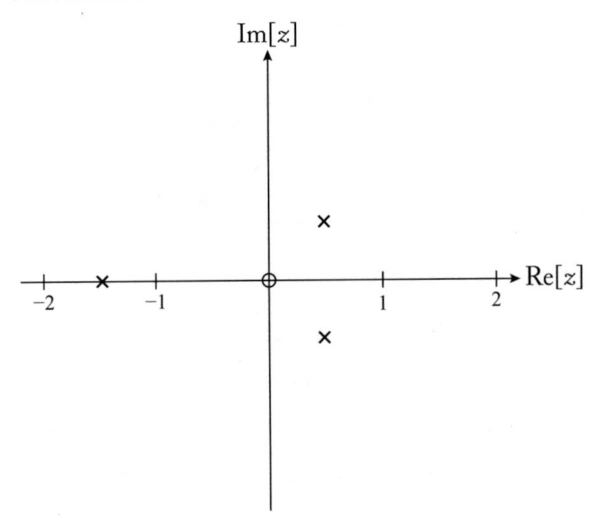

Figure 25.2: Poles and zero of a discrete time system.

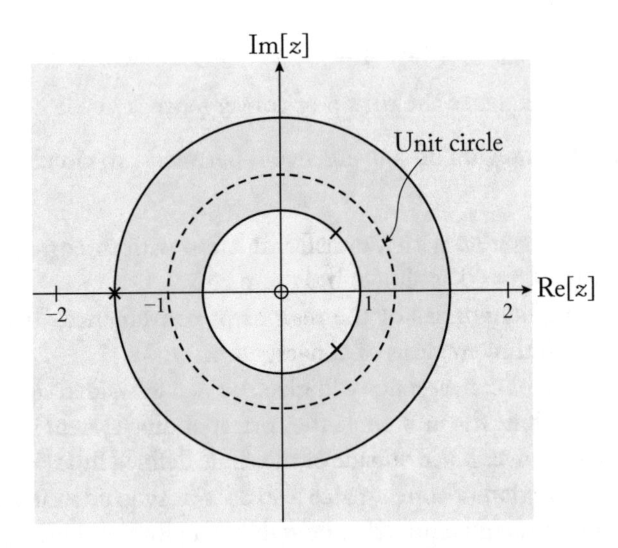

Figure 25.3: The regions of convergence for a three pole one zero system.

decrease; and we have an exponentially decreasing function. Furthermore, if $x[n]$ is the impulse response of a linear time-invariant system, it satisfies the requirement for BIBO stability.

$$\sum_{n=0}^{\infty} |x[n]| = \sum_{n=0}^{\infty} |a|^n = \frac{|a|^0 - |a|^{\rightarrow\infty}}{1-|a|} = \frac{1}{1-|a|}$$

However, if the magnitude of a is greater than one then as n increases the value of $|a|^n$ increases and

$$\sum_{n=0}^{\infty} |x[n]| = \sum_{n=0}^{\infty} |a|^n = \frac{|a|^0 - |a|^{\to\infty}}{1-|a|} \to \infty$$

and the system is not BIBO stable. Looking at Figure 24.1 we can see that in the first case where $|a| < 1$ the region of convergence includes the circle with a radius of one—the unit circle, while in the second case where $|a| > 1$ the region of convergence does not include the unit circle.

We can understand the importance of the inclusion of the unit circle in the region of convergence in a couple of different ways. The Z-transform is given by

$$X(z) = \sum_{n=-\infty}^{\infty} x[n]z^{-n}$$

If we write z in polar coordinates $z = re^{j\omega}$ the Z-transform becomes

$$\sum_{n=-\infty}^{\infty} x[n]r^{-n}e^{-jn\omega}$$

If we want to evaluate this transform on the unit circle we set $r = 1$ and this becomes

$$\sum_{n=-\infty}^{\infty} x[n]e^{-jn\omega}$$

which is simply the discrete time Fourier transform of $x[n]$. If $x[n]$ is the impulse response of a linear time-invariant discrete system the existence of the discrete time Fourier transform implies that the system is stable. Therefore, if we can evaluate the Z-transform of an impulse response on the unit circle then we can say that the system is stable. But we can only evaluate the Z-transform on the unit circle if the unit circle is contained in the region of convergence. Therefore, if the unit circle is contained in the region of convergence we can say that the associated system is stable.

We can also reach the same conclusion if we look at the transformation between the z-plane and the s-plane. The transformation between the two planes is given by

$$z = e^s = e^{\sigma + j\omega}$$

The transformation maps points from the s-plane to the z-plane. This means that the points along a line $\sigma = \sigma_o$ will be mapped into the points

$$z = e^{\sigma_o + j\omega} = r_o e^{j\omega}$$

where

$$r_o = e^{\sigma_o}$$

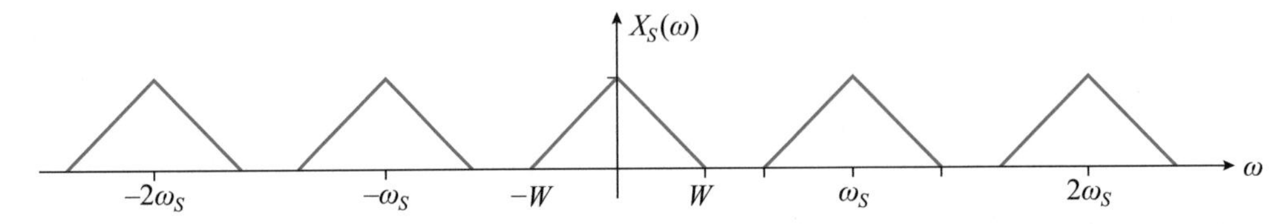

Figure 25.4: Spectral profile of a sampled signal.

The equation

$$z = r_o e^{j\omega}$$

is the equation of a circle with radius r_o. Thus, vertical lines in the s-plane map into circles in the z-plane. Therefore, as the boundaries of regions of convergence n the s-plane were vertical lines, the boundaries of regions of convergence in the z-plane will be circles. Let's look at the special case of the $j\omega$ axis. The $j\omega$ axis corresponds to the line $\sigma = 0$. Therefore, in the z-plane it corresponds to the circle with radius $r_o = e^0 = 1$—the unit circle. Just as in the Laplace domain inclusion of the $j\omega$ axis in the region of convergence indicated stability of continuous time systems, inclusion of the unit circle in the z-plane indicates stability of discrete time systems.

In the Laplace domain we were often concerned with whether a region was to the left or the right of a boundary. How does the notion of being to the left or right of a boundary translate to the z-domain? Consider two points s_1 and s_2 in the s-plain with $s_1 = \sigma_1 + j\omega$ and $s_2 = \sigma_2 + j\omega$. If $\sigma_2 < \sigma_1$ then the point s_2 will be to the left of the line defined by $\sigma = \sigma_1$. If we look at these points in the z-plane, the points corresponding to the line $\sigma = \sigma_1$ correspond to a circle with radius e^{σ_1}. A point in the s-plane with $\sigma = \sigma_2$ where $\sigma_2 < \sigma_1$ will have a radius of e^{σ_2}. As $\sigma_2 < \sigma_1$, $e^{\sigma_2} < e^{\sigma_1}$, therefore the point in the z-plane corresponding to s_2 will lie inside the circle of radius e^{σ_1}. Similarly if $\sigma_2 > \sigma_1$, s_2 will lie to the right of the line defined by $\sigma = \sigma_1$ in the s-plane and the corresponding point in the z-plane will lie outside of the circle of radius e^{σ_1}. Thus, points to the left of a line in the s-plane lie inside the corresponding circle in the z-plane and points to the right of a line in the s-plane lie outside of the corresponding circle in the z-plane.

Before we leave the topic of mapping between the s and z-planes we need to deal with one more issue. In the s-domain we obtained the frequency response of a system by traversing the $j\omega$ axis and we could do this for distinct values of ω from $-\infty$ to ∞. The $j\omega$ axis maps into the unit circle in the z-domain, but here once ω goes from 0 to 2π, or $-\pi$ to π we come back to the same point and repeat our traversal. We can see why this is so if we recall that a discrete time signal is generated from a continuous time signal by sampling in the time domain. And sampling in the time domain corresponds to replication in the frequency domain where the replicas are separated by the sampling frequency ω_s (or f_s in Hz). The spectral profile of a sampled signal is shown in Figure 25.4

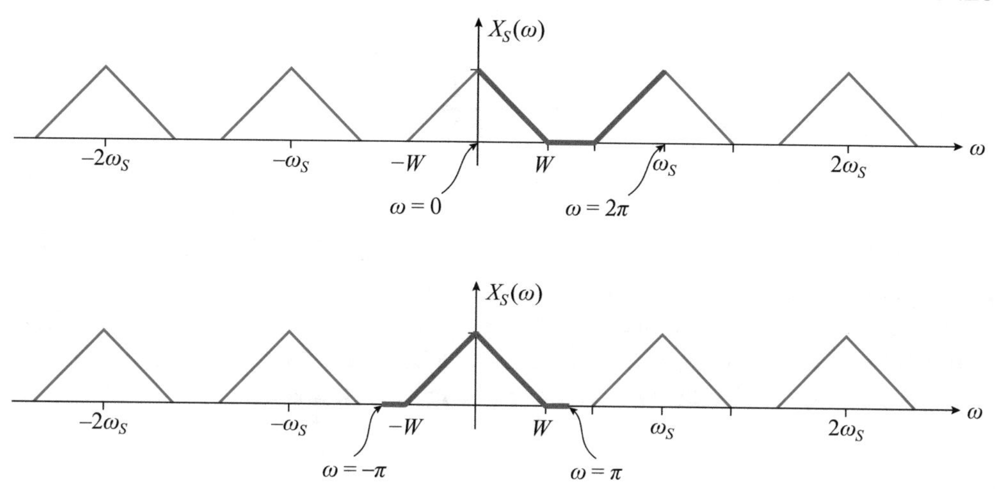

Figure 25.5: Spectral profile of a sampled signal.

Notice that as we traverse the frequency axis we keep encountering replicas of the original spectrum. When we traverse a circle we encounter the same value every 2π radians, here we encounter the same value ever ω_s radians. If we map $\omega = 0$ to $re^0 = 1$, and $\omega = \omega_s$ in the s-plane to $\omega = 2\pi$ in the z-plane we get the same effect. We can think of the mapping in one of two ways. We can think of the point $z = 1$ on the unit circle as corresponding to $\omega = 0$ in the s-plane and a full revolution around the circle brings us to ω_s, or we can think of the revolution going from $-\pi$ to π where $-\pi$ corresponds to $-\omega_s/2$ and π corresponding to $\omega_s/2$. The two ways of looking at the frequency response are shown in Figure 25.5. In practice when we are dealing with real signals the magnitude of the frequency response is even and phase is odd so knowing the frequency response from 0 to $\omega_s/2$ is sufficient and we map π to $\omega_s/2$.

Finally, we can use this transformation to explain why whereas for a continuous function which took on finite nonzero values over a finite interval the region of convergence for the Laplace transform was the entire s-plane, for a discrete time sequence which takes on finite nonzero values over a finite interval the region of convergence for the Z-transform is the entire z-plane *except* for $z = 0$. To see why this is so consider the Laplace transform $X(s) = e^{-s}$. This does not go to infinity for any finite value of s. However, if we let σ go to negative infinity, $X(s)$ will also go to infinity. Therefore, we actually do have a pole—it's just not in the finite s-plane. However when we go through the transformation

$$z = e^s = e^{\sigma + j\omega} = e^{\sigma} e^{j\omega}$$

when σ goes to negative infinity z goes to zero, and the pole at infinity in the s-plane gets mapped into a pole at $z = 0$ in the z-plane.

Example 25.2 Given a transfer function

$$H(s) = \frac{1}{s+1}$$

and a definition of the Laplace domain that includes infinity let's map the poles and zeros of this transfer function to the z-plane. There is clearly a pole at $s = -1$. Using the mapping $z = e^s$. This pole would map to $e^{-1} = 0.368$. Note that a pole in the left half plane in the s-domain maps to a point inside the unit circle in the z-domain.

At first sight it seems like that is it. There are no more poles and zeros. However if we let s go to infinity $H(s)$ will go to zero. Therefore, we also have a zero at $s = \pm\infty$. The zero at negative infinity in the s-domain will map to $z = 0$ in the z-domain.

25.2 SUMMARY

This module was a bit of something old something new. We saw that

1. A left sided discrete time sequence has a region of convergence inside a circular boundary where the circle is centered at the origin.

2. A right sided sequence has a region of convergence outside a circular boundary where the circle is centered at the origin.

3. A two sided sequence has a region of convergence between two circular boundaries where the circles are centered at the origin.

4. The Laplace domain and the z-domain are related through the mapping

$$z = e^s$$

5. If we evaluate the Z-transform around the unit circle we get the frequency profile of the discrete time sequence. The points 0 to π around the unit circle correspond to the frequencies from 0 Hz to $f_s/2$ Hz where f_s is the sampling frequency used to obtain the discrete time sequence from the continuous time sequence.

25.3 EXERCISES

(Answers on the following page)

1. Find the z-transform of

$$x[n] = (0.6)^n u[n] + (1.2)^n u[-n]$$

2. Find the z-transform of

$$x[n] = \left(\frac{1}{2}\right)^{|n|}$$

3. Find the z-transform of

$$x[n] = \left(\frac{1}{2}\right)^n u[n+1] + \left(\frac{1}{2}\right)^{-n} u[1-n]$$

4. $x[n]$ is the two sided impulse response of a stable system. Which of the following could be the poles of the system.
   (a) $-1/4, -1/2$
   (b) $-1/4, 1/2$
   (c) $1/2, -2$
   (d) $-1/2, 2$

5. In the Laplace transform module we constructed a band pass filter with a poles at $s = -2 \pm j12$ and a zero at $s = 0$. Where would the poles and zeros of the corresponding discrete time filter be in the z-domain.

6. If you look at the transfer function

$$H(s) = \frac{s}{s^2 + 4s + 148}$$

you can immediately identify a zero at $s = 0$ and poles at $s = -2 \pm j12$. There is another zero in this system. Can you identify that zero and find the corresponding zero in the z-plane.

7. A simple low pass filter has a transfer function in the Laplace domain given by

$$H(s) = \frac{1}{s+1}$$

Clearly we have a pole at $s = -1$. Also, notice that if we let σ go to infinity the transfer function will go to zero. Therefore, we also have a zero at infinity.
   (a) where would the corresponding pole be in the z-plane?
   (b) Where would the corresponding zero be in the z-plane?
   (c) What is the transfer function of the discrete time system?

25.4 ANSWERS

1.

$$X(z) = \frac{z^2 - 2.4z + 0.72}{(z-0.6)(z-1.2)} \quad 0.6 < |z| < 1.2$$

2.

$$X(z) = \frac{-\frac{3}{2}z}{\left(z-\frac{1}{2}\right)(z-2)} \quad \frac{1}{2} < |z| < 2$$

3.

$$X(z) = \frac{2z^4 - 4z^3 - z + \frac{1}{2}}{z\left(z-\frac{1}{2}\right)(z-2)} \quad \frac{1}{2} < |z| < 2$$

4. (c) and (d)

5. Zero at $z = 1$ and poles at $z = e^{-2}e^{\pm j12}$.

6. If we let σ go to infinity in either the positive or negative direction the transfer function will go to zero.

7. (a) The corresponding pole would be at $1/e = 0.368$.

   (b) In the finite z-plane $s \to -\infty$ corresponds to $z = 0$ so the corresponding zero is at 0.

   (c)

$$H(z) = \frac{z}{z-0.368} \quad |z| > 0.368$$

MODULE 26

Properties of the Z-Transform

In this module we will look at some of the properties of the Z-transform—a few of which we have been using already without explicitly identifying them. These properties match those of properties of the Laplace and Fourier transforms—properties we have already talked about.

26.1 LINEARITY

We begin, as always with the property of linearity. This a property we have already used to let us take the Z-transform of sums of sequences. The proof is very simple.

Let

$$\begin{aligned} \mathcal{Z}[x[n]] &= X(z) \;\; \text{ROC}: R_x \\ \mathcal{Z}[y[n]] &= Y(z) \;\; \text{ROC}: R_y \end{aligned}$$

Then

$$\begin{aligned} \mathcal{Z}[\alpha x[n] + \beta y[n]] &= \sum_{n=-\infty}^{\infty} [\alpha x[n] + \beta y[n]] \, z^{-n} \\ &= \alpha \sum_{n=-\infty}^{\infty} x[n] z^{-n} + \beta \sum_{n=-\infty}^{\infty} y[n] z^{-n} \\ &= \alpha X(z) + \beta Y(z) \;\; \text{ROC}: R_x \cap R_y \end{aligned}$$

The one thing we do need to be careful about though is the region of convergence.

Example 26.1 Recall the Z-transform of

$$x[n] = a^{|n|}$$

We can write $x[n]$ as the sum of two terms

$$x[n] = a^n u[n] + a^{-n} u[-n-1]$$

and then take the Z-transform of each term.

$$\mathcal{Z}[a^n u[n]] = \frac{z}{z-a}; \;\; |z| > |a|$$

$$\mathcal{Z}\left[a^{-n}u[-n-1]\right] = \frac{az}{1-az}; \quad |z| < \frac{1}{|a|}$$

In order to get the Z-transform of $x[n]$ we can use linearity to add these two terms together. However, we have to make sure that there is a region in the z-plane in which both terms converge. If $|a| < 1$ we can find such a region and

$$X(z) = \frac{(1-a^2)z}{(z-a)(1-az)}; \quad |a| < |z| < \frac{1}{|a|}$$

However, if $|a| > 1$ there is no intersection between the region outside the circle of radius $|a|$, $|z| > |a|$, and the region inside the circle of radius $\frac{1}{|a|}$, $|z| < \frac{1}{|a|}$, and despite the linearity property we cannot add the individual terms to obtain $X(z)$. In fact, $X(z)$ does not exist.

26.2 CONVOLUTION

$$\mathcal{Z}\left[x[n] \circledast h[n]\right] = X(z)H(z) \quad \text{ROC} : R_x \cap R_h$$

We can prove this exactly the same way as we have done previously for Fourier series, Fourier transforms, and Laplace transforms.

$$\begin{aligned}
\mathcal{Z}\left[x[n] \circledast h[n]\right] &= \mathcal{Z}\left[\sum_{k=-\infty}^{\infty} x[k]h[n-k]\right] \\
&= \sum_{n=-\infty}^{\infty}\left[\sum_{k=-\infty}^{\infty} x[k]h[n-k]\right] z^{-n}
\end{aligned}$$

If we set $m = n - k$, $n = m + k$ and

$$\begin{aligned}
\mathcal{Z}\left[x[n] \circledast h[n]\right] &= \sum_{m=-\infty}^{\infty}\sum_{k=-\infty}^{\infty} x[k]h[m]z^{-(m+k)} \\
&= \sum_{m=-\infty}^{\infty}\left[\sum_{k=-\infty}^{\infty} x[k]z^{-k}\right] h[m]z^{-m} \\
&= \sum_{m=-\infty}^{\infty} X(z)h[m]z^{-m} \\
&= X(z)\sum_{m=-\infty}^{\infty} h[m]z^{-m} \\
&= X(z)H(z)
\end{aligned}$$

Clearly if the region of convergence for $X(z)$ is R_x and the region of convergence for $Y(z)$ is R_y then the region of convergence for their product will either be $R_x \cap R_h$ or will contain $R_x \cap R_h$. The reason for the latter would be if a pole of either of the terms is canceled by a zero in the other.

Example 26.2 The step response of a discrete time linear time invariant system is given by

$$y[n] = 2u[n] - \left(\frac{1}{2}\right)^n u[n]$$

What is the transfer function of this system? What is the impulse response of this system?

We know that

$$y[n] = x[n] \circledast h[n]$$

By the convolution property

$$Y(z) = X(z)H(z)$$

Noting that the step response is the response of the system to an input which is the step function $u[n]$,

$$X(z) = \mathcal{Z}\left[u[n]\right] = \frac{z}{z-1}; \quad |z| > 1$$

If we now find $Y(z)$ we can obtain $H(z)$ as

$$H(z) = \frac{Y(z)}{X(z)}$$

So let's find $Y(z)$. We know that

$$\mathcal{Z}\left[2u[n]\right] = \frac{2z}{z-1}; \quad |z| > 1$$

and

$$\mathcal{Z}\left[\left(\frac{1}{2}\right)^n u[n]\right] = \frac{z}{z-1/2}; \quad |z| > 1/2$$

Using the linearity property

$$Y(z) = \frac{2z}{z-1} - \frac{z}{z-1/2}; \quad |z| > 1$$

or

$$Y(z) = \frac{z^2}{(z-1)(z-1/2)}; \quad |z| > 1$$

Dividing $Y(z)$ by $X(z)$ we get

$$H(z) = \frac{z}{z-1/2}; \quad |z| > 1/2$$

where because the pole at $z = 1$ got cancelled the region of convergence has expanded. Note the new region of convergence $|z| > 1/2$ includes the region of convergence $|z| > 1$.

By inspection we can see that

$$h[n] = \left(\frac{1}{2}\right)^n u[n]$$

26.3 TIME SHIFTING

$$\mathcal{Z}\left[x[n-n_0]\right] = z^{-n_0} X(z)$$

This is going to be a property that we will use often. Proving it is straightforward.

$$\begin{aligned}
\mathcal{Z}\left[x[n-n_0]\right] &= \sum_{n=-\infty}^{\infty} x[n-n_0] z^{-n} \\
&= \sum_{m=-\infty}^{\infty} x[m] z^{-(m+n_0)} \\
&= z^{-n_0} \sum_{m=-\infty}^{\infty} x[m] z^{-m} \\
&= z^{-n_0} X(z)
\end{aligned}$$

The region of convergence for the shifted sequence is the same as the region of convergence for the original sequence except perhaps for $z = 0$ as the only additional pole would be at $z = 0$.

Example 26.3 Let's see if we can use this property to implement the input output relationship for a system with a particular impulse response. As an example let's use

$$h[n] = a^n u[n]$$

We can find the transfer function

$$H(z) = \frac{1}{1 - az^{-1}}$$

Remember that

$$Y(z) = X(z)H(z)$$

we have

$$Y(z) = X(z)\frac{1}{1 - az^{-1}}$$

or

$$Y(z)\left(1 - az^{-1}\right) = X(z)$$

which is

$$Y(z) - az^{-1}Y(z) = X(z)$$

We can now use the linearity property of the Z-transform. The function $Y(z)$ is the transform of $y[n]$ and the function $X(z)$ is the Z-transform of $x[n]$. According to the time shifting property $z^{-1}Y(z)$ is the Z-transform of $y[n-1]$. Putting all of these together we get the input-output relationship of this discrete time system.

$$y[n] - ay[n-1] = x[n]$$

or

$$y[n] = x[n] + ay[n-1]$$

A couple of things to note in this example which will be useful to us when we try to implement discrete time systems. The factor z^{-1} can be implemented simply as a delay. This means that if we can write our input output relationship in terms of z^{-1} (and its powers) then we can implement discrete time systems using just delays and multipliers. The other thing to note is that this is a system with feedback. The output $y[n]$ is fed back after a delay of one time interval with a gain of a. The fact that the transfer function has a pole (at a location other than zero) is also indicative of the fact that the system contains feedback.

26.4 SCALING IN THE z-DOMAIN

If we multiply the time sequence $x[n]$ with powers of a complex number z_o, we get a scaling of the Z-transform.

$$\begin{aligned}
\mathcal{Z}\left[z_o^n x[n]\right] &= \sum_{n=-\infty}^{\infty} z_o^n x[n] z^{-n} \\
&= \sum_{n=-\infty}^{\infty} x[n]\left(\frac{z}{z_o}\right)^{-n}
\end{aligned}$$

In other words, if $X(z)$ is the Z-transform of $x[n]$

$$\mathcal{Z}\left[z_o^n x[n]\right] = X\left(\frac{z}{z_o}\right), \quad \text{ROC} = |z_o|R$$

We can see the reason for the change in the region of convergence if we consider a single pole in $X(z)$. Suppose that pole is at p_o. This means that the denominator of $X(z)$ has a root at $z = p_o$. Or we can say that the denominator has a factor $(z - p_o)$. Now consider $X(z/z_o)$. In this function we replace every z in $X(z)$ with z/z_o. That means the factor $(z - p_o)$ becomes $(z/z_o - p_o)$.

$$\frac{z}{z_o} - p_o = 0 \Rightarrow z = p_0 z_0$$

so the pole at p_o moves to $z_o p_o$. As in the case of the Laplace domain, scaling in the z-domain is a way to move poles around. We will see more of this when we look at feedback systems.

Example 26.4 Let's verify this property by finding the shifted poles and zeros of a system using this property and then see if we get the same poles and zeros when we substitute z/z_o for every z. Suppose we have a system with a zero at 1 and poles at 2 and 3.

$$H(z) = \frac{z-1}{(z-2)(z-3)}; \quad |z| > 3$$

If we multiply the impulse response with $(1/4)^n$ then using the property the new zero would be at $1/4$ and the new poles will be at 0.5 and 0.75.

If we replace z with $z/(1/4)$ or $4z$ we get

$$\begin{aligned}
H(4z) &= \frac{4z-1}{(4z-2)(4z-3)} \\
&= \frac{4(z-1/4)}{16(z-2/4)(z-3/4)} \\
&= \frac{z-1/4}{4(z-0.5)(z-0.75)}; \quad |z| > 0.75
\end{aligned}$$

where the change in the region of convergence is because the impulse response is still a right sided sequence and the outermost pole is at 0.75. Notice that this also makes the previously unstable system stable.

26.5 TIME REVERSAL

Reversal of a time sequence results in an inversion of its poles and zeros and its region of convergence. Let's first derive the property and then look at the consequences of time reversal.

$$\begin{aligned}
\mathcal{Z}[x[-n]] &= \sum_{n=-\infty}^{\infty} x[-n]z^{-n} \\
&= \sum_{m=-\infty}^{\infty} x[m]z^{m}
\end{aligned}$$

where we have used the variable substitution $m = -n$. We can rewrite this as

$$\mathcal{Z}[x[-n]] = \sum_{m=-\infty}^{\infty} x[m](z^{-1})^{-m}$$

of if $X(z) = \mathcal{Z}[x[n]]$ with region of convergence R then

$$\mathcal{Z}[x[-n]] = X(z^{-1}) = X\left(\frac{1}{z}\right) \quad \text{ROC} = \frac{1}{R}$$

If $X(z)$ has a pole at p_o, then the denominator of $X(z)$ has a root at $z = p_o$, and the denominator of $X(z)$ has a factor $(z - p_o)$. For $X(1/z)$ this factor becomes $1/z - p_o$, but

$$\frac{1}{z} - p_o = 0 \Rightarrow z = \frac{1}{p_o}$$

Therefore, if $X(z)$ has a pole at p_o, $X(1/z)$ will have a pole at $1/p_o$. If p_o has a magnitude less than one then $1/p_o$ will have a magnitude greater than one. Thus, all poles of $X(z)$ that are inside the unit circle will get mapped to poles outside the unit circle for $X(1/z)$. And similarly all poles of $X(z)$ that lie outside the unit circle will translate to poles inside the unit circle for $X(1/z)$.

Despite all this movement of poles a stable system stays a stable system and an unstable system stays an unstable system under this transformation. To see this remember that by replacing n with $-n$ we are changing a left sided sequence into a right sided sequence and vice versa. If we have a right sided sequence the region of convergence will be outside of a circle defined by the outermost pole. If the system is stable this outermost pole, and hence all poles, will be inside the unit circle. This will make the unit circle be in the region of convergence. When we replace n with $-n$ we get a left sided sequence with a region of convergence inside a circle with radius being equal to the innermost pole. As all the poles inside the unit circle have migrated to outside the unit circle this means that the innermost pole will also be outside the unit circle. Therefore, the region of convergence will again include the unit circle and the system will be again stable.

26.6 DIFFERENTIATION IN THE z-DOMAIN

The Z-transform of a sequence $x[n]$ weighted by the time instant n is given by

$$\mathcal{Z}[nx[n]] = -z\frac{dX(z)}{dz} \quad \text{ROC} = R$$

We can show this easily be setting up the expression for the Z-transform and then taking the derivative of the Z-transform with respect to Z.

$$
\begin{aligned}
\frac{d}{dz}X(z) &= \frac{d}{dz}\sum_{n=-\infty}^{\infty} x[n]z^{-n} \\
&= \sum_{n=-\infty}^{\infty} x[n]\frac{d}{dz}z^{-n} \\
&= \sum_{n=-\infty}^{\infty} x[n](-n)z^{-n-1} \\
&= -z^{-1}\sum_{n=-\infty}^{\infty} nx[n]z^{-n} \\
&= -z^{-1}\mathcal{Z}\left[nx[n]\right]
\end{aligned}
$$

Multiplying both sides by $-z$ we get

$$-z\frac{dX(z)}{dz} = \mathcal{Z}\left[nx[n]\right]$$

Example 26.5 This property is very helpful when we need to find the Z-transform of sequences that include factors of n^k. We can show this with a simple example. Lets find the Z-transform of

$$x[n] = n^2u[n]$$

We can rewrite this as

$$x[n] = n\left[nu[n]\right]$$

We know that

$$\mathcal{Z}\left[u[n]\right] = \frac{z}{z-1}$$

So

$$
\begin{aligned}
\mathcal{Z}\left[nu[n]\right] &= -z\frac{d}{dz}\frac{z}{z-1} \\
&= \frac{z}{(z-1)^2}
\end{aligned}
$$

and

$$
\begin{aligned}
\mathcal{Z}\left[n^2u[n]\right] &= -z\frac{d}{dz}\frac{z}{(z-1)^2} \\
&= \frac{z(z+1)}{(z-1)^3}
\end{aligned}
$$

26.7 SUMMARY

We have discussed a number of properties of the Z-transform in this module which will be helpful to us when analyzing discrete time systems.

1. Linearity: The Z-transform of a sum is the sum of the Z-transforms with the region of convergence being the intersection of the regions of convergence of the individual transforms.

2. Convolution: The Z-transform of the convolution of two discrete time sequences is the product of the Z-transform of the individual sequences.

3. Time shifting: The Z-transform of a sequence shifted by n_o time units is the Z-transform of the original sequence multiplied by z^{-n_o}. An implication of this is that the unit delay operation in the time domain is represented by z^{-1} in the z-domain.

4. Scaling in the z-domain: Multiplication with z_o^n in the time domain results in scaling by z_o in the z-domain. This results in the movement of poles and zeros.

5. Time reversal: Changing of a left sided sequence to a right sided sequence and vice verso results in all poles and zeros inside the unit circle being moved to outside the unit circle and all poles and zeros outside the unit circle being moved to inside the unit circle.

6. Differentiation in the z-domain. Multiplication of $x[n]$ by n in the time domain results in the a differentiation of $X(z)$ in the z-domain multiplied by $-z$.

26.8 EXERCISES

(Answers on the following page)

1. The sequence $y[n]$ is the step response of a system with impulse response $h[n]$ given by

$$h[n] = \left(\frac{1}{2}\right)^n u[n]$$

   That is $y[n]$ is the output when the input is the unit step function $u[n]$. Find $Y(z)$.

2. The input output relationship of a discrete linear time invariant system is given by

$$y[n] = x[n] + 0.5y[n-1]$$

   Find the transfer function $H(z)$.

3. Find the z-transform of

$$x[n] = n^2 u[n]$$

4. $X(z)$ has a pole at $z = 0.5$ and a zero at $z = 0$. Given that

$$y[n] = \left(\frac{1}{2}\right)^n x[n]$$

   where are the pole and zero for $Y(z)$?

5. What is the z-transform of

$$x[n] = n\left(\frac{1}{2}\right)^n u[n]$$

6. The impulse response of a discrete linear time invariant system is given by

$$h[n] = \left(\frac{1}{2}\right)^n u[n]$$

   If the input is $x[n] = \delta[n]$ find the z-transform of the output.

7. The input output relationship of a discrete linear time invariant system is given by

$$y[n] = x[n] - \frac{3}{4}y[n-1] - \frac{1}{8}y[n-2]$$

   Find the transfer function $H(z)$ of this system.

8. The input output relationship of a discrete linear time invariant system is given by

$$y[n] = x[n] - x[n-2] - 0.81y[n-2]$$

   Find the transfer function $H(z)$ of this system.

26.9 ANSWERS

1.

$$Y(z) = \frac{z^2}{(z-1)(z-\frac{1}{2})}; \quad |z| > 1$$

2.

$$H(z) = \frac{z}{z-0.5}; \quad |z| > 0.5$$

3.

$$X(z) = \frac{z(z+1)}{(z-1)^3}$$

4. Pole at $z = 0.25$, zero at $z = 0$.

5.

$$X(z) = \frac{0.5z}{(z-0.5)^2}$$

6.

$$Y(z) = \frac{z}{z-0.5}; \quad |z| > 0.5$$

7.

$$H(z) = \frac{z^2}{(z+\frac{1}{2})(z+\frac{1}{4})}$$

8.

$$H(z) = \frac{z^2-1}{z^2+0.81}; \quad |z| > 0.9$$

MODULE 27

The Inverse Z-Transform

The Z-transform provides us with an alternative view of discrete time sequences just as the Laplace and Fourier transforms provided us with alternative views of continuous time functions. We have seen that finding the Z-transform of sequences of interest to us is really quite simple. All we need is the geometric sum formula and some properties of the Z-transform. However, for us to be able to use the real power of the Z-transform we need to know how to move from the z domain to the discrete time sequence—in other words we need the inverse Z-transform.

The formal definition of the inverse Z-transform which we will denote as

$$\mathcal{Z}^{-1}[X(z)] = x[n]$$

is given by the following contour integral

$$x[n] = \frac{1}{2\pi j} \oint X(z) z^{n-1} dz$$

where the contour lies within the region of convergence of $X(z)$.

While a bit intimidating at first sight this integral is not that difficult to compute. In order to solve it we use a method based on the *Cauchy residue theorem*. Using this theorem we can show that as long as the contour is a simple curve (no loops) the integral is equal to the sum of the residues corresponding to the poles inside the contour. A residue R_i corresponding to a simple pole at p_i is given by

$$R_i = \lim_{z \to p_i} (z - p_i) X(z) z^{n-1}$$

(looks a bit like partial fraction expansion). For higher order poles the residue R_i for a pole of order k is given by

$$R_i = \frac{1}{(k-1)!} \lim_{z \to p_i} \frac{d^{k-1}}{dz^{k-1}} \left[(z - p_i)^k X(z) z^{n-1} \right]$$

Once we have found the residues the integral is simply the sum of these residues.

$$\frac{1}{2\pi j} \oint X(z) z^{n-1} dz = \sum_i R_i$$

Where the sidedness of the sequence is dictated by the region of convergence.

While the formal approach may not be that difficult we can make life even easier for ourselves if we use the same shortcut we used for the inverse Laplace transform. And being lazy (the best engineer is a lazy engineer) that is what we will do. As in the case of the Laplace transform the shortcut makes use of the fact that most of the Z-transforms we are interested in are rational polynomials in z. If we have a Z-transform

$$X(z) = \frac{N(z)}{D(z)} = \frac{(z-p_1)(z-p_2)\ldots(z-p_n)}{(z-z1)(z-z_2)\ldots(z-z_m)}$$

where the degree of the numerator polynomial is less than the degree of the denominator polynomial then, just as in the case of the Laplace transform, we can use partial fraction expansion to write this as

$$X(z) = \sum \frac{A_i}{z-z_i}$$

We run into a slight problem here. Recall that for the inverse Laplace transform we made use of the following facts:

1. the Laplace transform and the inverse Laplace transform are linear transforms and

2. we knew that the inverse Laplace transform of $1/(s-z_i)$ was either a right sided or a left sided exponential.

Is this also true for the Z-transform? Well the first point certainly holds; the Z-transform and its inverse are linear transforms. However, the second fact is only partially correct. The Z-transform of an exponential in the discrete world is $1/(1-z_i z^{-1})$ which is equal to $z/(z-z_i)$, not $1/(z-z_i)$. We can adjust for this by finding the partial fraction expansion of $\frac{X(z)}{z}$. If

$$\frac{X(z)}{z} = \sum \frac{A_i}{z-z_i}$$

then

$$X(z) = \sum A_i \frac{z}{z-z_i}$$

and now we can proceed as we did with the Laplace transform.

Let's look at some examples of finding the inverse Z-transform using this method.

Example 27.1 Let's find the inverse transform of

$$X(z) = \frac{3z^2 - \frac{5}{6}z}{z^2 - \frac{7}{12}z + \frac{1}{12}} \quad |z| > \frac{1}{3}$$

The first thing we do is write the denominator in terms of factors.

$$\begin{aligned} X(z) &= \frac{3z^2 - \frac{5}{6}z}{z^2 - \frac{7}{12}z + \frac{1}{12}} \quad |z| > \frac{1}{3} \\ &= \frac{z\left(3z - \frac{5}{6}\right)}{\left(z - \frac{1}{4}\right)\left(z - \frac{1}{3}\right)} \end{aligned}$$

and
$$\frac{X(z)}{z} = \frac{3z - \frac{5}{6}}{\left(z - \frac{1}{4}\right)\left(z - \frac{1}{3}\right)}$$

Now lets find the partial fraction expansion of $X(z)/z$.

$$\frac{3z - \frac{5}{6}}{\left(z - \frac{1}{4}\right)\left(z - \frac{1}{3}\right)} = \frac{A}{z - \frac{1}{4}} + \frac{B}{z - \frac{1}{3}}$$

and

$$\begin{aligned} A &= \left.\frac{3z - \frac{5}{6}}{z - \frac{1}{3}}\right|_{z=\frac{1}{4}} \\ &= \frac{\frac{3}{4} - \frac{5}{6}}{\frac{1}{4} - \frac{1}{3}} \\ &= \frac{\frac{9-10}{12}}{\frac{3-4}{12}} \\ &= 1 \end{aligned}$$

Similarly

$$B = \left.\frac{3z - \frac{5}{6}}{z - \frac{1}{4}}\right|_{z=\frac{1}{3}} = \frac{\frac{3}{3} - \frac{5}{6}}{\frac{1}{3} - \frac{1}{4}} = \frac{\frac{1}{6}}{\frac{1}{12}} = 2$$

Therefore,

$$\frac{X(z)}{z} = \frac{1}{z - \frac{1}{4}} + \frac{2}{z - \frac{1}{3}} \quad |z| > \frac{1}{3}$$

and

$$X(z) = \frac{z}{z - \frac{1}{4}} + \frac{2z}{z - \frac{1}{3}} \quad |z| > \frac{1}{3}$$

or dividing top and bottom by z,

$$X(z) = \frac{1}{1 - \frac{1}{4}z^{-1}} + \frac{2}{1 - \frac{1}{3}z^{-1}} \quad |z| > \frac{1}{3}$$

and we can find $x[n]$, the inverse transform of $X(z)$ as

$$x[n] = \left(\frac{1}{4}\right)^n u[n] + 2\left(\frac{1}{3}\right)^n u[n]$$

Example 27.2 What is the response of a filter with impulse response

$$h(n) = \left(\frac{1}{2}\right)^n u[n]$$

if the input is a unit step?

We know the Z-transform of

$$x[n] = a^n u[n]$$

to be

$$X(z) = \frac{1}{1 - az^{-1}} \quad |z| > |a|$$

Therefore, we know that

$$H(z) = \frac{1}{1 - \frac{1}{2}z^{-1}}$$

but what about the unit step $u[n]$? If we look at the equation for $x[n]$ above we can see that if we set $a = 1$, then $x[n] = u[n]$. Therefore,

$$\mathcal{Z}\left[u[n]\right] = \frac{1}{1 - z^{-1}} \quad |z| > 1$$

(As an aside, note that this means that a filter with impulse response $u[n]$ would not be BIBO stable as the region of convergence for the Z-transform does not include the unit circle). Using the convolution property we can write

$$Y[z] = \frac{1}{1 - \frac{1}{2}z^{-1}} \cdot \frac{1}{1 - z^{-1}} \quad |z| > 1$$

Multiplying top and bottom by z^2 we get

$$Y[z] = \frac{z^2}{\left(z - \frac{1}{2}\right)(z - 1)}$$

or

$$\frac{Y(z)}{z} = \frac{z}{\left(z - \frac{1}{2}\right)(z - 1)}$$

Using partial fraction expansion we can write $Y(z)/z$ as

$$\frac{Y(z)}{z} = \frac{A}{z - \frac{1}{2}} + \frac{B}{z - 1}$$

where

$$\begin{aligned} A &= \left.\frac{z}{z - 1}\right|_{z=\frac{1}{2}} \\ &= \frac{\frac{1}{2}}{\frac{1}{2} - 1} \\ &= -1 \end{aligned}$$

and

$$\begin{aligned} B &= \frac{z}{z-\frac{1}{2}}\Big|_{z=1} \\ &= \frac{1}{1-\frac{1}{2}} \\ &= 2 \end{aligned}$$

Therefore,

$$\begin{aligned} Y(z) &= \frac{2z}{z-1} - \frac{z}{z-\frac{1}{2}} \\ &= 2\frac{1}{1-z^{-1}} - \frac{1}{1-\frac{1}{2}z^{-1}} \end{aligned}$$

and therefore,

$$\begin{aligned} y[n] &= 2u[n] - \left(\frac{1}{2}\right)^n u[n] \\ &= \left(2 - \left(\frac{1}{2}\right)^n\right) u[n] \end{aligned}$$

We can confirm this last result using our favorite method—discrete time convolution.

Example 27.3 Using convolution find the output of a filter with impulse response

$$h[n] = \left(\frac{1}{2}\right)^n u[n]$$

when the input is $x[n] = u[n]$

$$\begin{aligned} y[n] &= x[n] \circledast h[n] \\ &= \sum_{k=-\infty}^{\infty} x[n-k]h[k] \\ &= \sum_{k=-\infty}^{\infty} u[n-k] \left(\frac{1}{2}\right)^k u[k] \end{aligned}$$

The term $u[k]$ will make the lower limit go to 0 while the term $u[n-k]$ will make the upper limit n as long as $n \geq 0$. For $n < 0$, $y[n] = 0$. For $n \geq 0$

$$\begin{aligned} y[n] &= \sum_{k=0}^{n} \left(\frac{1}{2}\right)^k \\ &= \frac{\left(\frac{1}{2}\right)^0 - \left(\frac{1}{2}\right)^{n+1}}{1-\frac{1}{2}} \\ &= 2\left(1-\left(\frac{1}{2}\right)^{n+1}\right) \\ &= 2-\left(\frac{1}{2}\right)^n \end{aligned}$$

or

$$y[n] = \left(2-\left(\frac{1}{2}\right)^n\right)u[n]$$

and we end up with the same result.

Let's take a look at one more example this time for a two sided sequence.

Example 27.4 Let's find the inverse transform of

$$X(z) = \frac{3z^2 - \frac{5}{6}z}{z^2 - \frac{7}{12}z + \frac{1}{12}} \quad \frac{1}{4} < |z| < \frac{1}{3}$$

The rational polynomial we are using here is the same as the one in the first example, but the region of convergence is different. In the first example we had used a region of convergence outside of a circle of radius 1/3 resulting in a right sided sequence. In this case we have an annular region of convergence which will result in a two sided sequence. We already know what the partial fraction expansion of this function will give us.

$$X(z) = \frac{z}{z-\frac{1}{4}} + \frac{2z}{z-\frac{1}{3}} \quad \frac{1}{4} < |z| < \frac{1}{3}$$

For the first term the pole is at $1/4$ and the region of convergence is outside of the circle with radius $1/4$. Therefore, the inverse Z-transform will be a right sided sequence.

$$\mathcal{Z}^{-1}\left[\frac{z}{z-\frac{1}{4}};\ |z|>\frac{1}{4}\right]=\left(\frac{1}{4}\right)^n u[n]$$

The second term has a pole at $z = 1/3$ and has a region of convergence inside a circle with radius 1/3. Therefore, the inverse transform will be a left sided sequence.

$$\mathcal{Z}^{-1}\left[\frac{2z}{z-\frac{1}{3}};\ |z|<\frac{1}{3}\right]=-2\left(\frac{1}{3}\right)^n u[-n-1]$$

Combining these together we get

$$\mathcal{Z}^{-1}\left[\frac{3z^2-\frac{5}{6}z}{z^2-\frac{7}{12}z+\frac{1}{12}}\quad \frac{1}{4}<|z|<\frac{1}{3}\right]=\left(\frac{1}{4}\right)^n u[n]-2\left(\frac{1}{3}\right)^n u[-n-1]$$

27.1 INVERTING BY DIVIDING

There is a third way of finding the inverse Z-transform which makes use of the form of the Z-transform equation. Remember that

$$X(z)=\sum_{n=-\infty}^{\infty} x[n]z^{-n}$$

For a right sided sequence the summation starts from 0 and we can expand the summation as

$$X(z)=\sum_{n=0}^{\infty} x[n]z^{-n}=x[0]+x[1]z^{-1}+x[2]z^{-2}+x[3]z^{-3}+\cdots$$

So if could write $X(z)$ as a polynomial in z^{-1} we could glean $x[n]$ from there. One way we can do that is by long division. Let's see how this work by example.

Example 27.5 Previously we have found that the inverse z-transform of

$$Y[z]=\frac{z^2}{\left(z-\frac{1}{2}\right)(z-1)}\qquad |z|>1$$

is given by

$$y[n] = \left(2 - \left(\frac{1}{2}\right)^n\right) u[n]$$

Or $y[0] = 2 - 1 = 1$, $y[1] = 2 - \frac{1}{2} = 1.5$, $y[2] = 2 - \left(\frac{1}{2}\right)^2 = 1.75$, etc.

Let's expand $Y(z)$ using long division.

$$\begin{array}{rllll}
 & 1 & +1.5z^{-1} & +1.75z^{-2} & \\
\hline
z^2 - 1.5z + .5 \surd & z^2 & & & \\
 & z^2 & -1.5z & +.5 & \\
\hline
 & & 1.5z & -0.5 & \\
 & & 1.5z & -2.25 & +.75z^{-1} \\
\hline
 & & & 1.75 & -0.75z^{-1} \\
 & & & 1.75 & -2.625z^{-1} \quad +0.875 \\
\hline
 & & & & 1.875z^{-1} \quad -0.875 \\
 & & & & \cdots
\end{array}$$

Comparing the results of the division, $1 + 1.5z^{-1} + 1.75z^{-2} + \cdots$, to $y[0] + y[1]z^{-1} + y[2]z^{-2} + \cdots$ we can see that $y[0] = 1$, $y[1] = 1.5$, $y[2] = 1.75$, etc.

How would we use long division if the we were dealing with a left sided sequence?

Example 27.6 Suppose we wanted to find the inverse Z-transform of

$$Y(z) = \frac{z^2}{\left(z - \frac{1}{2}\right)(z-1)} \qquad |z| < \frac{1}{2}$$

For a left sided sequence we know that the Z-transform will look something like

$$Y(z) = \sum_{n=-\infty}^{-1} y[n]z^{-n} = y[1]z + y[2]z^2 + y[3]z^4 + \cdots$$

Let's redo the previous division but let's flip the divisor

$$
\begin{array}{rllllll}
 & 2z^2 & +6z^3 & +14z^4 & & & \\
\cline{2-7}
0.5 - 1.5z + z^2 \surd & z^2 & & & & & \\
 & z^2 & -3z^3 & +2z^4 & & & \\
\cline{2-4}
 & & 3z^3 & -2z^4 & & & \\
 & & 3z^3 & -9z^4 & +6z^5 & & \\
\cline{3-5}
 & & & 7z^4 & -6z^5 & & \\
 & & & 7z^4 & -21z^5 & +14z^6 & \\
\cline{4-6}
 & & & & 15z^5 & -14z^6 & \\
 & & & & \cdots & &
\end{array}
$$

Comparing the quotient $2z^2 + 6z^3 + 10z^4 + \cdots$ to $y[1]z + y[2]z^2 + y[3]z^3 + y[4]z^4 + \cdots$ we can see that we have

$$
\begin{array}{rcl}
y[1] & = & 0 \\
y[2] & = & 2 \\
y[3] & = & 6 \\
y[4] & = & 14 \\
\vdots & & \vdots
\end{array}
$$

We can verify that this is indeed the right result by taking the inverse Z-transform using our favorite method of expanding $Y(z)/z$ using partial fraction expansion

27.2 SUMMARY

We have introduced three different ways of finding the inverse Z-transform.

1. We can evaluate the inverse Z-transform integral using the Cauchy residue theorem.

2. We can expand $X(z)/z$ using partial fraction expansion so that we can write $X(z)$ as a sum of terms of the form $z/(z - p_i)$ or $z/(z - p_1)^2$. Based on the region of convergence the corresponding discrete time sequences will be of the form $p_i^n u[n]$, $p_i^n u[-n-1]$,$np_i^n u[n]$, or $np_i^n u[-n-1]$.

3. We can carry out long division to get a polynomial in terms of z^{-n} where the way we arrange the divisor depends on the region of convergence.

Of these methods we will find that the second one is almost always the most convenient (your mileage though may differ).

27.3 EXERCISES

(Answers on the following page)

1. Find the inverse Z-transform of

$$X(z) = \frac{z/6}{z^2 - \frac{5}{6}z + 1/6}; \quad |z| > \frac{1}{2}$$

2. Find the inverse Z-transform of

$$X(z) = \frac{z}{6z^2 - 5z + 1}; \quad |z| > \frac{1}{2}$$

3. Find the inverse Z-transform of

$$X(z) = \frac{-1.5z}{z^2 - 2.5z + 1}; \quad \frac{1}{2} < |z| < 2$$

4. Find the inverse Z-transform of

$$X(z) = \frac{z/6}{z^2 - \frac{5}{6}z + 1/6}; \quad |z| < \frac{1}{3}$$

5. A causal linear time invariant system is characterized by the following input/output relationship.

$$y[n] = \frac{1}{6}x[n] + \frac{5}{6}y[n-1] - \frac{1}{6}y[n-2]$$

What is the impulse response of this system?

6. A causal linear time invariant system is characterized by the following input/output relationship.

$$y[n] = x[n] - x[n-2] - 0.25y[n-2]$$

What is the impulse response of this system?

7. If the transfer function for a linear time invariant system is given by

$$H(z) = \frac{z^2 - \frac{1}{2}z}{\left(z - \frac{1}{4}\right)^2}; \quad |z| > \frac{1}{2}$$

Find the impulse response $h[n]$.

27.4 ANSWERS

1.

$$x[n] = \left(\frac{1}{2}\right)^n u[n] - \left(\frac{1}{3}\right)^n u[n]$$

2.

$$x[n] = \left(\frac{1}{2}\right)^n u[n] - \left(\frac{1}{3}\right)^n u[n]$$

3.

$$x[n] = \left(\frac{1}{2}\right)^{|n|}$$

4.

$$x[n] = -\left(\frac{1}{2}\right)^n u[-n-1] + \left(\frac{1}{3}\right)^n u[-n-1]$$

5.

$$h[n] = \left(\frac{1}{2}\right)^n u[n] - \left(\frac{1}{3}\right)^n u[n]$$

6.

$$h[n] = -4\delta[n] + 5(0.5)^n \cos\left(\frac{n\pi}{2}\right) u[n]$$

7.

$$h[n] = \left(-\frac{1}{2}\right)^n u[n]$$

MODULE 28

Filters and Difference Equations

In the s-domain we could get an idea about the frequency domain behavior of a filter by going up and down the $j\omega$ axis. Depending on how close or how far we were from the poles and zeros of the transfer function we would get either an amplification of the response or an attenuation of the response. When we map from the Laplace domain to the z-domain the $j\omega$ axis maps into the unit circle. When we do the mapping we have to remember that we are looking at the frequency profile of a sampled signal. Recall that in order to be able to recover the frequency profile of a signal we had to keep the frequency profile free from aliasing we had to sample at more than twice the highest frequency. The other side of this is that given a sampled signal the only frequencies we can see are those that are less than half the sampling frequency. When we sampled a signal the frequency profile repeated every f_s Hz where f_s is the sampling frequency. So, as we go around the unit circle we are going from a frequency of zero Hz to a frequency of f_s Hz . Keeping this in mind if we want to look at the frequency domain behavior of a discrete time system we can do that by moving around the unit circle and seeing how close or how far we are from the poles and zeros of the transfer function.

Using the same logic we can also design systems to have certain frequency characteristics. We can place zeros close to or on the unit circle at points that correspond to frequencies we wish to attenuate or eliminate. We can put poles close to points on the unit circle that correspond to frequencies we wish to enhance. The closer the pole is to the unit circle the higher the level of amplification.

Let's see how this work using a simple example.

28.1 FREQUENCY RESPONSE AND POLES AND ZEROS

Consider the system with two complex conjugate poles and a zero at $z = 1$ shown in Figure 28.1.

In the following Figures 28.2–28.6 we plot the magnitude of the frequency response as we go around the unit circle. At a frequency of 0, because we have a zero at $z = 1$ the response is zero.

As we traverse the unit circle and come close to the pole the magnitude of the frequency response increases. It peaks at the value of ω for which we are closest to the pole and then de-

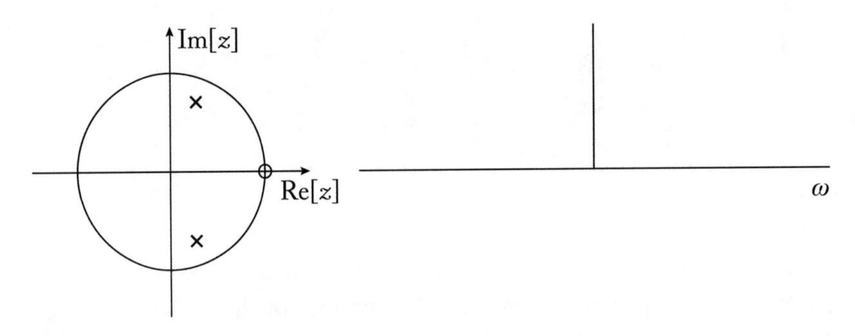

Figure 28.1: A two pole one zero system.

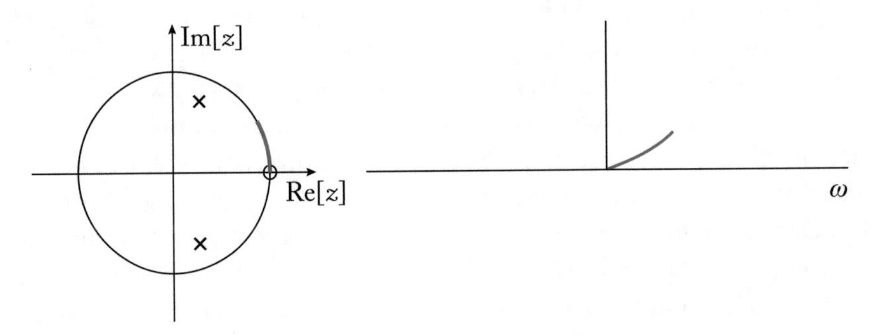

Figure 28.2: Beginning the traversal.

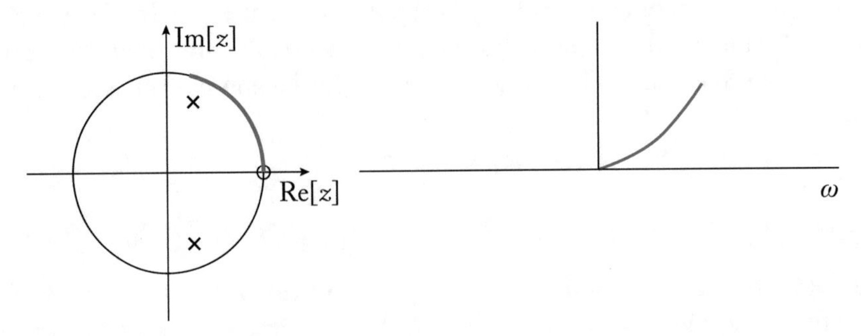

Figure 28.3: Approaching the pole.

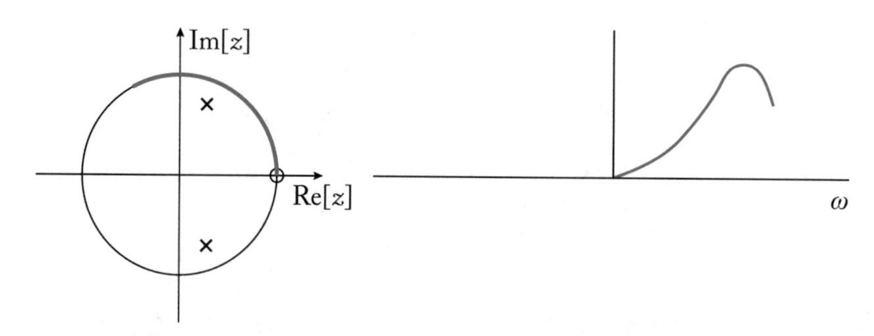

Figure 28.4: Past the pole.

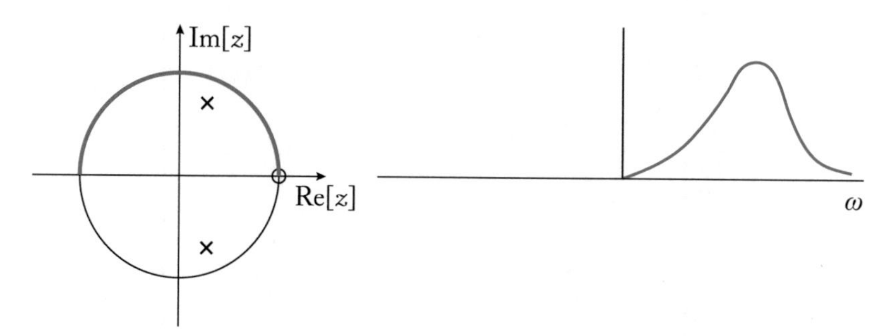

Figure 28.5: Far from the poles.

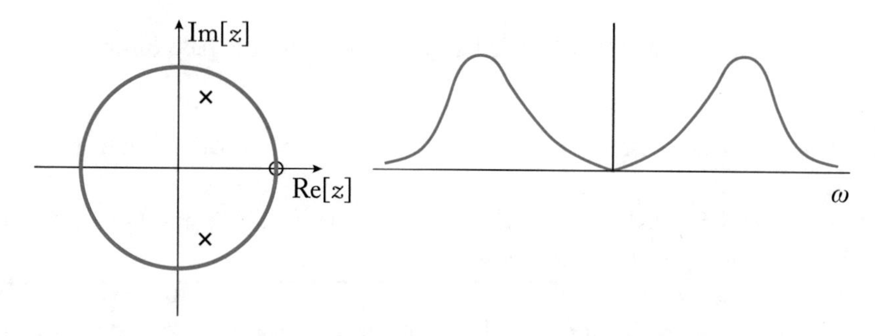

Figure 28.6: Traversing the circle in both directions.

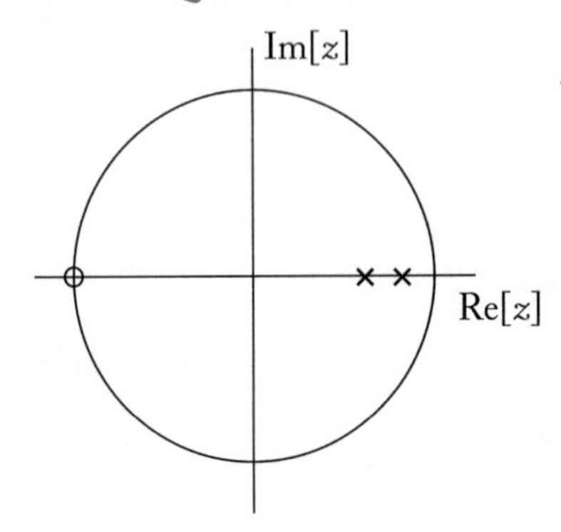

Figure 28.7: Pole and zero placement for a simple low pass filter.

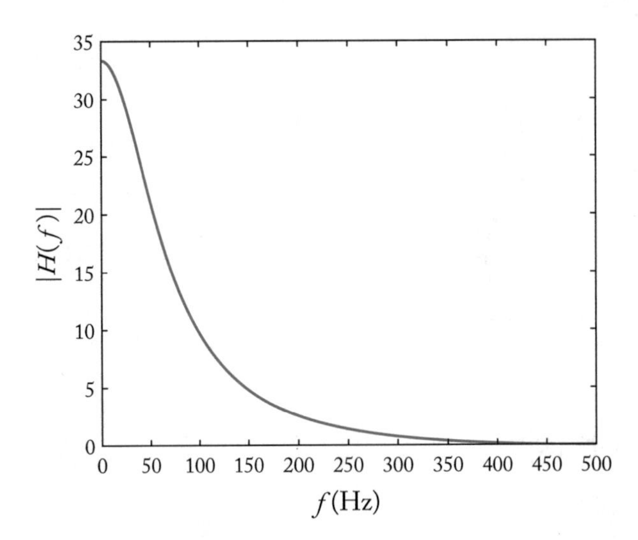

Figure 28.8: Frequency response of a simple two pole two zero low pass filter.

creases as we move away from the pole. The same happens as we move in the negative frequency direction from 0 to $-\pi$.

This filter would be called a bandpass filter—not a particularly good one but still a bandpass filter—as it lets through a band of frequencies.

If we place a pole on the positive real axis we will be closest to it when we begin our traversal of the unit circle so the signal will be enhanced at frequencies close to zero. A zero placed anywhere on the unit circle will result in the blocking of the frequencies at the zero. The placement of poles and zeros shown in Figure 28.7 results in only the lower frequencies getting through as shown in Figure 28.8, therefore, it is called a low pass filter.

In Figure 28.9 we show a system with two zeros at $z = 1$ and poles at $0.9e^{\pm j\frac{\pi}{10}}$.

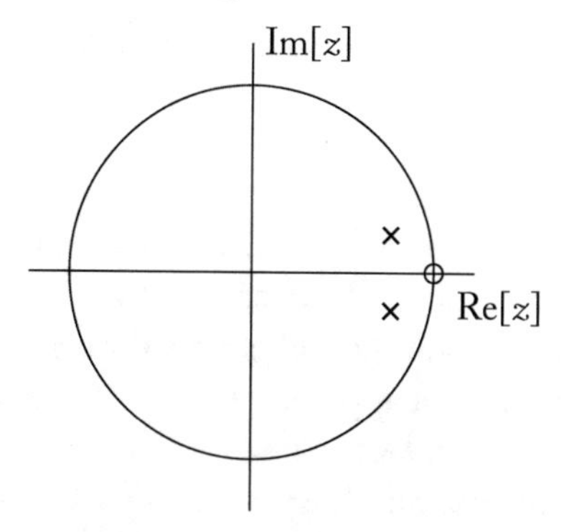

Figure 28.9: Pole and zero placement for a simple high pass filter.

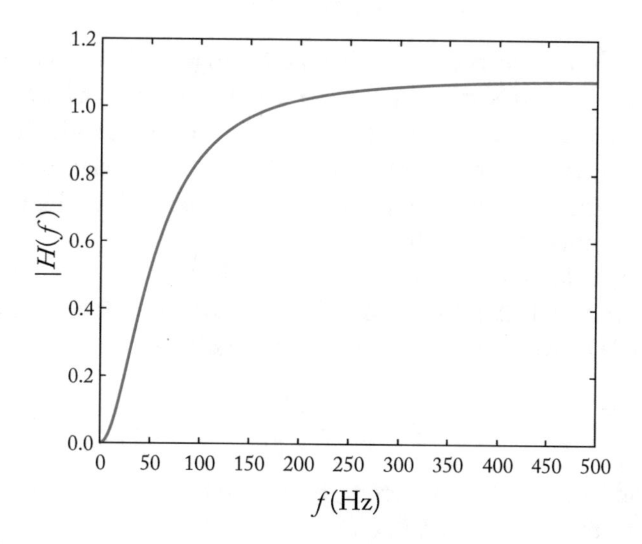

Figure 28.10: Frequency response of a simple two pole two zero high pass filter.

Now the zero frequency or dc component is blocked and the lower frequencies are attenuated. The attenuation is then canceled by the poles. This behavior can be seen in the frequency response plotted in Figure 28.10.

We have not said anything about the labels on the frequency axis in these figures. One reason is that the sampling frequency used to generate the discrete time signal from the continuous time signal has not been specified. If we know the sampling frequency ω_s in radians or f_s in Hz, we can map $\omega_s/2$ or $f_s/2$ to π and then divide up the span from $z = 1$ to $z = -1$ on the unit circle in a uniform manner between 0 and $\omega_s/2$ or $f_s/2$.

Example 28.1 Find the frequency response of a system with transfer function

$$H(z) = 1 - \frac{1}{z}; \quad |z| > 1$$

Would you classify this as a low-pass, a high-pass, or a band-pass filter?

We can do this in two different ways. The first approach is to figure out where the poles and zeros are and then qualitatively evaluate their effect for different frequency ranges. The second approach is a more exact computational approach where we actually evaluate the magnitude of the transfer function on the unit circle. Let us begin with the first approach.

We haven't been told what the sampling frequency associated with this system is so we will look at the behavior between $\omega = 0$ to $\omega = \pi$. To find the pole zero locations let's write $H(z)$ as a rational polynomial.

$$H(z) = \frac{z-1}{z}$$

This system has one pole and one zero; the pole is at the origin so it is equidistant from all points on the unit circle, and the zero is at $z = 1$. The zero at $z = 1$ will block the zero frequency and attenuate the frequencies close to zero. As we traverse the unit circle and ω increases the distance from the zero will increase and the effect of the zero will be canceled out by the effect of the pole. Thus, the low frequencies will be attenuated or blocked and the high frequencies will be unaffected. This means the system is a high-pass filter.

The second way is to evaluate the magnitude of the transfer function on the unit circle. For all points on the unit circle the magnitude of z is 1, so on the unit circle $z = e^{j\omega}$. To evaluate $H(z)$ on the unit circle all we need to do is to replace z with $e^{j\omega}$.

$$H(e^{j\omega}) = \frac{e^{j\omega} - 1}{e^{j\omega}}$$

To find the magnitude we take the product of this transfer function with its complex conjugate and then take the square root of the product.

$$\begin{aligned} |H(e^{j\omega})|^2 &= \frac{e^{j\omega} - 1}{e^{j\omega}} \cdot \frac{e^{-j\omega} - 1}{e^{-j\omega}} \\ &= \frac{1 - e^{j\omega} - e^{-j\omega} + 1}{1} \\ &= 2 - 2\cos(\omega) \end{aligned}$$

Therefore

$$|H(e^{j\omega})| = \sqrt{2 - 2\cos(\omega)}$$

We have plotted this in Figure 28.11. We can see that this is a high pass response even if the filter characteristics leave something to be desired. Compare this with the high-pass filter response shown in Figure 28.10. That was a two pole two zero system as oppose to this system which

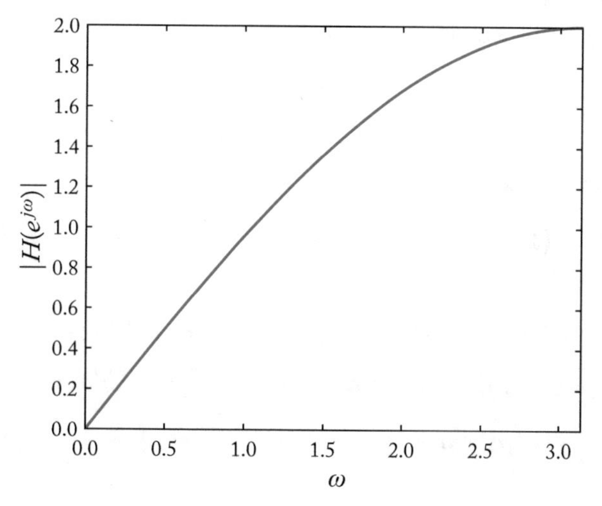

Figure 28.11: Frequency response of a simple single pole single zero high pass filter.

consists of only a single pole and a single zero. You can see how much better the magnitude response is when you increase the number of poles and zeros.

28.2 FINDING THE INPUT OUTPUT RELATIONSHIP

Given the pole-zero placement we cannot only obtain the frequency response, we can also obtain the difference equation for the system which has this particular frequency response. In this particular example the zero is at $z = 1$. Let's assume the poles are at p_1 and p_1^*. Then the transfer function is given by

$$H(z) = \frac{z-1}{(z-p_1)(z-p_1^*)} = \frac{z-1}{z^2 - p_1 z - z p_1^* + |p_1|^2}$$

or

$$H(z) = \frac{z-1}{z^2 - 2Re(p_1)z + |p_1|^2}$$

where we have used the fact that the sum of a complex number and its complex conjugate is equal to twice the real part of the complex number. If we now divide top and bottom by z^2 we get

$$H(z) = \frac{z^{-1} - z^{-2}}{1 - 2Re(p_1)z^{-1} + |p_1|^2 z^{-2}}$$

To get the difference equation we begin with the fact that

$$Y(z) = H(z)X(z)$$

Substituting for $H(z)$ in this equation we get

$$Y(z) = \frac{z^{-1} - z^{-2}}{1 - 2Re(p_1)z^{-1} + |p_1|^2 z^{-2}} X(z)$$

Multiplying both sides by the denominator of the right hand side

$$Y(z)\left(1 - 2Re(p_1)z^{-1} + |p_1|^2 z^{-2}\right) = X(z)\left(z^{-1} - z^{-2}\right)$$

or

$$Y(z) - 2Re(p_1)z^{-1}Y(z) + |p_1|^2 z^{-2}Y(z) = z^{-1}X(z) - z^{-2}X(z)$$

Using the time shift property the inverse Z-transform of both sides is

$$y[n] - 2Re(p_1)y[n-1] + |p_1|^2 y[n-2] = x[n-1] - x[n-2]$$

or

$$y[n] = x[n-1] - x[n-2] + 2Re(p_1)y[n-1] - |p_1|^2 y[n-2]$$

Example 28.2 Find the input output relationship for the system with transfer function

$$H(z) = \frac{z^3}{z^3 + 2z^2 - 3}$$

Dividing top and bottom by z^3 we get

$$H(z) = \frac{1}{1 + 2z^{-1} - 3z^{-3}}$$

Setting this equal to $Y(z)/X(z)$ and cross multiplying we get

$$Y(z)\left(1 + 2z^{-1} - 3z^{-3}\right) = X(z)$$

or

$$Y(z) + 2z^{-1}Y(z) - 3z^{-3}Y(z) = X(z)$$

Using linearity and the time shift property

$$\mathcal{Z}\left[x[n - n_o]\right] = z^{-n_o}X(z)$$

we get

$$y[n] + 2y[n-1] - 3y[n-3] = x[n]$$

or

$$y[n] = x[n] - 2y[n-1] + 3y[n-3]$$

28.3 DESIGNING A SIMPLE DISCRETE TIME FILTER

What if we are interested in specifying the frequency response of the system in terms of Hz? In order this we first have to know the sampling frequency f_s. This is because all locations around the unit circle (if we are thinking of frequency in Hz) is determined by the sampling frequency. Let's suppose we want our system to enhance the components of the signal at frequency f_p Hz and attenuate the signal at frequency f_z Hz. In order to place the complex conjugate poles to at f_p with magnitude r_p and the complex conjugate zeros at f_z with magnitude r_z our system transfer function will be.

$$H(z) = \frac{\left(z - r_z e^{j2\pi \frac{f_z}{f_s}}\right)\left(z - r_z e^{-j2\pi \frac{f_z}{f_s}}\right)}{\left(z - r_p e^{j2\pi \frac{f_p}{f_s}}\right)\left(z - r_p e^{-j2\pi \frac{f_p}{f_s}}\right)}$$

Note that the closer the magnitudes of the poles and zeros, r_p and r_z, are to 1 the more will be their impact. For notational convenience let's define

$$\begin{aligned} \alpha &= r_z \cos\left(2\pi \frac{f_z}{f_s}\right) \\ \beta &= r_p \cos\left(2\pi \frac{f_p}{f_s}\right) \end{aligned}$$

where α is the real part of the zero and β is the real part of the pole. Following the same procedure as above we can find the difference equation for this filter as

$$y[n] = x[n] - 2\alpha x[n-1] + r_z^2 x[n-2] + 2\beta y[n-1] - r_p^2 y[n-2]$$

A MATLAB implementation of this filter is shown below.

```
fs = 16000 % sampling frequency
duration = 2; % duration in seconds
Ns = duration*fs; % Number of samples
fz = 1000; %location of zero in Hz
fp = 500; % location of pole in Hz
rz = 1; % magnitude of zero
rp = .99; % magnitude of pole;
alpha  = rz*cos((fz/fs)*2*pi)
beta = rp*cos((fp/fs)*2*pi)
y = zeros(Ns,1);
t = [1:1:Ns]/fs;
```

```
while(1)

        fprobe = input('Enter probe frequency: ')
         x = cos(2*pi*fprobe*t);
sound(x,fs);
pause
for i= 3:1:Ns
    y(i) = x(i) - 2*alpha*x(i-1) + rz*rz*x(i-2) + 2*beta*y(i-1)
                 - rp*rp*y(i-2);
end

sound(y,fs);

% plot input and output
figure
subplot(2,1,1)
plot(t,x)
axis([0,.1,-2,2])
title('input waveform')
subplot(2,1,2)
plot(t,y)
axis([0,.1,-2,2])
title('output waveform')

end
```

We can see the frequency shaping effect of the filter if we excite the filter with random noise instead of different probe frequencies. In the MATLAB program this would mean replacing the line

```
          x = cos(2*pi*fprobe*t);
```

with

```
            x = randn(Ns,1);
```

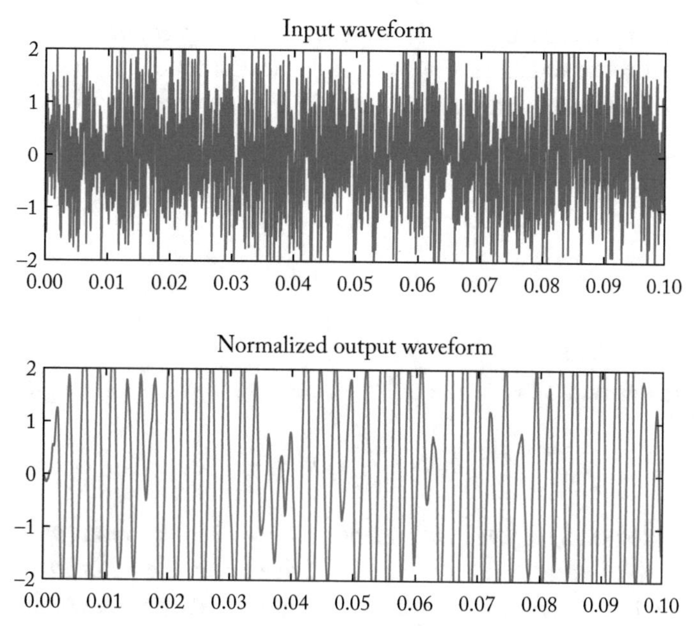

Figure 28.12: Input and output signals for a two pole two zero system with the zeros at the origin and the poles near the unit circle.

This will generate a Gauassian random number sequence of length `Ns`. Lets also move the zeros to the origin by setting `rz = 0` so we see just the effect of the poles. To magnify the effect of the poles we move them close to the unit circle by setting `rp = 0.99`. The input and output waveforms are shown in Figure 28.12. The output was normalized because as a result of the poles being close to the unit circle the magnitude of the response was significantly larger than the magnitude of the input signal. You can see how the filter is blocking most of the frequency content except for the frequency where the poles are located.

Example 28.3 Suppose you have a signal sampled at 480 samples per second which has been contaminated by the power line so that you have an unwanted 60 Hz component in the signal. You want to block that component. You can block or attenuate a frequency component using a pair of zeros in the system. Where would you place these zeros?

The zeros are going to be a complex conjugate pair. We want to block the signal not attenuate it so the zeros will be placed on the unit circle. This means the magnitude of the zeros will be unity. The angle of the zeros is given by

$$\theta_z = \frac{60}{480} \cdot 2\pi = \frac{\pi}{4}$$

Therefore the zeros will be at $e^{\pm j\frac{\pi}{4}}$, or in cartesian coordinates at $0.707 \pm j0.707$.

Example 28.4 Suppose you have a signal sampled at 480 samples per second with a component at 120 Hz that you want to enhance. Given a choice of the following pole pairs which should you pick if you want maximum enhancement of this component?

1. $0.6 \pm j0.6$
2. $\pm 0.9j$
3. 0.45 ± 0.78
4. $\pm 0.75j$

If the component is at 120 Hz the poles we need should have an angle of

$$\theta_p = \frac{120}{480} \cdot 2\pi = \frac{\pi}{2}$$

Choices 1 and 3 correspond to poles with angles clearly different from $\pi/2$. Choices 2 and 4 both correspond to poles with angles of $\pm\pi/2$. Of these two the poles in option 2, $\pm 0.9j$ are closer to the unit circle and hence with more of an impact. Therefore, we should place poles at $\pm 0.9j$.

We can also look at the impulse response of this filter. Here is the MATLAB code we will be using:

```
clear all
%%%%%%%%%%%%%%%%%%%%%%%%%%%%%%%%%%%%%%%%%%%%%%%%%%%%%%%%%%%%
%                                                          %
%                    Initialization                        %
%                                                          %
%%%%%%%%%%%%%%%%%%%%%%%%%%%%%%%%%%%%%%%%%%%%%%%%%%%%%%%%%%%%

rp = 1; % Magnitude of pole
fp = 10; % pole location in Hz
fs = 500; % sampling frequency in Hz
Ns = 500; % Number of samples
t = [1:1:Ns]/fs; % time points
alpha = rp * cos(2*pi*(fp/fs)); % real part of pole
y = zeros(Ns,1);
x = zeros(Ns,1);
```

```
%%%%%%%%%%%%%%%%%%%%%%%%%%%%%%%%%%%%%%%%%%%%%%%%%%%%%%%%%%%%%%%
%                                                             %
%                       Filtering                             %
%                                                             %
%%%%%%%%%%%%%%%%%%%%%%%%%%%%%%%%%%%%%%%%%%%%%%%%%%%%%%%%%%%%%%%

x(1) = 1;
y(1) = x(1);
y(2) = x(2) + 2*alpha*y(1);

for i= 3:1:Ns
    y(i) = x(i)+2*alpha*y(i-1)-rp*rp*y(i-2);
end

plot(t,y)
```

In this program we have set the input to be an impulse. The value of $x(1)$ is set to 1 and all other values of the input remain set to 0. As in the previous program α is the real part of the pole and rp is the magnitude of the pole. We have set the angle of the pole to be 10 Hz. As the sampling frequency is 500 Hz the poles are at $e^{\pm j \frac{20}{500}\pi}$.

Let's first look at what happens when we place the poles on the unit circle. This is the same as putting poles on the $j\omega$ axis in the Laplace domain. Recall that the Laplace transform of a pure sinusoid results in poles on the $j\omega$ axis. Therefore, we would expect the impulse response of this system to be a pure sinusoid. This is the case as we can see from the response shown in Figure 28.13.

If we move the poles back a bit from the origin by setting the magnitude to be 0.99 we obtain the impulse response shown in Figure 28.14.

We can see that after some oscillations the response dies down. If we pull the poles even further back with a magnitude of 0.9 the damping effect is much more pronounced as evident from the plot of the response in Figure 28.15.

28.4 SUMMARY

In this module we showed how we can figure out the frequency response of a system based on the location of its poles and zeros. We only examined stable systems so we kept our poles inside the unit circle.

1. Moving around the unit circle in the z-domain is like moving along the $j\omega$ axis in the Laplace domain.

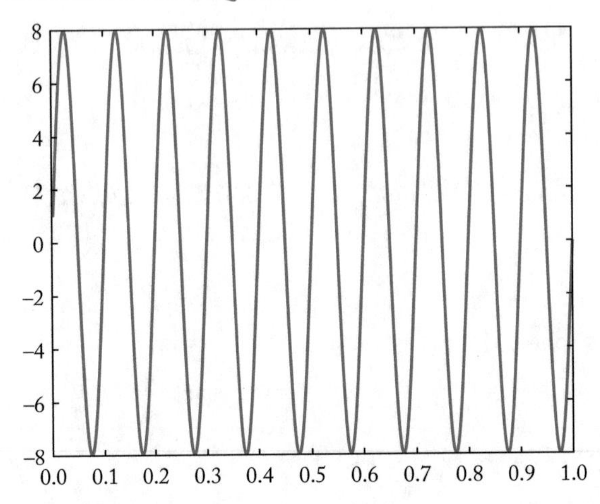

Figure 28.13: Response of a two pole two zero system with the zeros at the origin and the poles on the unit circle at an angle corresponding to 10 Hz.

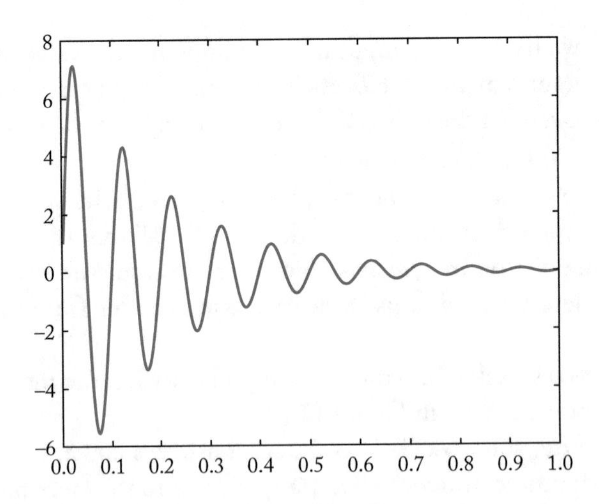

Figure 28.14: Response of a two pole two zero system with the zeros at the origin and the poles near the unit circle with a magnitude of 0.99 and an angle corresponding to 10 Hz.

2. The frequency demarcation depends on the sampling frequency with the angle of zero corresponding to zero and the angle π corresponding to half the sampling frequency.

3. The points on the unit circle which are closer to poles correspond to frequencies which are enhanced by the system. The closer the pole is to the unit circle the more pronounced the effect.

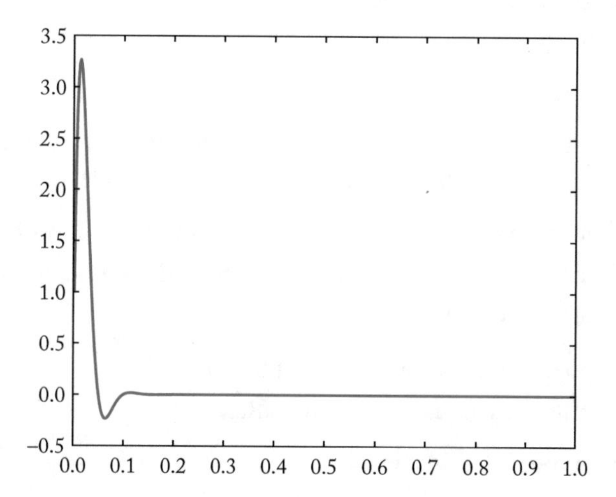

Figure 28.15: Response of a two pole two zero system with the zeros at the origin and the poles near the unit circle with a magnitude of 0.9 and an angle corresponding to 10 Hz.

4. The points on the unit circle which are closer to zeros correspond to frequencies which are attenuated by the system. The closer the zero is to the unit circle the more pronounced the effect.

5. We can design a simple system by placing poles near points of the circle corresponding to frequencies we want to enhance and zeros near points of the circle corresponding to frequencies we want to attenuate. From the pole zero plot we can obtain the transfer function $H(z)$.

6. Given a transfer function we can write it in terms of z^{-1} from which we can extract a difference equation which can be easily implemented.

28.5 EXERCISES

(Answers on the following page)

1. You have a real signal which has been sampled at 240 samples/second. You want to design a two pole, two zero discrete time filter with the following specifications.
   - The filter should block the component of the signal at 60 Hz.
   - The filter should enhance the component of the signal at 30 Hz.
   - The magnitudes of the poles should be 0.9.

   (a) Specify the pole zero locations for this filter.

   (b) Specify the difference equation for this filter.

2. Given the input output relationship

$$y[n] = x[n] - x[n-2]$$

   (a) Where are the poles and zeros of this filter?

   (b) What kind of a filter is it?

3. A biological signal has been sampled at 480 samples per second. There is a component of the signal around 120 Hz which is of particular interest and there is a 60 cycle interference from the power line. Specify the locations of the poles and zeros of a two-zero two pole discrete time filter which will remove this 60 cycle hum while enhancing the component of the signal at 120 Hz.

4. Given a system with a pole at $z = -0.9$ and a zero at $z = 0.9$, would the best description of this system be that it was a low pass filter or a high pass filter.

5. Given a system with a pole at $z = 0.9$ and a zero at $z = -0.9$, would the best description of this system be that it was a low pass filter or a high pass filter.

6. Is the system described by the difference equation

$$y[n] = x[n] + x[n-1]$$

   a low pass filter, a band pass filter, or a high pass filter?

7. Is the system described by the difference equation

$$y[n] = x[n] - x[n-1]$$

   a low pass filter, a band pass filter, or a high pass filter?

8. Is the system described by the difference equation

$$y[n] = x[n] - 0.5y[n-2]$$

   a low pass filter, a band pass filter, or a high pass filter?

28.6 ANSWERS

1. (a) Zero locations: $z = \pm j$, Pole locations: $z = 0.9e^{\pm j\frac{\pi}{4}}$.

   (b)

$$y[n] = x[n] + x[n-2] + \frac{1.8}{\sqrt{2}}y[n-1] - 0.81y[n-2]$$

2. (a) Zeros at $z = \pm 1$, poles at $z = 0$.

   (b) Band pass filter.

3. Poles at $z = \pm 0.9j$, Zeros at $\frac{1}{\sqrt{2}} \pm j\frac{1}{\sqrt{2}}$.

4. High pass filter.

5. Lowpass filter.

6. Low pass filter.

7. High pass filter.

8. Band pass filter.

MODULE 29

Discrete-Time Feedback Systems

As with continuous time systems complex discrete time systems depend on feedback for many of their characteristics. Just as in the case of continuous time systems we can use feedback to move the poles and zeros of a system to stabilize it or to shape its response. In practical terms there are however differences—some of the computational aspects are different enough to trip you up if you are not careful.

Let's begin with obtaining the general transfer function for a feedback system in the z domain. Consider the system shown in Figure 29.1.

As in the case of our analysis of the feedback system in the Laplace domain let's define $E(z)$ to be the output of the summer.

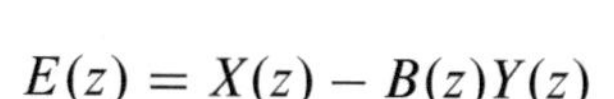

$$E(z) = X(z) - B(z)Y(z)$$

then

$$\begin{aligned} Y(z) &= E(z)A(z) \\ &= (X(z) - B(z)Y(z))A(z) \\ &= X(z)A(z) - A(z)B(z)Y(z) \end{aligned}$$

Therefore,

$$Y(z)\,(1 + A(z)B(z) = A(z)X(z))$$

and the transfer function is

$$H(z) = \frac{A(z)}{1 + A(z)B(z)}$$

29.1 STABILIZING A SYSTEM

Let's take a simple example of an unstable system and see how we can use feedback to move the poles and stabilize the system. Lets take $A(z)$ to be our favorite one pole system but with a pole outside the unit circle

$$A(z) = \frac{1}{1 - 2z^{-1}} = \frac{z}{z - 2}$$

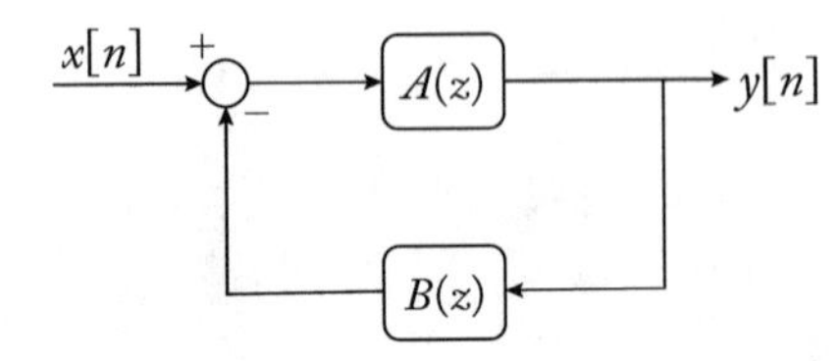

Figure 29.1: A feedback system.

and let's let $B(z)$ be a gain Kz^{-1}. Notice that unlike the situation in the continuous time case for this simple example we have used $B(z) = Kz^{-1}$ instead of $B(z) = K$. There is a reason for this. Usually in a discrete time system the feedback from the output is not instantaneous and there is a delay between a signal appearing at the output and being fed back to appear as the feedback component. Think in terms of the digital circuits you might have built. The z^{-1} in the feedback path accounts for this delay. The overall transfer function with this $B(z)$ is

$$H(z) = \frac{A(z)}{1 + Kz^{-1}A(z)}$$

Substituting the expression for $A(z)$ we get

$$H(z) = \frac{\frac{z}{z-2}}{1 + Kz^{-1} \cdot \frac{z}{z-2}} = \frac{z}{z - 2 + K}$$

from which we can see that we have a pole at $2 - K$.

It is at this point you have to be a bit careful. In the case of the continuous-time systems in order for a system to be causal the region of convergence is to the right of a boundary—where the boundary is defined by the right-most pole. For stability we want the $j\omega$ axis to be in the region of convergence. Therefore, for a causal system to be stable we need the poles to be in the left half plane. In this particular case that would mean picking $K < -1$. However, we are analyzing a discrete-time system. And for a discrete-time system to be causal the region of convergence is to the outside of a circular boundary—the radius of the circular boundary being defined by the poles furthest from the origin. For the system to be stable we want the region of the convergence to include the unit circle. Therefore, for a causal system to be stable we need the poles to be inside the unit circle. That is we need $|2 - K| < 1$. This will happen when $1 < K < 3$. This is a point which trips up a lot of people. Be aware of it.

Example 29.1 Consider the situation we have all probably seen or participated in at one time or another. Someone steps up to a microphone, begins speaking, and there is this annoying squeal that has everyone wincing. You turn down the volume a bit or move the speakers further away from the microphone and the squealing stops. Let's model this system and see why this is so.

Let's make the forward path simply be the gain of the amplifier. Therefore,

$$A(z) = A$$

In the feedback path we will have some gain or attenuation α and a delay—let's assume the delay is one time instant. So,

$$B(z) = \alpha z^{-1}$$

The feedback is positive so the output of the summer will be

$$E(z) = X(z) + \alpha z^{-1} Y(z)$$

and

$$Y(z) = AE(z) = AX(z) + A\alpha Y(z)$$

Therefore, the transfer function for the overall system is

$$\begin{aligned} H(z) &= \frac{A}{1 - A\alpha z^{-1}} \\ &= \frac{Az}{z - A\alpha} \end{aligned}$$

This system has a pole at $A\alpha$. If the value of $A\alpha$ is greater than one then the pole will be outside the unit circle and we will get the squealing. We can get rid of the squealing by making sure that $|A\alpha| < 1$. We can do this in a number of different ways. We can reduce the gain of the microphone to reduce A. Or we can move the speakers further away from the microphone so the value of α goes down and the pole moves inside the unit circle.

29.2 PRODUCING A DESIRED OUTPUT

As in the case of continuous-time systems we often want the discrete-time system to generate a desired output. We accomplish this by driving the error between the desired response and the actual response to zero. In the control systems literature the system whose response we are trying to control is often called the *plant* while the system we use to control the system is called the *controller*. The common setup for this is shown in Figure 29.2.

In this figure the block with transfer function $A(z)$ is the plant and the block marked $B(z)$ is the controller. Unlike the continuous time control system the output is not directly fed back. Instead there is a delay in the feedback path for the reason mentioned above. As in the case of continuous time systems the control signal can be proportional to the error signal, or it can be the sum of a term proportional to the error and the integral of the error, or it can have a term which is proportional to the error and a term which is the derivative of the error, or it can be the sum of terms proportional to the error the integral of the error and the derivative of the error. Let's take a look at the simplest case first.

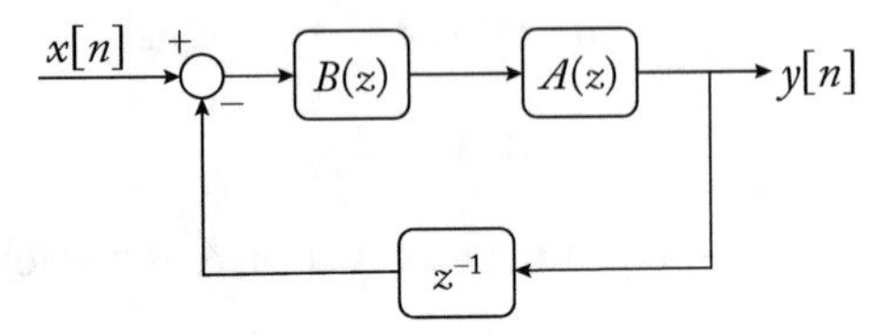

Figure 29.2: The control system setup.

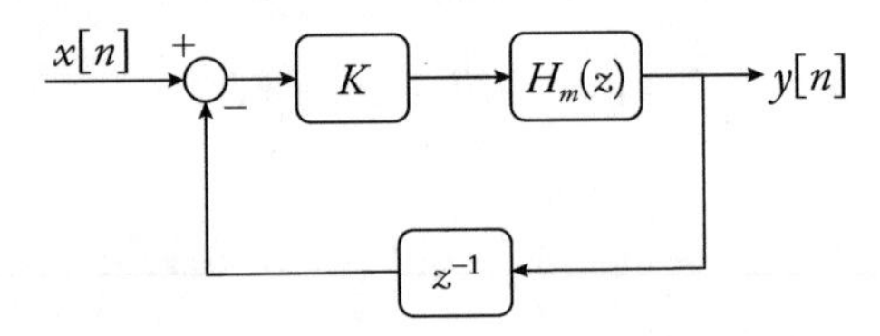

Figure 29.3: A simple motor controller.

29.3 PROPORTIONAL CONTROL

Suppose we want issue commands to move a robotic arm and we have a really simple model for the motor. At each time instant the new position $y[n]$ is the sum of the previous position and the input at that time.

$$y[n] = y[n-1] + x[n-1]$$

The transfer function for this simple motor model can be obtained by taking the Z transform of both sides.

$$Y(z) = z^{-1}Y(z) + z^{-1}X(z)$$

Moving the $Y(z)$ terms to the left hand side of the equation and finding the ratio of $Y(z)$ to $X(z)$ we get the transfer function for the motor as

$$H_m(z) = \frac{z^{-1}}{1 - z^{-1}}$$

We would like to control this motor so that we can give it a position and have the motor track the position. That is, given a desired position $x[n]$ we would like the difference between the measurement of the actual position $y[n]$ and the desired position $x[n]$ to be zero. Let's try to do this using proportional control. The block diagram for the overall system is shown in Figure 29.3.

Notice the z^{-1} in the feedback loop. We have included that because we would need at least one time instant to feed the measurement of the position back to the input. Let's now find the transfer function for the overall system.

$$
\begin{aligned}
E(z) &= X(z) - z^{-1}Y(z) \\
Y(z) &= E(z) \cdot KH_m(z) \\
&= \frac{Kz^{-1}}{1-z^{-1}}E(z) \\
&= \frac{Kz^{-1}\left(X(z) - z^{-1}Y(z)\right)}{1-z^{-1}} \\
&= \frac{Kz^{-1}X(z) - Kz^{-2}Y(z)}{1-z^{-1}}
\end{aligned}
$$

Multiplying both sides by $1 - z^{-1}$ and rearranging we can obtain the overall transfer function as

$$
H(z) = \frac{Y(z)}{X(z)} = \frac{Kz^{-1}}{Kz^{-2} - z^{-1} + 1}
$$

Multiplying top and bottom by z^2 we get

$$
H(z) = \frac{Kz}{z^2 - z + K}
$$

The poles for this system are at

$$
z_{1,2} = \frac{1}{2} \pm \frac{\sqrt{1-4K}}{2}
$$

We can see that we will get real poles as long as $K < 1/4$ and complex poles otherwise. Furthermore the poles migrate outside the unit circle for $K > 1$. So we can expect the system to go unstable for values of $K > 1$.

After all of this—to paraphrase Kipling[1]—we still really haven't answered that age old question "but does it work?" We can't build one of these right now but we could simulate it. We can use the version of the transfer function in terms of z^{-1} to generate the difference equation for this system and use that to simulate the system.

Given that

$$
H(z) = \frac{Y(z)}{X(z)} = \frac{Kz^{-1}}{Kz^{-2} - z^{-1} + 1}
$$

$$
Y(z)\left(Kz^{-2} - z^{-1} + 1\right) = Kz^{-1}X(z)
$$

or

$$
Kz^{-2}Y(z) - z^{-1}Y(z) + Y(z) = Kz^{-1}X(z)
$$

Taking the inverse Z transform we get the difference equation

$$
y[n] = Kx[n-1] + y[n-1] - Ky[n-2]
$$

[1]Google *The Conundrum of the Workshops.*

Here is some MATLAB code to implement this difference equation.

```
clear all
Ns = 500; % number of samples
y = zeros(Ns,1); % motor position
x = zeros(Ns,1); % desired position
K = 0.2; % Proportional controller

%%%%%%%%%%%%%%%%%%%%%%%%%%%%%%%%%%%%%%%%%%%%%%%%
%                                              %
%       Generate the desired signal            %
%                                              %
%%%%%%%%%%%%%%%%%%%%%%%%%%%%%%%%%%%%%%%%%%%%%%%%

for k= 1:1:100

  x(k) = 1;
end

%%%%%%%%%%%%%%%%%%%%%%%%%%%%%%%%%%%%%%%%%%%%%%%%
%                                              %
%            Simulate the system               %
%                                              %
%%%%%%%%%%%%%%%%%%%%%%%%%%%%%%%%%%%%%%%%%%%%%%%%
y(2) = K*x(1);
for k=3:1:Ns
    y(k) = K*x(k-1) + y(k-1) - K*y(k-2);
end

%%%%%%%%%%%%%%%%%%%%%%%%%%%%%%%%%%%%%%%%%%%%%%%%
%                                              %
%            Plot                              %
%                                              %
%%%%%%%%%%%%%%%%%%%%%%%%%%%%%%%%%%%%%%%%%%%%%%%%


figure
subplot(2,1,1)
plot(x)
```

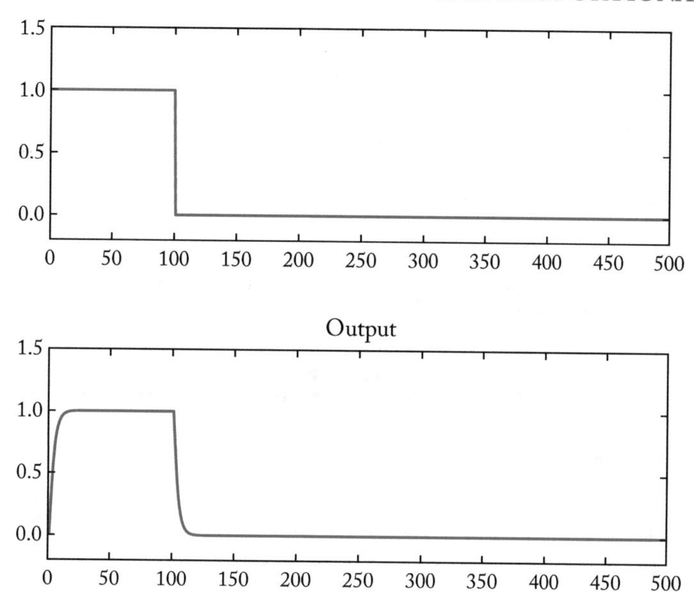

Figure 29.4: A simple motor controller, with gain $K = 0.2$.

```
axis([0 Ns -0.2 1.5])
subplot(2,1,2)
title('Control Signal')
plot(y)
axis([0 Ns -0.2 1.5])
title('Output')
```

Let's run the simulation for different values of K. Let's begin with $K = 0.2$. We want the output to be 1 for a while and then go to 0. The result of the simulation is shown in Figure 29.4.

Notice that it takes a few samples—about sixteen—for the system output to track the input. The response is smooth as we would expect for a system with only real poles.

Lets increase the gain to $K = 0.5$. The resulting performance is shown in Figure 29.5.

Notice a couple of things about this response. First the output follows the input much more rapidly than in the previous case. The bad news is that we have some ringing during sharp transitions in the control signal. This is a result of the poles becoming complex. This ringing may or may not be acceptable depending on the application. However, what is definitely not acceptable is what happens when the gain is increased beyond 1. Figure 29.6 shows what happens when we set $K = 1.001$.

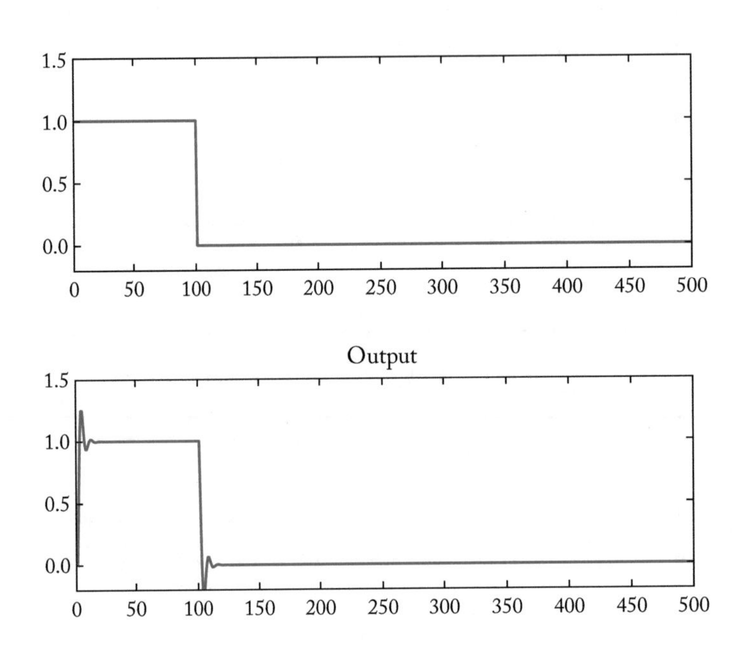

Figure 29.5: A simple motor controller, with gain $K = 0.5$.

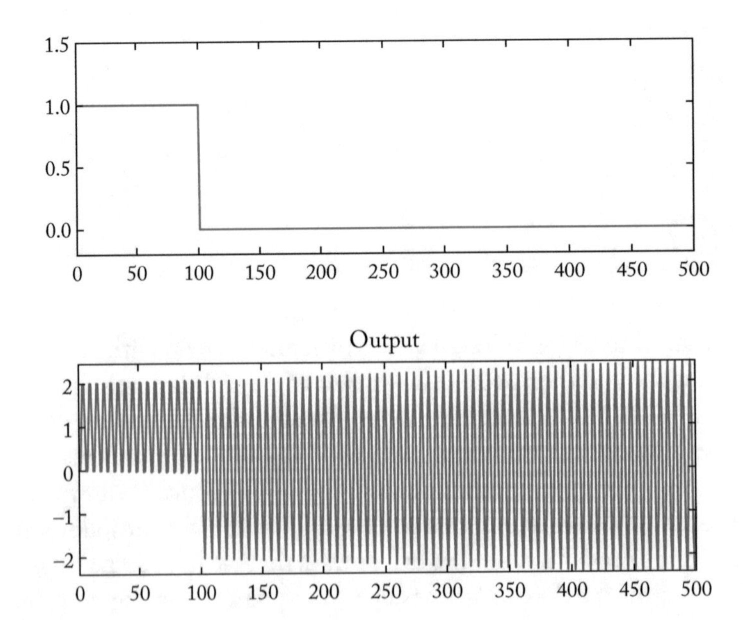

Figure 29.6: A simple motor controller, with gain $K = 1.001$.

Clearly not a nice situation.

29.4 DISCRETE TIME PID CONTROLLERS

Just as in the continuous time case, in the discrete-time case we may want to get rid of small steady state errors using integral control or use some level of proactive control using a derivative term in the control. To see what some possible equivalents are in the discrete world let's take a look at the equivalent systems for integration and differentiation in the discrete world. The analog to integration in continuous time is summation in discrete time. So the analog to an integrator in continuous time with input-output relationship

$$y(t) = \int_{-\infty}^{t} x(\tau)d\tau$$

is the system in discrete time

$$y[n] = \sum_{k=-\infty}^{n} x[k]$$

The impulse response of this discrete-time integrator is

$$h[n] = \sum_{k=-\infty}^{n} \delta[k] = \begin{cases} 1 & n \geq 0 \\ 0 & \text{otherwise} \end{cases} = u[n]$$

The Z-transform of this impulse response is

$$H(z) = \frac{1}{1 - z^{-1}}; \quad |z| > 1$$

Therefore if we wanted to use a PI controller, the controller instead of simply being K would be given by $K_p + K_i/(1 - z^{-1})$.

Similarly a derivative in continuous time can be approximated with a difference in discrete time

$$y[n] = x[n] - x[n-1]$$

We can see that the Z-transform of the derivative operator would be $1 - z^{-1}$. Therefore the PID controller would have the form $K_p + K_i/(1 - z^{-1}) + K_d(1 - z^{-1})$.

As in the continuous time case the integrator in the controller allows you to reduce steady state error and the derivative allows you to anticipate and therefore speeds up your response.

Here is some MATLAB code for a PID controller for the simple motor.

```
clear all
Ns = 500; % number of samples
```

```
y = zeros(Ns,1); % motor position
x = zeros(Ns,1); % desired position
Kp = 0.3; % Proportional controller
Ki = 0.001; % Integral controller
Kd = 0.2; % derivative term
a = Kp + Ki + Kd;
b = Kp + 2*Kd;


%%%%%%%%%%%%%%%%%%%%%%%%%%%%%%%%%%%%%%%%%%%%
%                                          %
%     Generate the desired signal          %
%                                          %
%%%%%%%%%%%%%%%%%%%%%%%%%%%%%%%%%%%%%%%%%%%%

for k= 1:1:100

  x(k) = 1;
end
%%%%%%%%%%%%%%%%%%%%%%%%%%%%%%%%%%%%%%%%%%%%
%                                          %
%         Simulate the system              %
%                                          %
%%%%%%%%%%%%%%%%%%%%%%%%%%%%%%%%%%%%%%%%%%%%
y(2) = a*x(1);
y(3) = a*x(2) -b*x(1) + 2*y(2);
y(4) = a*x(3) -b*x(2) + Kd*x(1) +2*y(3) -(a+1)*y(2);
for k=5:1:Ns
    y(k) = a*x(k-1) - b*x(k-2) + Kd*x(k-3) + 2*y(k-1) - (a+1)*y(k-2)
          + b*y(k-3) - Kd*y(k-4);
end

%%%%%%%%%%%%%%%%%%%%%%%%%%%%%%%%%%%%%%%%%%%%
%                                          %
%         Plot                             %
%                                          %
%%%%%%%%%%%%%%%%%%%%%%%%%%%%%%%%%%%%%%%%%%%%
```

```
figure
subplot(2,1,1)
plot(x,'-k','LineWidth',2)
axis([0 Ns -0.2 1.5])
title('Control Signal')
subplot(2,1,2)
plot(x,':k','LineWidth',2)
hold on
plot(y,'-k','LineWidth',2)
axis([0 Ns -0.2 1.5])
title('Output')
hold off
```

29.5 SUMMARY

In this module we introduced the idea of feedback and showed how we can use feedback to stabilize a system, to make it more responsive, to obtain the inverse of a system, and to control the output of a system in order to get a desired response.

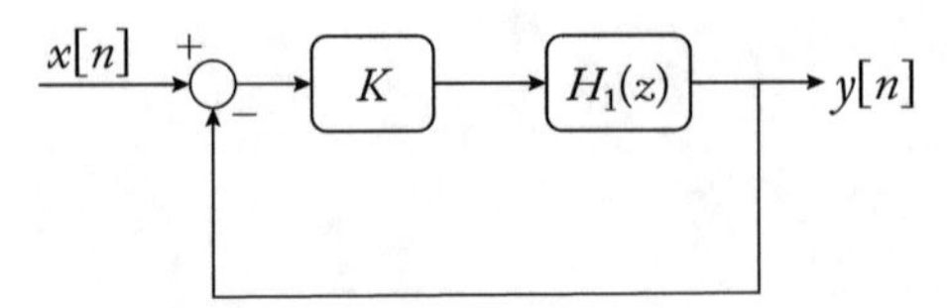

Figure 29.7: A simple discrete time feedback system.

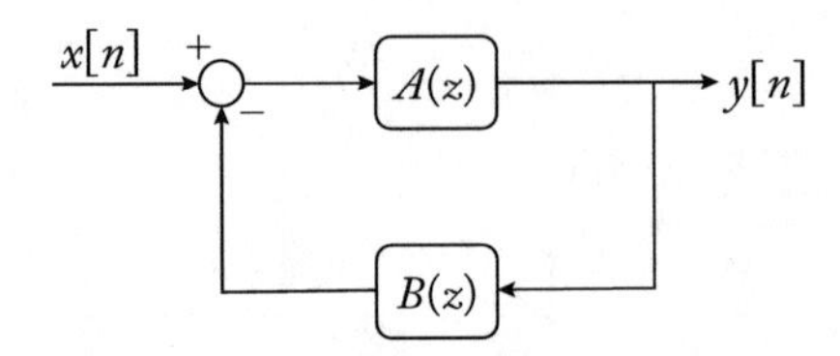

Figure 29.8: Another simple feedback system.

29.6 EXERCISES

(Answers on the following page)

1. The difference equation describing the system with transfer function $H_1(z)$ (Figure 29.7) is given by

$$y[n] = x[n] + 2y[n-1]$$

   (a) Find the difference equation relating the input and output of the overall system.

   (b) Find the values of K for which the system is stable.

2. In the system shown in Figure 29.8

$$\begin{aligned} A(z) &= 1 - z^{-1} \\ B(z) &= z^{-1} \end{aligned}$$

   Find the difference equation relating $x[n]$ and $y[n]$.

3. In Figure 29.7

$$H_1(z) = \frac{z}{z - 0.9}$$

   If $K = 0.8$

   (a) Is the system stable?

   (b) What is the difference equation relating $x[n]$ and $y[n]$?

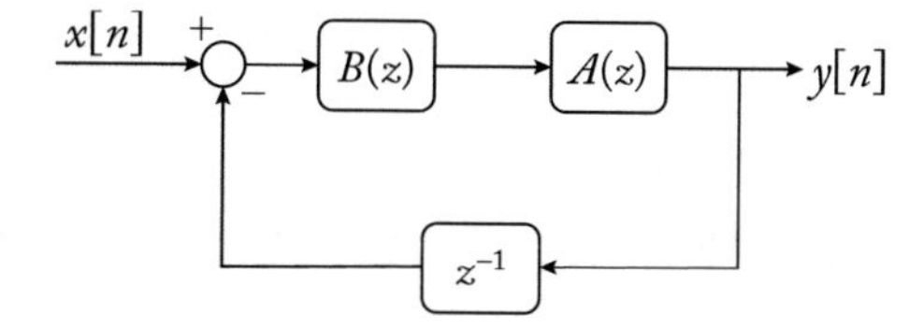

Figure 29.9: Feedback system using a control system formulation.

4. In the system shown in Figure 29.9 the plant $A(z)$ and the PI controller B(z) are given by

$$\begin{aligned} A(z) &= \frac{z^{-1}}{1-z^{-1}} \\ B(z) &= K_p + \frac{K_i}{1-z^{-1}} \end{aligned}$$

Find the difference equation relating $x[n]$ and $y[n]$.

29.7 ANSWERS

1. (a)

$$y[n] = \frac{K}{1+K}x[n] + \frac{2}{1+K}y[n-1]$$

   (b) $K > 1$ or $K < -3$.

2.

$$y[n] = x[n] - x[n-1] - y[n-1] + y[n-2]$$

3. (a) Yes it is stable for $K = 0.8$

   (b)

$$y[n] = \frac{4}{9}x[n] - \frac{1}{2}y[n-1]$$

4.

$$\begin{aligned} y[n] =& (K_p + K_i)x[n-1] - K_p x[n-2] - 2y[n-1] \\ & - (K - p + K_i + 1)y[n-2] + K_p y[n-3] \end{aligned}$$

Author's Biography

KHALID SAYOOD

Khalid Sayood received the B.S. and M.S. degrees in electrical engineering from the University of Rochester, Rochester, NY, in 1977 and 1979, respectively, and the Ph.D. degree in electrical engineering from Texas A&M University, College Station, in 1982. From 1995 to 1996, he served as the Founding Head of the Computer Vision and Image Processing Group at the Turkish National Research Council Informatics Institute. From 1996 to 1997, he was a Visiting Professor with Bogazici University, Istanbul, Turkey. Since 1982, he has been with the University of Nebraska–Lincoln, where he is currently serving as a Professor in the Department of Electrical and Computer Engineering. He is the author of *Introduction to Data Compression*, 5th ed., Morgan Kaufmann, 2017, *Understanding Circuits: Learning Problem Solving Using Circuit Analysis*, Morgan & Claypool, 2005, *Learning Programming Using MATLAB*, Morgan& Claypool, 2006, *Joint Source Channel Coding Using Arithmetic Codes*, Morgan & Claypool, 2010, and *Computational Genomic Signatures*, Morgan & Claypool, 2011, and the Editor of *Lossless Compression Handbook*, Academic Press, 2002. His research interests include bioinformatics, data compression, and biological signal processing.

Printed in the United States
by Baker & Taylor Publisher Services